T0255904

Lecture Notes in Computer Science

Lecture Notes in Artificial Intelligence   14645

Founding Editor

Jörg Siekmann

Series Editors

Randy Goebel, *University of Alberta, Edmonton, Canada*
Wolfgang Wahlster, *DFKI, Berlin, Germany*
Zhi-Hua Zhou, *Nanjing University, Nanjing, China*

The series Lecture Notes in Artificial Intelligence (LNAI) was established in 1988 as a topical subseries of LNCS devoted to artificial intelligence.

The series publishes state-of-the-art research results at a high level. As with the LNCS mother series, the mission of the series is to serve the international R & D community by providing an invaluable service, mainly focused on the publication of conference and workshop proceedings and postproceedings.

De-Nian Yang · Xing Xie · Vincent S. Tseng ·
Jian Pei · Jen-Wei Huang · Jerry Chun-Wei Lin
Editors

Advances in Knowledge Discovery and Data Mining

28th Pacific-Asia Conference
on Knowledge Discovery and Data Mining, PAKDD 2024
Taipei, Taiwan, May 7–10, 2024
Proceedings, Part I

Springer

Editors
De-Nian Yang (ID)
Academia Sinica
Taipei, Taiwan

Vincent S. Tseng (ID)
National Yang Ming Chiao Tung University
Hsinchu, Taiwan

Jen-Wei Huang (ID)
National Cheng Kung University
Tainan, Taiwan

Xing Xie (ID)
Microsoft Research Asia
Beijing, China

Jian Pei (ID)
Duke University
Durham, NC, USA

Jerry Chun-Wei Lin (ID)
Silesian University of Technology
Gliwice, Poland

ISSN 0302-9743   ISSN 1611-3349 (electronic)
Lecture Notes in Artificial Intelligence
ISBN 978-981-97-2241-9   ISBN 978-981-97-2242-6 (eBook)
https://doi.org/10.1007/978-981-97-2242-6

LNCS Sublibrary: SL7 – Artificial Intelligence

© The Editor(s) (if applicable) and The Author(s), under exclusive license
to Springer Nature Singapore Pte Ltd. 2024

This work is subject to copyright. All rights are solely and exclusively licensed by the Publisher, whether
the whole or part of the material is concerned, specifically the rights of translation, reprinting, reuse of
illustrations, recitation, broadcasting, reproduction on microfilms or in any other physical way, and transmission
or information storage and retrieval, electronic adaptation, computer software, or by similar or dissimilar
methodology now known or hereafter developed.
The use of general descriptive names, registered names, trademarks, service marks, etc. in this publication
does not imply, even in the absence of a specific statement, that such names are exempt from the relevant
protective laws and regulations and therefore free for general use.
The publisher, the authors and the editors are safe to assume that the advice and information in this book
are believed to be true and accurate at the date of publication. Neither the publisher nor the authors or the
editors give a warranty, expressed or implied, with respect to the material contained herein or for any errors
or omissions that may have been made. The publisher remains neutral with regard to jurisdictional claims in
published maps and institutional affiliations.

This Springer imprint is published by the registered company Springer Nature Singapore Pte Ltd.
The registered company address is: 152 Beach Road, #21-01/04 Gateway East, Singapore 189721, Singapore

Paper in this product is recyclable.

General Chairs' Preface

On behalf of the Organizing Committee, we were delighted to welcome attendees to the 28th Pacific-Asia Conference on Knowledge Discovery and Data Mining (PAKDD 2024). Since its inception in 1997, PAKDD has long established itself as one of the leading international conferences on data mining and knowledge discovery. PAKDD provides an international forum for researchers and industry practitioners to share their new ideas, original research results, and practical development experiences across all areas of Knowledge Discovery and Data Mining (KDD). This year, after its two previous editions in Taipei (2002) and Tainan (2014), PAKDD was held in Taiwan for the third time in the fascinating city of Taipei, during May 7–10, 2024. Moreover, PAKDD 2024 was held as a fully physical conference since the COVID-19 pandemic was contained.

We extend our sincere gratitude to the researchers who submitted their work to the PAKDD 2024 main conference, high-quality tutorials, and workshops on cutting-edge topics. The conference program was further enriched with seven high-quality tutorials and five workshops on cutting-edge topics. We would like to deliver our sincere thanks for their efforts in research, as well as in preparing high-quality presentations. We also express our appreciation to all the collaborators and sponsors for their trust and cooperation. We were honored to have three distinguished keynote speakers joining the conference: Ed H. Chi (Google DeepMind), Vipin Kumar (University of Minnesota), and Huan Liu (Arizona State University), each with high reputations in their respective areas. We enjoyed their participation and talks, which made the conference one of the best academic platforms for knowledge discovery and data mining. We would like to express our sincere gratitude for the contributions of the Steering Committee members, Organizing Committee members, Program Committee members, and anonymous reviewers, led by Program Committee Chairs De-Nian Yang and Xing Xie. It is through their untiring efforts that the conference had an excellent technical program. We are also thankful to the other Organizing Committee members: Workshop Chairs, Chuan-Kang Ting and Xiaoli Li; Tutorial Chairs, Jiun-Long Huang and Philippe Fournier-Viger; Publicity Chairs, Mi-Yen Yeh and Rage Uday Kiran; Industrial Chairs, Kun-Ta Chuang, Wei-Chao Chen and Richie Tsai; Proceedings Chairs, Jen-Wei Huang and Jerry Chun-Wei Lin; Registration Chairs, Chih-Ya Shen and Hong-Han Shuai; Web and Content Chairs, Cheng-Te Li and Shan-Hung Wu; Local Arrangement Chairs, Yi-Ling Chen, Kuan-Ting Lai, Yi-Ting Chen, and Ya-Wen Teng. We feel indebted to the PAKDD Steering Committee for their constant guidance and sponsorship of manuscripts. We are also grateful to the hosting organizations, National Yang Ming Chiao Tung University and Academia Sinica, and all our sponsors for continuously providing institutional and financial support to PAKDD 2024.

May 2024

Vincent S. Tseng
Jian Pei

PC Chairs' Preface

It is our great pleasure to present the 28th Pacific-Asia Conference on Knowledge Discovery and Data Mining (PAKDD 2024) as Program Committee Chairs. PAKDD is one of the longest-established and leading international conferences in the areas of data mining and knowledge discovery. It provides an international forum for researchers and industry practitioners to share their new ideas, original research results, and practical development experiences in all KDD-related areas, including data mining, data warehousing, machine learning, artificial intelligence, databases, statistics, knowledge engineering, big data technologies, and foundations.

This year, PAKDD received a record number of 720 submissions, among which 86 submissions were rejected at a preliminary stage due to policy violations. There were 595 Program Committee members and 101 Senior Program Committee members involved in the double-blind reviewing process. For submissions entering the double-blind review process, each one received at least three quality reviews from PC members. Furthermore, each valid submission received one meta-review from the assigned SPC member, who also led the discussion with the PC members. The PC Co-chairs then considered the recommendations and meta-reviews from SPC members and looked into each submission as well as its reviews and PC discussions to make the final decision.

As a result of the highly competitive selection process, 175 submissions were accepted and recommended to be published, with 133 oral-presentation papers and 42 poster-presentation papers. We would like to thank all SPC and PC members whose diligence produced a high-quality program for PAKDD 2024. The conference program also featured three keynote speeches from distinguished data mining researchers, eight invited industrial talks, five cutting-edge workshops, and seven comprehensive tutorials.

We wish to sincerely thank all SPC members, PC members, and external reviewers for their invaluable efforts in ensuring a timely, fair, and highly effective paper review and selection procedure. We hope that readers of the proceedings will find the PAKDD 2024 technical program both interesting and rewarding.

May 2024

De-Nian Yang
Xing Xie

Organization

Organizing Committee

Honorary Chairs

Philip S. Yu — University of Illinois at Chicago, USA
Ming-Syan Chen — National Taiwan University, Taiwan

General Chairs

Vincent S. Tseng — National Yang Ming Chiao Tung University, Taiwan
Jian Pei — Duke University, USA

Program Committee Chairs

De-Nian Yang — Academia Sinica, Taiwan
Xing Xie — Microsoft Research Asia, China

Workshop Chairs

Chuan-Kang Ting — National Tsing Hua University, Taiwan
Xiaoli Li — A*STAR, Singapore

Tutorial Chairs

Jiun-Long Huang — National Yang Ming Chiao Tung University, Taiwan
Philippe Fournier-Viger — Shenzhen University, China

Publicity Chairs

Mi-Yen Yeh — Academia Sinica, Taiwan
Rage Uday Kiran — University of Aizu, Japan

Industrial Chairs

Kun-Ta Chuang   National Cheng Kung University, Taiwan
Wei-Chao Chen   Inventec Corp./Skywatch Innovation, Taiwan
Richie Tsai   Taiwan AI Academy, Taiwan

Proceedings Chairs

Jen-Wei Huang   National Cheng Kung University, Taiwan
Jerry Chun-Wei Lin   Silesian University of Technology, Poland

Registration Chairs

Chih-Ya Shen   National Tsing Hua University, Taiwan
Hong-Han Shuai   National Yang Ming Chiao Tung University,
                                       Taiwan

Web and Content Chairs

Shan-Hung Wu   National Tsing Hua University, Taiwan
Cheng-Te Li   National Cheng Kung University, Taiwan

Local Arrangement Chairs

Yi-Ling Chen   National Taiwan University of Science and
                                       Technology, Taiwan
Kuan-Ting Lai   National Taipei University of Technology, Taiwan
Yi-Ting Chen   National Yang Ming Chiao Tung University,
                                       Taiwan
Ya-Wen Teng   Academia Sinica, Taiwan

Steering Committee

Chair

Longbing Cao   Macquarie University, Australia

Vice Chair

Gill Dobbie   University of Auckland, New Zealand

Treasurer

Longbing Cao   Macquarie University, Australia

Members

Ramesh Agrawal   Jawaharlal Nehru University, India
Gill Dobbie   University of Auckland, New Zealand
João Gama   University of Porto, Portugal
Zhiguo Gong   University of Macau, Macau SAR
Hisashi Kashima   Kyoto University, Japan
Hady W. Lauw   Singapore Management University, Singapore
Jae-Gil Lee   KAIST, Korea
Dinh Phung   Monash University, Australia
Kyuseok Shim   Seoul National University, Korea
Geoff Webb   Monash University, Australia
Raymond Chi-Wing Wong   Hong Kong University of Science and
                                          Technology, Hong Kong SAR
Min-Ling Zhang   Southeast University, China

Life Members

Longbing Cao   Macquarie University, Australia
Ming-Syan Chen   National Taiwan University, Taiwan
David Cheung   University of Hong Kong, China
Joshua Z. Huang   Chinese Academy of Sciences, China
Masaru Kitsuregawa   Tokyo University, Japan
Rao Kotagiri   University of Melbourne, Australia
Ee-Peng Lim   Singapore Management University, Singapore
Huan Liu   Arizona State University, USA
Hiroshi Motoda   AFOSR/AOARD and Osaka University, Japan
Jian Pei   Duke University, USA
P. Krishna Reddy   IIIT Hyderabad, India
Jaideep Srivastava   University of Minnesota, USA
Thanaruk Theeramunkong   Thammasat University, Thailand
Tu-Bao Ho   JAIST, Japan
Vincent S. Tseng   National Yang Ming Chiao Tung University,
                                          Taiwan
Takashi Washio   Osaka University, Japan
Kyu-Young Whang   KAIST, Korea
Graham Williams   Australian National University, Australia
Chengqi Zhang   University of Technology Sydney, Australia

Ning Zhong   Maebashi Institute of Technology, Japan
Zhi-Hua Zhou   Nanjing University, China

Past Members

Arbee L. P. Chen   Asia University, Taiwan
Hongjun Lu   Hong Kong University of Science and
                                   Technology, Hong Kong SAR
Takao Terano   Tokyo Institute of Technology, Japan

Senior Program Committee

Aijun An   York University, Canada
Aris Anagnostopoulos   Sapienza Università di Roma, Italy
Ting Bai   Beijing University of Posts and
                                   Telecommunications, China
Elisa Bertino   Purdue University, USA
Arnab Bhattacharya   IIT Kanpur, India
Albert Bifet   Université Paris-Saclay, France
Ludovico Boratto   Università degli Studi di Cagliari, Italy
Ricardo Campello   University of Southern Denmark, Denmark
Longbing Cao   University of Technology Sydney, Australia
Tru Cao   UTHealth, USA
Tanmoy Chakraborty   IIT Delhi, India
Jeffrey Chan   RMIT University, Australia
Pin-Yu Chen   IBM T. J. Watson Research Center, USA
Bin Cui   Peking University, China
Anirban Dasgupta   IIT Gandhinagar, India
Wei Ding   University of Massachusetts Boston, USA
Eibe Frank   University of Waikato, New Zealand
Chen Gong   Nanjing University of Science and Technology,
                                   China
Jingrui He   UIUC, USA
Tzung-Pei Hong   National University of Kaohsiung, Taiwan
Qinghua Hu   Tianjin University, China
Hong Huang   Huazhong University of Science and Technology,
                                   China
Jen-Wei Huang   National Cheng Kung University, Taiwan
Tsuyoshi Ide   IBM T. J. Watson Research Center, USA
Xiaowei Jia   University of Pittsburgh, USA
Zhe Jiang   University of Florida, USA

Toshihiro Kamishima	National Institute of Advanced Industrial Science and Technology, Japan
Murat Kantarcioglu	University of Texas at Dallas, USA
Hung-Yu Kao	National Cheng Kung University, Taiwan
Kamalakar Karlapalem	IIIT Hyderabad, India
Anuj Karpatne	Virginia Tech, USA
Hisashi Kashima	Kyoto University, Japan
Sang-Wook Kim	Hanyang University, Korea
Yun Sing Koh	University of Auckland, New Zealand
Hady Lauw	Singapore Management University, Singapore
Byung Suk Lee	University of Vermont, USA
Jae-Gil Lee	KAIST, Korea
Wang-Chien Lee	Pennsylvania State University, USA
Chaozhuo Li	Microsoft Research Asia, China
Gang Li	Deakin University, Australia
Jiuyong Li	University of South Australia, Australia
Jundong Li	University of Virginia, USA
Ming Li	Nanjing University, China
Sheng Li	University of Virginia, USA
Ying Li	AwanTunai, Singapore
Yu-Feng Li	Nanjing University, China
Hao Liao	Shenzhen University, China
Ee-peng Lim	Singapore Management University, Singapore
Jerry Chun-Wei Lin	Silesian University of Technology, Poland
Shou-De Lin	National Taiwan University, Taiwan
Hongyan Liu	Tsinghua University, China
Wei Liu	University of Technology Sydney, Australia
Chang-Tien Lu	Virginia Tech, USA
Yuan Luo	Northwestern University, USA
Wagner Meira Jr.	UFMG, Brazil
Alexandros Ntoulas	University of Athens, Greece
Satoshi Oyama	Nagoya City University, Japan
Guansong Pang	Singapore Management University, Singapore
Panagiotis Papapetrou	Stockholm University, Sweden
Wen-Chih Peng	National Yang Ming Chiao Tung University, Taiwan
Dzung Phan	IBM T. J. Watson Research Center, USA
Uday Rage	University of Aizu, Japan
Rajeev Raman	University of Leicester, UK
P. Krishna Reddy	IIIT Hyderabad, India
Thomas Seidl	LMU München, Germany
Neil Shah	Snap Inc., USA

Yingxia Shao	Beijing University of Posts and Telecommunications, China
Victor S. Sheng	Texas Tech University, USA
Kyuseok Shim	Seoul National University, Korea
Arlei Silva	Rice University, USA
Jaideep Srivastava	University of Minnesota, USA
Masashi Sugiyama	RIKEN/University of Tokyo, Japan
Ju Sun	University of Minnesota, USA
Jiliang Tang	Michigan State University, USA
Hanghang Tong	UIUC, USA
Ranga Raju Vatsavai	North Carolina State University, USA
Hao Wang	Nanyang Technological University, Singapore
Hao Wang	Xidian University, China
Jianyong Wang	Tsinghua University, China
Tim Weninger	University of Notre Dame, USA
Raymond Chi-Wing Wong	Hong Kong University of Science and Technology, Hong Kong SAR
Jia Wu	Macquarie University, Australia
Xindong Wu	Hefei University of Technology, China
Xintao Wu	University of Arkansas, USA
Yiqun Xie	University of Maryland, USA
Yue Xu	Queensland University of Technology, Australia
Lina Yao	University of New South Wales, Australia
Han-Jia Ye	Nanjing University, China
Mi-Yen Yeh	Academia Sinica, Taiwan
Hongzhi Yin	University of Queensland, Australia
Min-Ling Zhang	Southeast University, China
Ping Zhang	Ohio State University, USA
Zhao Zhang	Hefei University of Technology, China
Zhongfei Zhang	Binghamton University, USA
Xiangyu Zhao	City University of Hong Kong, Hong Kong SAR
Yanchang Zhao	CSIRO, Australia
Jiayu Zhou	Michigan State University, USA
Xiao Zhou	Renmin University of China, China
Xiaofang Zhou	Hong Kong University of Science and Technology, Hong Kong SAR
Feida Zhu	Singapore Management University, Singapore
Fuzhen Zhuang	Beihang University, China

Program Committee

Zubin Abraham	Robert Bosch, USA
Pedro Henriques Abreu	CISUC, Portugal
Muhammad Abulaish	South Asian University, India
Bijaya Adhikari	University of Iowa, USA
Karan Aggarwal	Amazon, USA
Chowdhury Farhan Ahmed	University of Dhaka, Bangladesh
Ulrich Aïvodji	ÉTS Montréal, Canada
Esra Akbas	Georgia State University, USA
Shafiq Alam	Massey University Auckland, New Zealand
Giuseppe Albi	Università degli Studi di Pavia, Italy
David Anastasiu	Santa Clara University, USA
Xiang Ao	Chinese Academy of Sciences, China
Elena-Simona Apostol	Uppsala University, Sweden
Sunil Aryal	Deakin University, Australia
Jees Augustine	Microsoft, USA
Konstantin Avrachenkov	Inria, France
Goonmeet Bajaj	Ohio State University, USA
Jean Paul Barddal	PUCPR, Brazil
Srikanta Bedathur	IIT Delhi, India
Sadok Ben Yahia	University of Southern Denmark, Denmark
Alessandro Berti	Università di Pisa, Italy
Siddhartha Bhattacharyya	University of Illinois at Chicago, USA
Ranran Bian	University of Sydney, Australia
Song Bian	Chinese University of Hong Kong, Hong Kong SAR
Giovanni Maria Biancofiore	Politecnico di Bari, Italy
Fernando Bobillo	University of Zaragoza, Spain
Adrian M. P. Brasoveanu	Modul Technology GmbH, Austria
Krisztian Buza	Budapest University of Technology and Economics, Hungary
Luca Cagliero	Politecnico di Torino, Italy
Jean-Paul Calbimonte	University of Applied Sciences and Arts Western Switzerland, Switzerland
K. Selçuk Candan	Arizona State University, USA
Fuyuan Cao	Shanxi University, China
Huiping Cao	New Mexico State University, USA
Jian Cao	Shanghai Jiao Tong University, China
Yan Cao	University of Texas at Dallas, USA
Yang Cao	Hokkaido University, Japan
Yuanjiang Cao	Macquarie University, Australia

Sharma Chakravarthy	University of Texas at Arlington, USA
Harry Kai-Ho Chan	University of Sheffield, UK
Zhangming Chan	Alibaba Group, China
Snigdhansu Chatterjee	University of Minnesota, USA
Mandar Chaudhary	eBay, USA
Chen Chen	University of Virginia, USA
Chun-Hao Chen	National Kaohsiung University of Science and Technology, Taiwan
Enhong Chen	University of Science and Technology of China, China
Fanglan Chen	Virginia Tech, USA
Feng Chen	University of Texas at Dallas, USA
Hongyang Chen	Zhejiang Lab, China
Jia Chen	University of California Riverside, USA
Jinjun Chen	Swinburne University of Technology, Australia
Lingwei Chen	Wright State University, USA
Ping Chen	University of Massachusetts Boston, USA
Shang-Tse Chen	National Taiwan University, Taiwan
Shengyu Chen	University of Pittsburgh, USA
Songcan Chen	Nanjing University of Aeronautics and Astronautics, China
Tao Chen	China University of Geosciences, China
Tianwen Chen	Hong Kong University of Science and Technology, Hong Kong SAR
Tong Chen	University of Queensland, Australia
Weitong Chen	University of Adelaide, Australia
Yi-Hui Chen	Chang Gung University, Taiwan
Yile Chen	Nanyang Technological University, Singapore
Yi-Ling Chen	National Taiwan University of Science and Technology, Taiwan
Yi-Shin Chen	National Tsing Hua University, Taiwan
Yi-Ting Chen	National Yang Ming Chiao Tung University, Taiwan
Zheng Chen	Osaka University, Japan
Zhengzhang Chen	NEC Laboratories America, USA
Zhiyuan Chen	UMBC, USA
Zhong Chen	Southern Illinois University, USA
Peng Cheng	East China Normal University, China
Abdelghani Chibani	Université Paris-Est Créteil, France
Jingyuan Chou	University of Virginia, USA
Lingyang Chu	McMaster University, Canada
Kun-Ta Chuang	National Cheng Kung University, Taiwan

Robert Churchill	Georgetown University, USA
Chaoran Cui	Shandong University of Finance and Economics, China
Alfredo Cuzzocrea	Università della Calabria, Italy
Bi-Ru Dai	National Taiwan University of Science and Technology, Taiwan
Honghua Dai	Zhengzhou University, China
Claudia d'Amato	University of Bari, Italy
Chuangyin Dang	City University of Hong Kong, China
Mrinal Das	IIT Palakkad, India
Debanjan Datta	Virginia Tech, USA
Cyril de Runz	Université de Tours, France
Jeremiah Deng	University of Otago, New Zealand
Ke Deng	RMIT University, Australia
Zhaohong Deng	Jiangnan University, China
Anne Denton	North Dakota State University, USA
Shridhar Devamane	KLE Institute of Technology, India
Djellel Difallah	New York University, USA
Ling Ding	Tianjin University, China
Shifei Ding	China University of Mining and Technology, China
Yao-Xiang Ding	Zhejiang University, China
Yifan Ding	University of Notre Dame, USA
Ying Ding	University of Texas at Austin, USA
Lamine Diop	EPITA, France
Nemanja Djuric	Aurora Innovation, USA
Gillian Dobbie	University of Auckland, New Zealand
Josep Domingo-Ferrer	Universitat Rovira i Virgili, Spain
Bo Dong	Amazon, USA
Yushun Dong	University of Virginia, USA
Bo Du	Wuhan University, China
Silin Du	Tsinghua University, China
Jiuding Duan	Allianz Global Investors, Japan
Lei Duan	Sichuan University, China
Walid Durani	LMU München, Germany
Sourav Dutta	Huawei Research Centre, Ireland
Mohamad El-Hajj	MacEwan University, Canada
Ya Ju Fan	Lawrence Livermore National Laboratory, USA
Zipei Fan	Jilin University, China
Majid Farhadloo	University of Minnesota, USA
Fabio Fassetti	Università della Calabria, Italy
Zhiquan Feng	National Cheng Kung University, Taiwan

Len Feremans	Universiteit Antwerpen, Belgium
Edouard Fouché	Karlsruher Institut für Technologie, Germany
Dongqi Fu	UIUC, USA
Yanjie Fu	University of Central Florida, USA
Ken-ichi Fukui	Osaka University, Japan
Matjaž Gams	Jožef Stefan Institute, Slovenia
Amir Gandomi	University of Technology Sydney, Australia
Aryya Gangopadhyay	UMBC, USA
Dashan Gao	Hong Kong University of Science and Technology, China
Wei Gao	Nanjing University, China
Yifeng Gao	University of Texas Rio Grande Valley, USA
Yunjun Gao	Zhejiang University, China
Paolo Garza	Politecnico di Torino, Italy
Chang Ge	University of Minnesota, USA
Xin Geng	Southeast University, China
Flavio Giobergia	Politecnico di Torino, Italy
Rosalba Giugno	Università degli Studi di Verona, Italy
Aris Gkoulalas-Divanis	Merative, USA
Djordje Gligorijevic	Temple University, USA
Daniela Godoy	UNICEN, Argentina
Heitor Gomes	Victoria University of Wellington, New Zealand
Maciej Grzenda	Warsaw University of Technology, Poland
Lei Gu	Nanjing University of Posts and Telecommunications, China
Yong Guan	Iowa State University, USA
Riccardo Guidotti	Università di Pisa, Italy
Ekta Gujral	University of California Riverside, USA
Guimu Guo	Rowan University, USA
Ting Guo	University of Technology Sydney, Australia
Xingzhi Guo	Stony Brook University, USA
Ch. Md. Rakin Haider	Purdue University, USA
Benjamin Halstead	University of Auckland, New Zealand
Jinkun Han	Georgia State University, USA
Lu Han	Nanjing University, China
Yufei Han	Inria, France
Daisuke Hatano	RIKEN, Japan
Kohei Hatano	Kyushu University/RIKEN AIP, Japan
Shogo Hayashi	BizReach, Japan
Erhu He	University of Pittsburgh, USA
Guoliang He	Wuhan University, China
Pengfei He	Michigan State University, USA

Yi He	Old Dominion University, USA
Shen-Shyang Ho	Rowan University, USA
William Hsu	Kansas State University, USA
Haoji Hu	University of Minnesota, USA
Hongsheng Hu	CSIRO, Australia
Liang Hu	Tongji University, China
Shizhe Hu	Zhengzhou University, China
Wei Hu	Nanjing University, China
Mengdi Huai	Iowa State University, USA
Chao Huang	University of Hong Kong, Hong Kong SAR
Congrui Huang	Microsoft, China
Guangyan Huang	Deakin University, Australia
Jimmy Huang	York University, Canada
Jinbin Huang	Hong Kong Baptist University, Hong Kong SAR
Kai Huang	Hong Kong University of Science and Technology, China
Ling Huang	South China Agricultural University, China
Ting-Ji Huang	Nanjing University, China
Xin Huang	Hong Kong Baptist University, Hong Kong SAR
Zhenya Huang	University of Science and Technology of China, China
Chih-Chieh Hung	National Chung Hsing University, Taiwan
Hui-Ju Hung	Pennsylvania State University, USA
Nam Huynh	JAIST, Japan
Akihiro Inokuchi	Kwansei Gakuin University, Japan
Atsushi Inoue	Eastern Washington University, USA
Nevo Itzhak	Ben-Gurion University, Israel
Tomoya Iwakura	Fujitsu Laboratories Ltd., Japan
Divyesh Jadav	IBM T. J. Watson Research Center, USA
Shubham Jain	Visa Research, USA
Bijay Prasad Jaysawal	National Cheng Kung University, Taiwan
Kishlay Jha	University of Iowa, USA
Taoran Ji	Texas A&M University - Corpus Christi, USA
Songlei Jian	NUDT, China
Gaoxia Jiang	Shanxi University, China
Hansi Jiang	SAS Institute Inc., USA
Jiaxin Jiang	National University of Singapore, Singapore
Min Jiang	Xiamen University, China
Renhe Jiang	University of Tokyo, Japan
Yuli Jiang	Chinese University of Hong Kong, Hong Kong SAR
Bo Jin	Dalian University of Technology, China

Ming Jin	Monash University, Australia
Ruoming Jin	Kent State University, USA
Wei Jin	University of North Texas, USA
Mingxuan Ju	University of Notre Dame, USA
Wei Ju	Peking University, China
Vana Kalogeraki	Athens University of Economics and Business, Greece
Bo Kang	Ghent University, Belgium
Jian Kang	University of Rochester, USA
Ashwin Viswanathan Kannan	Amazon, USA
Tomi Kauppinen	Aalto University School of Science, Finland
Jungeun Kim	Kongju National University, Korea
Kyoung-Sook Kim	National Institute of Advanced Industrial Science and Technology, Japan
Primož Kocbek	University of Maribor, Slovenia
Aritra Konar	Katholieke Universiteit Leuven, Belgium
Youyong Kong	Southeast University, China
Olivera Kotevska	Oak Ridge National Laboratory, USA
P. Radha Krishna	NIT Warangal, India
Adit Krishnan	UIUC, USA
Gokul Krishnan	IIT Madras, India
Peer Kröger	CAU, Germany
Marzena Kryszkiewicz	Warsaw University of Technology, Poland
Chuan-Wei Kuo	National Yang Ming Chiao Tung University, Taiwan
Kuan-Ting Lai	National Taipei University of Technology, Taiwan
Long Lan	NUDT, China
Duc-Trong Le	Vietnam National University, Vietnam
Tuan Le	New Mexico State University, USA
Chul-Ho Lee	Texas State University, USA
Ickjai Lee	James Cook University, Australia
Ki Yong Lee	Sookmyung Women's University, Korea
Ki-Hoon Lee	Kwangwoon University, Korea
Roy Ka-Wei Lee	Singapore University of Technology and Design, Singapore
Yue-Shi Lee	Ming Chuan University, Taiwan
Dino Lenco	INRAE, France
Carson Leung	University of Manitoba, Canada
Boyu Li	University of Technology Sydney, Australia
Chaojie Li	University of New South Wales, Australia
Cheng-Te Li	National Cheng Kung University, Taiwan
Chongshou Li	Southwest Jiaotong University, China

Fengxin Li	Renmin University of China, China
Guozhong Li	King Abdullah University of Science and Technology, Saudi Arabia
Huaxiong Li	Nanjing University, China
Jianxin Li	Beihang University, China
Lei Li	Hong Kong University of Science and Technology (Guangzhou), China
Peipei Li	Hefei University of Technology, China
Qian Li	Curtin University, Australia
Rong-Hua Li	Beijing Institute of Technology, China
Shao-Yuan Li	Nanjing University of Aeronautics and Astronautics, China
Shuai Li	Cambridge University, UK
Shuang Li	Beijing Institute of Technology, China
Tianrui Li	Southwest Jiaotong University, China
Wengen Li	Tongji University, China
Wentao Li	Hong Kong University of Science and Technology (Guangzhou), China
Xin-Ye Li	Bytedance, China
Xiucheng Li	Harbin Institute of Technology, China
Xuelong Li	Northwestern Polytechnical University, China
Yidong Li	Beijing Jiaotong University, China
Yinxiao Li	Meta Platforms, USA
Yuefeng Li	Queensland University of Technology, Australia
Yun Li	Nanjing University of Posts and Telecommunications, China
Panagiotis Liakos	University of Athens, Greece
Xiang Lian	Kent State University, USA
Shen Liang	Université Paris Cité, France
Qing Liao	Harbin Institute of Technology (Shenzhen), China
Sungsu Lim	Chungnam National University, Korea
Dandan Lin	Shenzhen Institute of Computing Sciences, China
Yijun Lin	University of Minnesota, USA
Ying-Jia Lin	National Cheng Kung University, Taiwan
Baodi Liu	China University of Petroleum (East China), China
Chien-Liang Liu	National Yang Ming Chiao Tung University, Taiwan
Guiquan Liu	University of Science and Technology of China, China
Jin Liu	Shanghai Maritime University, China
Jinfei Liu	Emory University, USA
Kunpeng Liu	Portland State University, USA

Ning Liu	Shandong University, China
Qi Liu	University of Science and Technology of China, China
Qing Liu	Zhejiang University, China
Qun Liu	Louisiana State University, USA
Shenghua Liu	Chinese Academy of Sciences, China
Weifeng Liu	China University of Petroleum (East China), China
Yang Liu	Wilfrid Laurier University, Canada
Yao Liu	University of New South Wales, Australia
Yixin Liu	Monash University, Australia
Zheng Liu	Nanjing University of Posts and Telecommunications, China
Cheng Long	Nanyang Technological University, Singapore
Haibing Lu	Santa Clara University, USA
Wenpeng Lu	Qilu University of Technology, China
Simone Ludwig	North Dakota State University, USA
Dongsheng Luo	Florida International University, USA
Ping Luo	Chinese Academy of Sciences, China
Wei Luo	Deakin University, Australia
Xiao Luo	UCLA, USA
Xin Luo	Shandong University, China
Yong Luo	Wuhan University, China
Fenglong Ma	Pennsylvania State University, USA
Huifang Ma	Northwest Normal University, China
Jing Ma	Hong Kong Baptist University, Hong Kong SAR
Qianli Ma	South China University of Technology, China
Yi-Fan Ma	Nanjing University, China
Rich Maclin	University of Minnesota, USA
Son Mai	Queen's University Belfast, UK
Arun Maiya	Institute for Defense Analyses, USA
Bradley Malin	Vanderbilt University Medical Center, USA
Giuseppe Manco	Consiglio Nazionale delle Ricerche, Italy
Naresh Manwani	IIIT Hyderabad, India
Francesco Marcelloni	Università di Pisa, Italy
Leandro Marinho	UFCG, Brazil
Koji Maruhashi	Fujitsu Laboratories Ltd., Japan
Florent Masseglia	Inria, France
Mohammad Masud	United Arab Emirates University, United Arab Emirates
Sarah Masud	IIIT Delhi, India
Costas Mavromatis	University of Minnesota, USA

Maxwell McNeil	University at Albany SUNY, USA
Massimo Melucci	Università degli Studi di Padova, Italy
Alex Memory	Johns Hopkins University, USA
Ernestina Menasalvas	Universidad Politécnica de Madrid, Spain
Xupeng Miao	Carnegie Mellon University, USA
Matej Miheli	University of Zagreb, Croatia
Fan Min	Southwest Petroleum University, China
Jun-Ki Min	Korea University of Technology and Education, Korea
Tsunenori Mine	Kyushu University, Japan
Nguyen Le Minh	JAIST, Japan
Shuichi Miyazawa	Graduate University for Advanced Studies, Japan
Songsong Mo	Nanyang Technological University, Singapore
Jacob Montiel	Amazon, USA
Yang-Sae Moon	Kangwon National University, Korea
Sebastian Moreno	Universidad Adolfo Ibáñez, Chile
Daisuke Moriwaki	CyberAgent, Inc., Japan
Tsuyoshi Murata	Tokyo Institute of Technology, Japan
Charini Nanayakkara	Australian National University, Australia
Mirco Nanni	Consiglio Nazionale delle Ricerche, Italy
Wilfred Ng	Hong Kong University of Science and Technology, Hong Kong SAR
Cam-Tu Nguyen	Nanjing University, China
Canh Hao Nguyen	Kyoto University, Japan
Hoang Long Nguyen	Meharry Medical College, USA
Shiwen Ni	Chinese Academy of Sciences, China
Jian-Yun Nie	Université de Montréal, Canada
Tadashi Nomoto	National Institute of Japanese Literature, Japan
Tim Oates	UMBC, USA
Eduardo Ogasawara	CEFET-RJ, Brazil
Kouzou Ohara	Aoyama Gakuin University, Japan
Kok-Leong Ong	RMIT University, Australia
Riccardo Ortale	Consiglio Nazionale delle Ricerche, Italy
Arindam Pal	CSIRO, Australia
Eliana Pastor	Politecnico di Torino, Italy
Dhaval Patel	IBM T. J. Watson Research Center, USA
Martin Pavlovski	Yahoo Inc., USA
Le Peng	University of Minnesota, USA
Nhan Pham	IBM T. J. Watson Research Center, USA
Thai-Hoang Pham	Ohio State University, USA
Chengzhi Piao	Hong Kong Baptist University, Hong Kong SAR
Marc Plantevit	EPITA, France

Mario Prado-Romero	Gran Sasso Science Institute, Italy
Bardh Prenkaj	Sapienza Università di Roma, Italy
Jianzhong Qi	University of Melbourne, Australia
Buyue Qian	Xi'an Jiaotong University, China
Huajie Qian	Columbia University, USA
Hezhe Qiao	Singapore Management University, Singapore
Biao Qin	Renmin University of China, China
Zengchang Qin	Beihang University, China
Tho Quan	Ho Chi Minh City University of Technology, Vietnam
Miloš Radovanović	University of Novi Sad, Serbia
Thilina Ranbaduge	Australian National University, Australia
Chotirat Ratanamahatana	Chulalongkorn University, Thailand
Chandra Reddy	IBM T. J. Watson Research Center, USA
Ryan Rossi	Adobe Research, USA
Morteza Saberi	University of Technology Sydney, Australia
Akira Sakai	Fujitsu Laboratories Ltd., Japan
David Sánchez	Universitat Rovira i Virgili, Spain
Maria Luisa Sapino	Università degli Studi di Torino, Italy
Hernan Sarmiento	UChile & IMFD, Chile
Badrul Sarwar	CloudAEye, USA
Nader Shakibay Senobari	University of California Riverside, USA
Nasrin Shabani	Macquarie University, Australia
Ankit Sharma	University of Minnesota, USA
Chandra N. Shekar	RGUKT RK Valley, India
Chih-Ya Shen	National Tsing Hua University, Taiwan
Wei Shen	Nankai University, China
Yu Shen	Peking University, China
Zhi-Yu Shen	Nanjing University, China
Chuan Shi	Beijing University of Posts and Telecommunications, China
Yue Shi	Meta Platforms, USA
Zhenwei Shi	Beihang University, China
Motoki Shiga	Tohoku University, Japan
Kijung Shin	KAIST, Korea
Kai Shu	Illinois Institute of Technology, USA
Hong-Han Shuai	National Yang Ming Chiao Tung University, Taiwan
Zeren Shui	University of Minnesota, USA
Satyaki Sikdar	Indiana University, USA
Dan Simovici	University of Massachusetts Boston, USA
Apoorva Singh	IIT Patna, India

Bikash Chandra Singh	Islamic University, Bangladesh
Stavros Sintos	University of Illinois at Chicago, USA
Krishnamoorthy Sivakumar	Washington State University, USA
Andrzej Skowron	University of Warsaw, Poland
Andy Song	RMIT University, Australia
Dongjin Song	University of Connecticut, USA
Arnaud Soulet	Université de Tours, France
Ja-Hwung Su	National University of Kaohsiung, Taiwan
Victor Suciu	University of Wisconsin, USA
Liang Sun	Alibaba Group, USA
Xin Sun	Technische Universität München, Germany
Yuqing Sun	Shandong University, China
Hirofumi Suzuki	Fujitsu Laboratories Ltd., Japan
Anika Tabassum	Oak Ridge National Laboratory, USA
Yasuo Tabei	RIKEN, Japan
Chih-Hua Tai	National Taipei University, Taiwan
Hiroshi Takahashi	NTT, Japan
Atsuhiro Takasu	National Institute of Informatics, Japan
Yanchao Tan	Fuzhou University, China
Chang Tang	China University of Geosciences, China
Lu-An Tang	NEC Laboratories America, USA
Qiang Tang	Luxembourg Institute of Science and Technology, Luxembourg
Yiming Tang	Hefei University of Technology, China
Ying-Peng Tang	Nanjing University of Aeronautics and Astronautics, China
Xiaohui (Daniel) Tao	University of Southern Queensland, Australia
Vahid Taslimitehrani	PhysioSigns Inc., USA
Maguelonne Teisseire	INRAE, France
Ya-Wen Teng	Academia Sinica, Taiwan
Masahiro Terabe	Chugai Pharmaceutical Co. Ltd., Japan
Kia Teymourian	University of Texas at Austin, USA
Qing Tian	Nanjing University of Information Science and Technology, China
Yijun Tian	University of Notre Dame, USA
Maksim Tkachenko	Singapore Management University, Singapore
Yongxin Tong	Beihang University, China
Vicenç Torra	University of Umeå, Sweden
Nhu-Thuat Tran	Singapore Management University, Singapore
Yash Travadi	University of Minnesota, USA
Quoc-Tuan Truong	Amazon, USA

Yi-Ju Tseng	National Yang Ming Chiao Tung University, Taiwan
Turki Turki	King Abdulaziz University, Saudi Arabia
Ruo-Chun Tzeng	KTH Royal Institute of Technology, Sweden
Leong Hou U	University of Macau, Macau SAR
Jeffrey Ullman	Stanford University, USA
Rohini Uppuluri	Glassdoor, USA
Satya Valluri	Databricks, USA
Dinusha Vatsalan	Macquarie University, Australia
Bruno Veloso	FEP - University of Porto and INESC TEC, Portugal
Anushka Vidanage	Australian National University, Australia
Herna Viktor	University of Ottawa, Canada
Michalis Vlachos	University of Lausanne, Switzerland
Sheng Wan	Nanjing University of Science and Technology, China
Beilun Wang	Southeast University, China
Changdong Wang	Sun Yat-sen University, China
Chih-Hang Wang	Academia Sinica, Taiwan
Chuan-Ju Wang	Academia Sinica, Taiwan
Guoyin Wang	Chongqing University of Posts and Telecommunications, China
Hongjun Wang	Southwest Jiaotong University, China
Hongtao Wang	North China Electric Power University, China
Jianwu Wang	UMBC, USA
Jie Wang	Southwest Jiaotong University, China
Jin Wang	Megagon Labs, USA
Jingyuan Wang	Beihang University, China
Jun Wang	Shandong University, China
Lizhen Wang	Yunnan University, China
Peng Wang	Southeast University, China
Pengyang Wang	University of Macau, Macau SAR
Sen Wang	University of Queensland, Australia
Senzhang Wang	Central South University, China
Shoujin Wang	Macquarie University, Australia
Sibo Wang	Chinese University of Hong Kong, Hong Kong SAR
Suhang Wang	Pennsylvania State University, USA
Wei Wang	Fudan University, China
Wei Wang	Hong Kong University of Science and Technology (Guangzhou), China
Weicheng Wang	Hong Kong University of Science and Technology, Hong Kong SAR

Wei-Yao Wang	National Yang Ming Chiao Tung University, Taiwan
Wendy Hui Wang	Stevens Institute of Technology, USA
Xiao Wang	Beihang University, China
Xiaoyang Wang	University of New South Wales, Australia
Xin Wang	University of Calgary, Canada
Xinyuan Wang	George Mason University, USA
Yanhao Wang	East China Normal University, China
Yuanlong Wang	Ohio State University, USA
Yuping Wang	Xidian University, China
Yuxiang Wang	Hangzhou Dianzi University, China
Hua Wei	Arizona State University, USA
Zhewei Wei	Renmin University of China, China
Yimin Wen	Guilin University of Electronic Technology, China
Brendon Woodford	University of Otago, New Zealand
Cheng-Wei Wu	National Ilan University, Taiwan
Fan Wu	Central South University, China
Fangzhao Wu	Microsoft Research Asia, China
Jiansheng Wu	Nanjing University of Posts and Telecommunications, China
Jin-Hui Wu	Nanjing University, China
Jun Wu	UIUC, USA
Ou Wu	Tianjin University, China
Shan-Hung Wu	National Tsing Hua University, Taiwan
Shu Wu	Chinese Academy of Sciences, China
Wensheng Wu	University of Southern California, USA
Yun-Ang Wu	National Taiwan University, Taiwan
Wenjie Xi	George Mason University, USA
Lingyun Xiang	Changsha University of Science and Technology, China
Ruliang Xiao	Fujian Normal University, China
Yanghua Xiao	Fudan University, China
Sihong Xie	Lehigh University, USA
Zheng Xie	Nanjing University, China
Bo Xiong	Universität Stuttgart, Germany
Haoyi Xiong	Baidu, Inc., China
Bo Xu	Donghua University, China
Bo Xu	Dalian University of Technology, China
Guandong Xu	University of Technology Sydney, Australia
Hongzuo Xu	NUDT, China
Ji Xu	Guizhou University, China

Tong Xu	University of Science and Technology of China, China
Yuanbo Xu	Jilin University, China
Hui Xue	Southeast University, China
Qiao Xue	Nanjing University of Aeronautics and Astronautics, China
Akihiro Yamaguchi	Toshiba Corporation, Japan
Bo Yang	Jilin University, China
Liangwei Yang	University of Illinois at Chicago, USA
Liu Yang	Tianjin University, China
Shaofu Yang	Southeast University, China
Shiyu Yang	Guangzhou University, China
Wanqi Yang	Nanjing Normal University, China
Xiaoling Yang	Southwest Jiaotong University, China
Xiaowei Yang	South China University of Technology, China
Yan Yang	Southwest Jiaotong University, China
Yiyang Yang	Guangdong University of Technology, China
Yu Yang	City University of Hong Kong, Hong Kong SAR
Yu-Bin Yang	Nanjing University, China
Junjie Yao	East China Normal University, China
Wei Ye	Tongji University, China
Yanfang Ye	University of Notre Dame, USA
Kalidas Yeturu	IIT Tirupati, India
Ilkay Yildiz Potter	BioSensics LLC, USA
Minghao Yin	Northeast Normal University, China
Ziqi Yin	Nanyang Technological University, Singapore
Jia-Ching Ying	National Chung Hsing University, Taiwan
Tetsuya Yoshida	Nara Women's University, Japan
Hang Yu	Shanghai University, China
Jifan Yu	Tsinghua University, China
Yanwei Yu	Ocean University of China, China
Yongsheng Yu	Macquarie University, Australia
Long Yuan	Nanjing University of Science and Technology, China
Lin Yue	University of Newcastle, Australia
Xiaodong Yue	Shanghai University, China
Nayyar Zaidi	Monash University, Australia
Chengxi Zang	Cornell University, USA
Alexey Zaytsev	Skoltech, Russia
Yifeng Zeng	Northumbria University, UK
Petros Zerfos	IBM T. J. Watson Research Center, USA
De-Chuan Zhan	Nanjing University, China

Huixin Zhan	Texas Tech University, USA
Daokun Zhang	Monash University, Australia
Dongxiang Zhang	Zhejiang University, China
Guoxi Zhang	Beijing Institute of General Artificial Intelligence, China
Hao Zhang	Chinese University of Hong Kong, Hong Kong SAR
Huaxiang Zhang	Shandong Normal University, China
Ji Zhang	University of Southern Queensland, Australia
Jianfei Zhang	Université de Sherbrooke, Canada
Lei Zhang	Anhui University, China
Li Zhang	University of Texas Rio Grande Valley, USA
Lin Zhang	IDEA Education, China
Mengjie Zhang	Victoria University of Wellington, New Zealand
Nan Zhang	Wenzhou University, China
Quangui Zhang	Liaoning Technical University, China
Shichao Zhang	Central South University, China
Tianlin Zhang	University of Manchester, UK
Wei Emma Zhang	University of Adelaide, Australia
Wenbin Zhang	Florida International University, USA
Wentao Zhang	Mila, Canada
Xiaobo Zhang	Southwest Jiaotong University, China
Xuyun Zhang	Macquarie University, Australia
Yaqian Zhang	University of Waikato, New Zealand
Yikai Zhang	Guangzhou University, China
Yiqun Zhang	Guangdong University of Technology, China
Yudong Zhang	Nanjing Normal University, China
Zhiwei Zhang	Beijing Institute of Technology, China
Zike Zhang	Hangzhou Normal University, China
Zili Zhang	Southwest University, China
Chen Zhao	Baylor University, USA
Jiaqi Zhao	China University of Mining and Technology, China
Kaiqi Zhao	University of Auckland, New Zealand
Pengfei Zhao	BNU-HKBU United International College, China
Pengpeng Zhao	Soochow University, China
Ying Zhao	Tsinghua University, China
Zhongying Zhao	Shandong University of Science and Technology, China
Guanjie Zheng	Shanghai Jiao Tong University, China
Lecheng Zheng	UIUC, USA
Weiguo Zheng	Fudan University, China

Aoying Zhou	East China Normal University, China
Bing Zhou	Sam Houston State University, USA
Nianjun Zhou	IBM T. J. Watson Research Center, USA
Qinghai Zhou	UIUC, USA
Xiangmin Zhou	RMIT University, Australia
Xiaoping Zhou	Beijing University of Civil Engineering and Architecture, China
Xun Zhou	University of Iowa, USA
Jonathan Zhu	Wheaton College, USA
Ronghang Zhu	University of Georgia, China
Xingquan Zhu	Florida Atlantic University, USA
Ye Zhu	Deakin University, Australia
Yihang Zhu	University of Leicester, UK
Yuanyuan Zhu	Wuhan University, China
Ziwei Zhu	George Mason University, USA

External Reviewers

Zihan Li	University of Massachusetts Boston, USA
Ting Yu	Zhejiang Lab, China

Sponsoring Organizations

Accton

ACSI

Appier

Chunghwa Telecom Co., Ltd

DOIT, Taipei

ISCOM

Metaage

NSTC

Pegatron

Quanta Computer

TWS

Wavenet Co., Ltd

Contents – Part I

Clustering

Data Mining Processes and Pipelines

Anomaly and Outlier Detection

Spatial-Temporal Transformer with Error-Restricted Variance Estimation for Time Series Anomaly Detection

Yuye Feng[✉], Wei Zhang, Haiming Sun, and Weihao Jiang

Hikvision Institute, Hangzhou 310051, China
fengyuye@hikvision.com

Abstract. Due to the intricate dynamics of multivariate time series in cyber-physical system, unsupervised anomaly detection has always been a research hotspot. Common methods are mainly based on reducing reconstruction error or maximizing estimated probability for normal data, however, both of them may be sensitive to particular fluctuations in data. Meanwhile, these methods tend to model temporal dependency or spatial correlation individually, which is insufficient to detect diverse anomalies. In this paper, we propose an error-restricted framework with variance estimation, namely Spatial-Temporal Anomaly Transformer (S-TAR), which can provide a corresponding confidence for each reconstruction. First, it presents Error-Restricted Probability (ERP) loss by restricting the reconstruction error and its estimated probability skillfully, further improving the capability to distinguish outliers from normal data. Second, we adopt Spatial-Temporal Transformer with distinct attention modules to detect diverse anomalies. Extensive experiments on five real-world datasets are conducted, the results show that our method is superior to existing state-of-the-art approaches.

1 Introduction

With the accelerated development of Industrial 4.0 era, it is imperative to conduct proper management and control of cyber-physical system (CPS) [14]. To monitor the status of CPS effectively, complex multivariate time series (MTS) are collected by multiple sensors in real time [9]. Generally, MTS displays complex non-linear dynamics, the deviations from normal patterns may imply operational failures (i.e. anomalies) or even potential risks. To prevent these risks, anomaly detection has been studied for a long time [18,19].

As malfunctions are infrequent and the data labeling is costly, anomaly detection is mainly required at unsupervised settings. Among past methods, autoencoder (AE) and its variants have been the foci in recent researches [3,15,23]. The core of such modules is to learn the generalized patterns from normal data, then isolating anomalies which deviate from regular patterns.

It is noteworthy that there are still two intractable issues for existing methods. 1) Considering that the sources and dynamics of different sensors could be

© The Author(s), under exclusive license to Springer Nature Singapore Pte Ltd. 2024
D.-N. Yang et al. (Eds.): PAKDD 2024, LNAI 14645, pp. 3–14, 2024.
https://doi.org/10.1007/978-981-97-2242-6_1

distinct, the difficulty of reconstruction may differ significantly. For instance, the measurement of some actuators are digital while others are analog, and there are ineluctable uncertainties and noises in the measurements of some sensors. These characteristics can be easily ignored by most methods [8], resulting in some misjudgments consequentially. 2) The entity of MTS characterizes the temporal and spatial dependency natively. For each sensor, the dynamics usually present long-term or short-term periodic trend, and the interactions among different sensors reveal the inherent property of CPS. In practice, those deviations from normal patterns temporally or spatially may cause diverse anomalies, so it is insufficient for most methods to model from a single perspective. Some approaches attempt to extract fused spatiotemporal information, but the results are not ideal enough [6,15]. Furthermore, it has been pointed recently that, existing evaluation protocols could not provide an impartial assessment of its performance [11]. More rigorous evaluation should be considered.

To overcome the above flaws, we propose variance-based Spatial-Temporal Anomaly Transformer (S-TAR), which achieves reconstruction (or prediction) and variance estimation simultaneously. The estimated variance can quantify the degree of confidence for the residual error of reconstruction. First, a novel Error-Restricted Probability (ERP) loss is elaborately designed. It maximizes the estimated probability density of the reconstruction error while adaptively restricting the error to a certain range. In this way, the estimation of reconstruction difficulty shall be mainly considered, which is conducive to the discrimination of anomalies. ERP provides an unified perspective of the aforementioned training objectives, and it can be applied to different frameworks. Second, to extract the contextual information in terms of time and correlations among sensors, we present two separate Transformer-based models with distinct attention modules [26], and similarity limitation is utilized to reduce the overlapped information obtained from the two views.

Our main contributions are summarized as follows.

- We present a novel two-view framework S-TAR based on Transformer structure. With different attention modules, it can effectively extract the temporal and spatial dependencies in MTS.
- We propose Error-Restricted Probability loss to unsupervised anomaly detection. It can evidently improve the performance of different frameworks.
- Extensive experiments are performed on five public real-world datasets. Through a more rigorous evaluation, the results demonstrate the effectiveness of our model. S-TAR is superior in both F1 with PA and without it than previous methods.

2   Related Work

Anomaly detection is vital to various CPS systems. Many approaches have been applied to model the intra characteristics of MTS using classical techniques [24], here we highlight recent deep learning researches at unsupervised settings.

They are classified into two categories: error-based methods and probabilistic approaches.

Error-based methods focus on reducing the reconstruction or forecasting errors, which is utilized to quantify the degree of abnormality. To model the temporal or spatial associations, RNNs and GNNs have been leveraged to acquire useful representations, like NSIBF and GDN [7,8,22]. As the errors are sensitive to noises, some researchers argued to amplify such errors of anomalies. For instance, with one shared encoder and two different decoders, USAD and TranAD are both trained to enhance the reconstruction error in an adversarial manner [3]. However, the improvements are limited to a certain degree.

Probabilistic methods maximize the observed probability of input series, and the measurements with low estimated probability values are treated as anomalies. Some methods combine AE and statistical models to derive the probability estimation, like DAGMM [28]. Recently, stochastic models like Variational Autoencoders (VAE) [13] and Normalizing Flow (NF) [21] have attracted aboard attention. LSTM-VAE integrates LSTM with VAE to make variational inference for reconstruction [20]. It is also extended by OmniAnomaly [23], which employs VAE and Planar NF to acquire more robust representations. Besides, to learn both temporal and spatial embeddings, InterFusion applies the updated hierarchical VAE to capture normal patterns of data [15], but it still cannot address the issue caused by imbalanced data distribution.

In our work, we combine the tasks of reconstruction and probability observation by introducing adaptive variance estimation, which enable more robust estimated probability values for anomaly detection.

3 Approach

3.1 Problem Formulation

MTS are composed of multiple univariate sequences, which are continuously collected from specific sensors. Throughout the paper, denote s and S as the serial number and number of sensors respectively, t and T denote the serial number and number of timestamps respectively. From temporal perspective, MTS has the form as:

$$X = \{x_1, \cdots, x_T\}, \tag{1}$$

where $x_t \in \mathbb{R}^S$ is the data from all sensors at timestamp t. To achieve point-by-point anomaly detection, anomaly scores will be assigned to all x_t. If the score of x_t is greater than a threshold, then it is considered anomalous.

3.2 Proposed Architecture

As shown in Fig. 1, the proposed S-TAR consists of two Transformer-based networks: Spatial Transformer and Temporal Transformer, which share a similar architecture of an encoder(called Temporal Encoder and Spatial Encoder, respectively) and two decoders. By modeling the temporal and spatial correlations

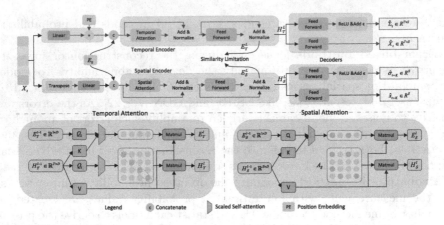

Fig. 1. Overall framework of the proposed variance-based Spatial-Temporal Anomaly Transformer (S-TAR).

respectively, the contextual information H_T^L and H_S^L can be acquired. Further, a shared non-contextual vector E_0 is employed to obtain the final aggregate representation E_T^L and E_S^L from distinct perspectives. To reduce the information overlap of the two encoders, we propose a similarity limitation technique skillfully. For the decoder parts, one aims at either achieving reconstruction (for Temporal Transformer) or prediction (for Spatial Transformer), while the other outputs the corresponding variance, which shall reflect the confidence of the former result.

Temporal Transformer. Similar to vanilla Transformer, the temporal encoder stacks the multi-head scaled self-attention and feed-forward layers alternately. For the hidden variable $H_T^l \in \mathbb{R}^{T \times D}$ of the l_{th} layer, the formulation of attention for each head is as follows.

$$\mathcal{Q}, \mathcal{K}, \mathcal{V} = H_T^l W_{\mathcal{Q}}^l, H_T^l W_{\mathcal{K}}^l, H_T^l W_{\mathcal{V}}^l,$$
$$\mathcal{A}_T(\mathcal{Q}, \mathcal{K}, \mathcal{V}) = \text{Softmax}(\text{Mask}(\frac{\mathcal{Q}\mathcal{K}^*}{\sqrt{D}}))\mathcal{V}, \tag{2}$$

where $\mathcal{Q}, \mathcal{K}, \mathcal{V}$ denote the query, key and value respectively, $W_{\mathcal{Q}}^l, W_{\mathcal{K}}^l, W_{\mathcal{V}}^l$ are learnable parameters. Mask will set the diagonal elements of a matrix to $-\infty$. In this way, for the input variable at each timestamp, the attention calculation is performed by covering up its own information of the input timestamp, which enables the representation to only fuse bidirectional contextual information. In addition, a non-contextual vector $E_0 \in \mathbb{R}^{1 \times D}$ is fed into the temporal encoder:

$$\mathcal{A}_E = \text{Softmax}(\frac{E_T^l W_E^l \mathcal{K}^*}{\sqrt{D}})\mathcal{V}, \tag{3}$$

where $E_T^0 = E_0$, W_E^l is a learnable parameter. Associated with the temporal features, the output variable E_T^L is expected as the aggregate representation of the input time window. Afterwards, the other output of temporal encoder H_T^L will be sent to two different decoders for reconstruction and variance estimation. It is formulated as

$$\hat{X}_t = \text{FeedForward}(H_T^L),$$
$$\hat{\Sigma}_t = \text{ReLU}(\text{FeedForward}(H_T^L)) + \epsilon.$$

(4)

where $0 < \epsilon \ll 1$ is a small value.

Spatial Transformer. Unlike temporal correlations, the spatial dependencies among sensors tend to be relatively fixed. Thus, we directly use an adaptive graph generation module to replace the original attention. Specifically, we utilize the Hermitian and skew-Hermitian splitting (HSS) method [5] to model the bidirectional and unidirectional correlations simultaneously. It defines that

$$A_S = \text{ReLU}(P + P^*) + \text{ReLU}(P - P^*),$$

(5)

where $P \in \mathbb{R}^{S \times S}$. As $H(P) = P + P^*$ is a Hermitian matrix, $\text{ReLU}(H(P))$ sets the negative elements to zero and resumes its Hermitian form, so it can represents the bidirectional dependency between two sensors. Besides, $S(P) = P - P^*$ is a skew-Hermitian matrix, then $\text{ReLU}(S(P))$ transforms half the elements of $S(P)$ to zero, which only reserves the unidirectional correlations. To reduce computational complexity, we define $P = MN^*$ in this work, where $M, N \in \mathbb{R}^{S \times F_0}$ ($F_0 \ll S$) are learnable parameters. The generated adjacency matrix A_S can be directly used in the Spatial Attention mechanism as follows.

$$\mathcal{A}_S(A_S, \mathcal{V}) = \text{Softmax}(A_S)\mathcal{V}.$$

(6)

The process of E_0 in Spatial Encoder is the same as that in Temporal Encoder, and we can get another aggregate representation E_S^L of the input window from spatial persepective.

Similarity Limitation. To reduce the overlap of temporal and spatial information extracted from the two transformer-based modules, we propose to minimize the absolute value of cosine similarity between the aggregate representations E_T^L and E_S^L. It is called the similarity loss:

$$\mathcal{L}_{sim} = \frac{|(E_T^L)^* \ E_S^L|}{||E_T^L|| \cdot ||E_S^L||}.$$

(7)

3.3   Error-Restricted Probability (ERP) Loss

In this section, we introduce our Error-Restricted Probability loss, which restricts the reconstruction or forecasting error (called error for short) and its estimated probability density in an inter-limitation manner.

Denote $\hat{x}_{ts}$ as the reconstruction or prediction result of sensor s at timestamp t, $\ddot{x}_{ts} = \hat{x}_{ts} - x_{ts}$, the error-based loss can be generally formulated by Mean Squared Error (MSE):

$$\mathcal{L}_{error} = \sum_{t=1}^{T} \sum_{s=1}^{S} (\ddot{x}_{ts})^2.$$

(8)

From probabilistic perspective, it is equivalent to maximize the probability density value of

$$\ddot{x}_{ts} \sim \mathcal{N}(0, a^2), \quad t = 1, \cdots T; s = 1, \cdots, S, \tag{9}$$

where $a > 0$ is a constant, $\mathcal{N}$ denotes the Gaussian distribution. This ignores a fact that, due to noises and fluctuations, the difficulty for reconstruction or prediction may vary significantly for different t and s. Probability density loss can alleviate this problem by adaptively assigning a.

Probability Density Loss. We assume that

$$\ddot{x}_{ts} \sim \mathcal{N}(0, \sigma_{ts}^2), \quad t = 1, \cdots, T; s = 1, \cdots, S. \tag{10}$$

Here, the optimization target is to maximize the probability density value $f(\ddot{x}_{ts})$. Set $\hat{\sigma}_{ts} = \frac{1}{\sigma_{ts}^2}$, by taking the negative logarithm, the probability density loss $\mathcal{L}_p = -\sum_{t=1}^{T} \sum_{s=1}^{S} f(\ddot{x}_{ts})$ can be transformed to

$$\mathcal{L}_p = \sum_{t=1}^{T} \sum_{s=1}^{S} (\hat{\sigma}_{ts} \cdot (\ddot{x}_{ts})^2 - \ln \hat{\sigma}_{ts}). \tag{11}$$

Intuitively, minimizing $\mathcal{L}_p$ includes two targets: reducing the reconstruction error $\ddot{x}_{ts}$, and providing an optimal value $\hat{\sigma}_{ts}$ to enhance the probability density of $\ddot{x}_{ts}$. However, due to the characteristic of Gaussian distribution, $f(\ddot{x}_{ts})$ goes to infinity when σ_{ts} and $\ddot{x}_{ts}$ both approach to 0. Therefore, for minimizing the probability density loss (11), the model will be biased towards reducing the errors $\ddot{x}_{ts}$ and deriving corresponding variances σ_{ts} close to 0 inevitably. Then, for some particular fluctuations that are hard to reconstruct or predict, the estimated variances loss its role to reflect this difficulty.

To tackle the mentioned issues, we argue to put more emphasis on providing proper variance $\hat{\sigma}_{ts}$ for each $\ddot{x}_{ts}$, rather than relying primarily on reducing reconstruction error. Specifically, a detached strategy is proposed to make the estimated variance $\hat{\sigma}_{ts}$ more credible. We derive a detached form of the probability density loss as

$$\mathcal{L}_{dp} = \sum_{t=1}^{T} \sum_{s=1}^{S} (\hat{\sigma}_{ts} \cdot (\ddot{x}^{ts}.detach)^2 - \ln \hat{\sigma}_{ts}). \tag{12}$$

where detach denotes stop-gradient. Obviously, $\mathcal{L}_{dp}$ only focuses on the estimation of $\hat{\sigma}_{ts}$, thus completely ignoring the effects of residual error, resulting in a loss of the reconstruction or prediction capability. To enable this capability of S-TAR simultaneously, based on $\mathcal{L}_{dp}$ and $\mathcal{L}_{error}$, we provide Error-Restricted probability loss that combining both the two objectives.

ERP Loss. The ERP loss function is expressed as

$$\mathcal{L}_{erp} = |\mathcal{L}_{dp} + \lambda * \mathcal{L}_{error}| + 2 * \mathcal{L}_{dp}. \tag{13}$$

The first half of $\mathcal{L}_{erp}$ includes reconstruction error and the detached probability density value of it, which aims at achieving a balance between the two relevant components. It practically makes $\mathcal{L}_{dp} + \lambda * \mathcal{L}_{error}$ close to 0, and then keeps the overall reconstruction error within a certain range. For the second half of $\mathcal{L}_{erp}$, it is designed to maximize the detached probability density value of reconstruction error.

According to the value of parameter λ, it can be divided into three cases: i) when $\lambda > 0$, $\mathcal{L}_{error}$ will not be continuously reduced but stabilized gradually, making the probability density value of each residual error more comparable; ii) when $\lambda = 0$, $\mathcal{L}_{erp}$ is equal to $\mathcal{L}_{dp}$; iii) when $\lambda < 0$, it is approximate to $\mathcal{L}_{error}$. In our experiment, we set λ as a positive integer.

Overall, the interaction of the two parts in $\mathcal{L}_{erp}$ promotes the reconstruction capability of model, and can better match the optimal variance of the reconstruction error. Combining the practice condition, we design an additive objective to train Spatial transformer and Temporal transformer concurrently but independently.

Final Loss. Denote $\mathcal{L}_{temporal}$ and $\mathcal{L}_{spatial}$ as the ERP loss for the Temporal Transformer and Spatial Transformer respectively, the final loss can be formulated as:

$$\mathcal{L}_{final} = \mathcal{L}_{temporal} + \mathcal{L}_{spatial} + \mathcal{L}_{sim}. \tag{14}$$

3.4   Anomaly Score

First, the estimated probability density is used to compute anomaly score for each timestamp t and sensor s:

$$S(t,s) = \hat{\sigma}_{ts} \cdot (\ddot{x}_{ts})^2 - \ln \hat{\sigma}_{ts}, \tag{15}$$

then the anomaly scores are rescaled by a robust normalization for each sensor.

$$\widetilde{S}(t,s) = \frac{S(t,s) - \text{Median}(s)}{\text{IQR}(s)}, \tag{16}$$

where $\text{Median}(s)$ and $\text{IQR}(s)$ denote the median and inter-quartile range of Score$(\cdot, s)$ respectively. Further, to capture the deviations from different perspectives, we adopt a weighted-sum to combine the anomaly scores derived from two Transformer-based modules:

$$\text{Score}(t,s) = \alpha \widetilde{S}_T(t,s) + (1 - \alpha)\widetilde{S}_S(t,s), \tag{17}$$

where $\alpha \in [0,1]$ is a hyperparameter, $\widetilde{S}_T$ and $\widetilde{S}_S$ are scores from Temporal Transformer and Spatial Transformer respectively. Finally, the anomalous degree at each timestamp are quantified by the maximum of Score$(t, \cdot)$.

4   Experiments

4.1   Dataset Description

Five public datasets for anomaly detection are adopted to evaluate the performance of S-TAR: 1) **SWaT** (Secure Water Treatment) [17] dataset consists of

Table 1. Performance of different models. F1 and F1$_{PA}$ is the best F1-Score before and after point adjustment, respectively. Bold indicates the best results, and the last column shows the number of best results for all models on five datasets.

Model	SWaT		WADI		PSM		MSL		SMD		count
	F1	F1$_{PA}$	F1	F1$_{PA}$	F1	F1$_{PA}$	F1	F1$_{PA}$	F1	F1$_{PA}$	
IsolationForest	0.7457	0.8619	0.3004	0.6938	0.5030	0.9219	0.3118	0.6748	0.3150	0.8962	0
LSTM-VAE	0.7091	0.9013	0.3619	0.7364	0.5317	0.9499	0.3666	0.9248	0.3674	0.9361	0
DAGMM	0.5047	0.8569	0.1596	0.6165	0.4616	0.9549	0.3669	0.7608	0.2685	0.8547	0
BeatGAN	0.6773	0.8219	0.2545	0.3865	**0.6382**	0.7805	**0.4422**	0.9400	0.3886	0.8713	2
OmniAnomaly	0.7822	0.8283	0.1151	0.5021	0.4345	0.8759	0.3554	0.8953	0.2860	0.9415	0
USAD	0.7492	0.8227	0.2502	0.3465	0.4778	0.7048	0.4189	0.8536	0.3107	0.8044	0
NSIBF	0.7778	0.9190	0.1316	0.2364	0.5726	0.9218	0.3608	0.7503	**0.4527**	0.8182	1
GDN	0.8000	0.9256	**0.5071**	0.9001	0.5498	0.9079	0.3044	0.8404	0.4281	0.9153	1
TranAD	0.7461	0.8279	0.2639	0.3804	0.6048	0.8066	0.4204	0.8331	0.3865	0.8964	0
Anomaly Transformer	0.0710	**0.9407**	0.1021	0.5297	0.0912	0.9789	0.1203	0.9359	0.9022	0.9233	1
S-TAR	**0.8380**	0.9309	0.4816	**0.9101**	0.5595	**0.9853**	0.4150	**0.9535**	0.4224	**0.9491**	5

the recordings from 51 sensors in a modern industry control system (ICS). It is collected by continuous sampling at the frequency of 1 s. 2) **WADI** (Water Distribution) [2] testbed is similar to SWaT. It is acquired from a larger city system with 123 sensors which operates for about half a month. 3) **PSM** (Pooled Server Metrics) [1] expresses the properties of multiple application server nodes from eBay, including CPU utilization, memory and etc. 4) **MSL** (Mars Science Laboratory) [10] rover dataset is released by NASA. It has 27 different entities in total, and each of them contains anomaly ground truths derived from the Incident Surprise Anomaly (ISA) reports. 5) **SMD** (Server Machine Dataset) [23] is built by a large Internet company, which is analogous to PSM.

4.2   Baseline Methods

Ten prevalent methods for anomaly detection are selected for comparison. BeatGAN [27], USAD [3], GDN [7] and TranAD [25] belong to error-based approaches, LSTM-VAE [20], DAGMM [28] and OmniAnomaly [23] are probabilistic models. Besides, TranAD, LSTM-VAE, OmniAnomaly, NSIBF and Anomaly Transformer mainly aim to capture temporal associations of MTS data, while GDN is a GNN-based framework to model spatial dependencies. A classical method called Isolation Forest [16] is also compared in this paper.

4.3   Experimental Settings

All of the experiments are executed by Pytorch 1.7 with one NVIDIA TITAN X Pascal 12GB GPU. During training, Adam [12] with an initial learning rate of $5e^{-4}$ is leveraged as the optimizer. For the encoders in Temporal Transformer and Spatial Transformer, the number of layer and head are set as 1 and 4 respectively. The sliding window size is 12 for SWaT and WADI, while it is 60 for other datasets. The value of λ is dictated by grid search in range 1 to 10. For all models, the best threshold is obtained through equispaced search.

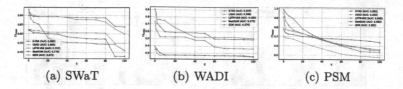

<table>
<tr><td></td><td>(a) SWaT</td><td>(b) WADI</td><td>(c) PSM</td></tr>
</table>

Fig. 2. F1 score with PA%K with varying K for three datasets. AUC is the area under the corresponding curve.

Table 2. The performance of three base frameworks with different training objective. Here, $* +$ Error indicates that it is trained by $\mathcal{L}_{error}$, $* +$ Pd corresponds to $\mathcal{L}_p$, and $* +$ ERP corresponds to our proposed training objective.

Model	SWaT		WADI	
	F1	F1$_{PA}$	F1	F1$_{PA}$
LSTM + Error	0.8049	0.8873	**0.4694**	0.6165
LSTM + Pd	0.8014	0.8995	0.3801	0.6288
LSTM + ERP	**0.8233**	**0.9371**	0.4217	**0.7545**
TCN + Error	0.8151	0.8896	0.1754	0.6384
TCN + Pd	0.8201	0.9034	0.1470	0.2247
TCN + ERP	**0.8246**	**0.9227**	**0.4743**	**0.7229**
Transformer + Error	0.7551	0.8682	0.3637	0.5829
Transformer + Pd	0.7888	0.8921	0.2976	**0.6804**
Transformer + ERP	**0.8243**	**0.9111**	**0.4627**	0.6480

For robust comparison, we adopt F1-Score before and after point adjustment (PA) as the evaluation metrics. PA can be illustrated as: if at least one timestamp in a successive anomalous segment S_k is detected, the detection results of all the timestamps in this segment are set as the ground truth. Along with them, the AUC metric based on PA%K [11] is also employed, which can be formulated as:

$$\hat{y}_t = \begin{cases} 1, & \text{if } t \in S_k \text{ and } \frac{|\{t'|t' \in S_k, \text{Score}(x_{t'}) > \delta\}|}{|S_k|} > K\%, \\ 0, & \text{otherwise}, \end{cases} \quad (18)$$

where $|\mathcal{C}|$ indicates the size of $\mathcal{C}$, $K \in [0, 100)$ is the detection accuracy in S_k.

4.4 Overall Results

The performance comparisons are shown in Table 1. For 5 public datasets, S-TAR achieves the best F1 score for SWaT and the best F1$_{PA}$ for the others, which confirms its superiority. Our proposed method is more capable to achieve satisfactory results under both of F1 and F1$_{PA}$ than others. On average, the F1 and F1$_{PA}$ of S-TAR are 0.5433 and 0.9458 respectively, which outperforms the other tested models (the second are 0.5275 for GDN and 0.8897 for LSTM-VAE).

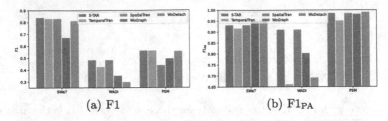

(a) F1   (b) F1$_{PA}$

Fig. 3. Performance comparison of different variants. SpatialTran and TemporalTran denotes the Spatial Transformer and Temporal Transformer respectively, WoDetach indicates that detach in the proposed training objective is removed, and WoGraph implies that the HSS method is replaced by the multi-head attention.

Figure 2 shows the curves of F1$_{PA\%K}$ with varying K for different methods and datasets. When K = 0 and K = 100, it is equal to F1$_{PA}$ and F1 respectively. AUC denotes the area under the curve, and it is the balanced result between F1 and F1$_{PA}$. Apparently, S-TAR achieves the best AUC for SWaT and PSM, and the second best for WADI. The curves also reveal the unreliability using only one of the F1 and F1$_{PA}$. For example, LSTM-VAE can acquire a relatively high F1$_{PA}$ on SWaT and PSM datasets, but it does not perform well in AUC.

4.5 Ablation Study

Generality of ERP Loss. Three base frameworks are tested to evaluate the universality of our proposed ERP loss, including LSTM, Temporal Convolutional Network (TCN) [4] and Transformer (with temporal attention). Concretely, reconstruction error (error for short) $\mathcal{L}_{error}$, probability density loss (Pd for short) $\mathcal{L}_p$ and the ERP loss $\mathcal{L}_{erp}$ are utilized as the training objective for comparison. As shown in Table 2, the ERP loss significantly improves the performance of different frameworks on SWaT and WADI datasets, especially for Transformer. Compared to the reconstruction error, $\mathcal{L}_{erp}$ can yield at most about 30% improvement for F1, and 14% for F1$_{PA}$. This strongly demonstrates the superiority and generality of our proposed training objective.

Effect of Submodules. We further investigate the influence of some special submodules and modifications on performance. To show the necessity of considering both temporal and spatial associations, the results achieved by Temporal Transformer and Spatial Transformer are given respectively. In Spatial Transformer, we compare the HSS method with the original multi-head attention of Transformer. The influence of detach in the proposed ERP loss is also presented in Fig. 3.

5   Conclusion

In this paper, we propose Spatial-Temporal Anomaly Transformer (S-TAR) with error-restricted variance estimation for unsupervised anomaly detection. The robustness of our proposed ERP loss is verified for different frameworks and various CPS systems, and we combine two Transformer-based modules technically to detect diverse anomalies from different perspectives. Experimental results on five benchmark datasets demonstrate the superiority of S-TAR compared with previous state-of-the-art methods.

References

1. Abdulaal, A., Liu, Z., Lancewicki, T.: Practical approach to asynchronous multivariate time series anomaly detection and localization, pp. 2485–2494. Association for Computing Machinery (2021)
2. Ahmed, C.M., Palleti, V.R., Mathur, A.P.: Wadi: a water distribution testbed for research in the design of secure cyber physical systems, pp. 25–28. Association for Computing Machinery (2017)
3. Audibert, J., Michiardi, P., Guyard, F., Marti, S., Zuluaga, M.A.: Usad: unsupervised anomaly detection on multivariate time series. In: Proceedings of the 26th ACM SIGKDD Conference on Knowledge Discovery & Data Mining, pp. 3395–3404 (2020)
4. Bai, S., Kolter, J.Z., Koltun, V.: An empirical evaluation of generic convolutional and recurrent networks for sequence modeling (2018)
5. Bai, Z.Z., Golub, G.H., Ng, M.K.: Hermitian and skew-hermitian splitting methods for non-hermitian positive definite linear systems. SIAM J. Matrix Anal. Appl. **24**(3), 603–626 (2003)
6. Dai, E., Chen, J.: Graph-augmented normalizing flows for anomaly detection of multiple time series. In: Proceedings of the 10th International Conference on Learning Representations (2022)
7. Deng, A., Hooi, B.: Graph neural network-based anomaly detection in multivariate time series. In: Proceedings of the AAAI Conference on Artificial Intelligence, vol. 35, pp. 4027–4035 (2021)
8. Feng, C., Tian, P.: Time series anomaly detection for cyber-physical systems via neural system identification and bayesian filtering. In: Proceedings of the 27th ACM SIGKDD Conference on Knowledge Discovery & Data Mining, pp. 2858–2867 (2021)
9. Goh, J., Adepu, S., Tan, M., Lee, Z.S.: Anomaly detection in cyber physical systems using recurrent neural networks. In: Proceedings of the 18th International Symposium on High Assurance Systems Engineering (HASE), pp. 140–145 (2017)
10. Hundman, K., Constantinou, V., Laporte, C., Colwell, I., Soderstrom, T.: Detecting spacecraft anomalies using lstms and nonparametric dynamic thresholding, pp. 387–395. Association for Computing Machinery (2018)
11. Kim, S., Choi, K., Choi, H.S., Lee, B., Yoon, S.: Towards a rigorous evaluation of time-series anomaly detection (2022)
12. Kingma, D.P., Ba, J.: Adam: a method for stochastic optimization (2014)
13. Kingma, D.P., Welling, M.: Auto-encoding variational bayes (2013)
14. Lee, J., Bagheri, B., Kao, H.A.: A cyber-physical systems architecture for industry 4.0-based manufacturing systems. Manufact. Lett. **3**, 18–23 (2015)
15. Li, Z., et al.: Multivariate time series anomaly detection and interpretation using hierarchical inter-metric and temporal embedding. In: Proceedings of the 27th ACM SIGKDD Conference on Knowledge Discovery & Data Mining, pp. 3220–3230. Association for Computing Machinery (2021)

16. Liu, F.T., Ting, K.M., Zhou, Z.H.: Isolation forest. In: Proceedings of the Eighth IEEE International Conference on Data Mining (ICDM)m pp. 413–422 (2008)
17. Mathur, A.P., Tippenhauer, N.O.: SWaT: a water treatment testbed for research and training on ics security. In: Proceedings of 2016 International Workshop on Cyber-physical Systems for Smart Water Networks, pp. 31–36 (2016)
18. Omar, S., Ngadi, A., Jebur, H.H.: Article: machine learning techniques for anomaly detection: an overview. Inter. J. Comput. Appli. **79**(2), 33–41 (2013)
19. Pang, G., Shen, C., Cao, L., Hengel, A.V.D.: Deep learning for anomaly detection: a review, vol. 54(2) (mar 2021)
20. Park, D., Hoshi, Y., Kemp, C.C.: A multimodal anomaly detector for robot-assisted feeding using an lstm-based variational autoencoder. IEEE Robot. Automat. Lett. **3**(3), 1544–1551 (2018)
21. Rezende, D., Mohamed, S.: Variational inference with normalizing flows. In: Proceedings of the 32nd International Conference on Machine Learning, pp. 1530–1538. PMLR, Lille, France (2015)
22. Shen, L., Yu, Z., Ma, Q., Kwok, J.T.: Time series anomaly detection with multiresolution ensemble decoding. In: Proceedings of the AAAI Conference on Artificial Intelligence, vol. 35, pp. 9567–9575 (2021)
23. Su, Y., Zhao, Y., Niu, C., Liu, R., Sun, W., Pei, D.: Robust anomaly detection for multivariate time series through stochastic recurrent neural network. In: Proceedings of the 25th ACM SIGKDD Conference on Knowledge Discovery & Data Mining, pp. 2828–2837 (2019)
24. Tax, D.M.J., Duin, R.P.W.: Support vector data description, vol. 54, pp. 45–66 (2004)
25. Tuli, S., Casale, G., Jennings, N.R.: TranAD: deep transformer networks for anomaly detection in multivariate time series data. In: Proceedings of Very Large Data Bases, vol. 15, pp. 1201–1214 (2022)
26. Vaswani, A., et al.: Attention is all you need. In: Proceedings of the 31st International Conference on Neural Information Processing Systems, vol. 30, pp. 6000–6010. Curran Associates, Inc. (2017)
27. Zhou, B., Liu, S.H., Hooi, B., Cheng, X., Ye, J.: BeatGAN: anomalous rhythm detection using adversarially generated time series. In: International Joint Conference on Artificial Intelligence (2019)
28. Zong, B., et al.: Deep autoencoding gaussian mixture model for unsupervised anomaly detection. In: International Conference on Learning Representations (2018)

Multi-task Contrastive Learning for Anomaly Detection on Attributed Networks

Junjie Zhang and Yuxin Ding[✉]

Harbin Institute of Technology (Shenzhen), Shenzhen, China
yxding@hit.edu.cn

Abstract. Anomaly detection on attributed networks is a vital task in graph data mining and has been widely applied in many real-world scenarios. Despite the promising performance, existing contrastive learning-based anomaly detection models still suffer from a limitation: the lack of fine-grained contrastive tasks tailored for different anomaly types, which hinders their capability to capture diverse anomaly patterns effectively. To address this issue, we propose a novel multi-task contrastive learning framework that jointly optimizes two well-designed contrastive tasks: context matching and link prediction. The context matching task identifies contextual anomalies by measuring the congruence of the target node with its local context. The link prediction task fully exploits self-supervised information from the network structure and identifies structural anomalies by assessing the rationality of the local structure surrounding target nodes. By integrating these two complementary tasks, our framework can more precisely identify anomalies. Extensive experiments on four benchmark datasets demonstrate that our method achieves considerable improvement compared to state-of-the-art baselines.

Keywords: Anomaly detection · Graph neural networks · Contrastive learning

1 Introduction

Attributed networks, as graph-structured data, are prevalent in various real-world scenarios, such as social networks and financial transaction networks. Anomaly detection on attributed networks is a crucial task that aims to identify nodes that significantly deviate from the normal behavior or patterns of the majority. Recently, this topic has attracted considerable attention from scholars for its broad applications in various areas, such as social spam detection [3], financial fraud detection [18], and log anomaly detection [21]. According to real-world anomaly patterns, anomalies can be categorized into two major types: *contextual anomalies* and *structural anomalies* [8]. Contextual anomalies refer to nodes whose attributes are significantly different from their neighboring nodes. Structural anomalies are a cluster of densely connected nodes or nodes that are connected to multiple different communities [8,10].

© The Author(s), under exclusive license to Springer Nature Singapore Pte Ltd. 2024
D.-N. Yang et al. (Eds.): PAKDD 2024, LNAI 14645, pp. 15–26, 2024.
https://doi.org/10.1007/978-981-97-2242-6_2

Considering the intricacy and mutability of anomalies in real-world networks, manually labeling nodes is labor-intensive. Consequently, the majority of existing anomaly detection methods [1,7,13,20] are unsupervised. These methods can be broadly categorized into traditional methods and deep learning-based methods. Traditional methods typically employ machine learning techniques for anomaly detection, such as density estimation [1], structural clustering [20], residual analysis [6], and matrix decomposition techniques [12]. However, these methods struggle to effectively incorporate both structural and attribute information for anomaly detection. Recently, Graph Neural Networks (GNNs) have emerged as the preferred tool for deep anomaly detection methods [2,4,9,22], owing to their strong capabilities in learning node representations. Specifically, the current deep methods can be classified into autoencoder-based methods and contrastive learning-based methods. DOMINANT [2] is a representative autoencoder-based model that detects anomalies by assessing errors in attribute and network structure reconstruction. Nevertheless, the process of reconstructing the entire network consumes substantial memory, restricting its applicability to large-scale networks. Additionally, its reconstruction objective is not specifically tailored for anomaly detection tasks. In contrast, contrastive learning-based models [4,9,22] have the potential to address these issues. CoLA [9] initially introduces the paradigm of contrastive learning into the task of anomaly detection, identifying anomalies by analyzing the relationship between target nodes and their corresponding local subgraphs. This model samples small-scale subgraphs for batch training, which reduces memory consumption while maintaining good scalability. Subsequent works [4,22] have further enhanced the performance of contrastive learning-based models through the implementation of multi-scale or multi-view contrastive learning approaches.

Although existing contrastive learning-based models [4,9,22] have demonstrated promising performance, several issues merit further exploration. First, these models generally lack contrastive tasks specifically designed for different types of anomalies, limiting their ability to capture diverse anomaly patterns effectively. Second, the potential of leveraging self-supervised information from topological structures to enhance anomaly detection is underexplored. Third, these models often employ the random walk with restart (RWR) [15] to sample local subgraphs around target nodes. However, this approach can lead to the sampling of many duplicate nodes, resulting in redundant contextual information and an increased risk of information leakage about target nodes. These models initially sample an excessive number of nodes before removing duplicates, consequently increasing both time and space complexity.

To address these issues, we propose a Subgraph-based multi-task contrastive learning framework, which consists of two well-designed contrastive learning modules (corresponding to two tasks): Context matching module and Link prediction module (SCL for convenience). In our framework, the original graph is augmented using a novel subgraph sampling strategy, generating contrastive views for the two contrastive modules. In the context matching module, an innovative attention-based center generator is utilized to adaptively capture the contextual information of target nodes. Furthermore, anomalies can be identified by assessing the match between nodes and their surrounding context. Simultaneously, the link prediction module fully leverages self-supervised information from the network structure and identifies anomalies by evaluating the rationality of the target nodes' local structure. By jointly optimizing these modules, SCL can effectively capture diverse anomaly patterns. Finally, the two complementary modules are integrated for anomaly detection.

Our main contributions are summarized as follows: (1) We propose a novel multi-task contrastive learning framework for anomaly detection on attributed networks that can effectively capture multiple abnormal patterns. (2) We design two specialized contrastive modules: the context matching module and the link prediction module, which focus on detecting contextual and structural anomalies, respectively. (3) We propose a novel subgraph sampling strategy for generating high-quality contrastive views. (4) We conduct comprehensive experiments on standard datasets to evaluate our proposed framework. The results demonstrate that our framework outperforms existing state-of-the-art methods.

2   Problem Definition

Notations. We use bold uppercase letters (e.g., $\mathbf{A}$), bold lowercase letters (e.g., $\mathbf{x}$), and standard lowercase letters (e.g., α) to denote matrices, vectors, and scalars, respectively. Let $\mathcal{G} = (\mathbf{A}, \mathbf{X})$ be an input attributed network with a node set $\mathcal{V} = \{v_1, \ldots, v_N\}$, where N is the number of the nodes in the graph. The binary adjacency matrix is denoted by $\mathbf{A} \in \mathbb{R}^{N \times N}$ and the attribute matrix is represented as $\mathbf{X} \in \mathbb{R}^{N \times F}$, with $\mathbf{x}_i \in \mathbb{R}^{F}(i = 1, \ldots, N)$ indicating the attribute vector of v_i. In this paper, our focus is on the problem of node-level anomaly detection in attributed networks, as defined below.

Definition 1. *Node-level Anomaly Detection on Attributed Networks. Given an attributed network $\mathcal{G} = (\mathbf{A}, \mathbf{X})$, the node-level anomaly detection aims to identify nodes that deviate significantly from the majority in both attribute and structure. The objective is to learn an anomaly score evaluator $score(\cdot)$: $\mathbb{R}^{N \times N} \times \mathbb{R}^{N \times F} \rightarrow \mathbb{R}^{N}$ to measure the degree of the abnormality for each node. Anomalous nodes are expected to obtain higher scores compared to normal nodes.*

3   The Proposed Framework

In this section, we introduce each component of the proposed framework SCL. As shown in Fig. 1, SCL consists of three major modules: the data augmentation module, the context matching contrastive learning module, and the link prediction contrastive learning module. Initially, we employ a subgraph sampling-based data augmentation strategy to generate instance pairs for contrastive learning. The context matching module is designed to identify contextual anomalies by evaluating the degree of match between a node and its context. In this module, we introduce an attention-based center generator that adaptively captures the context information of the target nodes. By leveraging attention mechanism, the model can effectively identify anomalies based on contextual information. In the link prediction module, we focus on identifying structural anomalies. This module assesses the rationality of the local structure surrounding the target nodes, leveraging the self-supervised learning signals provided by the network structure. Ultimately, the collaborative interplay of these three modules substantially enhances the detection of anomalies.

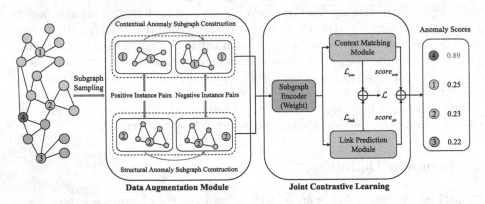

Fig. 1. Overall architecture of SCL. The symbol $\oplus$ represents the weighted summation of the outputs from both the context matching and the link prediction modules.

3.1   Subgraph Sampling Based Data Augmentation

The generation of contrastive views via data augmentation strategy is a key element in the design of contrastive learning-based models. Recent studies [4, 9, 22] have shown that exploring the relationships between nodes and their local subgraphs is promising for anomaly detection. These works commonly adopt random walk with restart (RWR) [15] to generate the local subgraph of each target node. However, a key issue with RWR is that the sampled subgraphs may contain duplicate nodes, which leads to redundant contextual information and the potential exposure of target node details. To address these issues, we propose a deduplicated random walk with restart (DRWR) strategy. Let $\mathcal{N}_t$ be the set of unsampled neighbors of the current node t, and $\mathcal{N}_c$ be the set of unsampled neighbors of the start node c, the sampling process can be described as follows:

$$
\pi(t, x) = \begin{cases} p \times \frac{1}{|\mathcal{N}_t|}, & \text{if } x \in \mathcal{N}_t \\ (1-p) \times \frac{1}{|\mathcal{N}_c|}, & \text{if } x \in \mathcal{N}_c \\ 0, & \text{otherwise,} \end{cases} \tag{1}
$$

where $\pi(t, x)$ is the probability transition function, x is the potential next node, p represents the random walk probability and $1-p$ controls the restart probability. As p increases in the DRWR algorithm, distant nodes are more likely to be sampled, enriching the subgraph with higher-order neighbors; conversely, a lower p focuses on local neighbors.

In our framework, given the input attributed network $\mathcal{G}$, we initially apply DRWR to sample a local subgraph $\mathcal{G}_i = (\mathbf{A}_i, \mathbf{X}_i)$ for each target node v_i. Here, $\mathbf{A}_i \in \mathbb{R}^{K \times K}$ is the adjacency matrix of $\mathcal{G}_i$, which contains K nodes including v_i, and $\mathbf{X}_i \in \mathbb{R}^{K \times F}$ is the corresponding attribute matrix. Notably, DRWR inherently avoids including duplicate nodes during the sampling process, eliminating the need for additional deduplication steps.

3.2 Context Matching Contrastive Learning

According to the homophily assumption, nodes in a network tend to connect with others that share similar features [11]. We refer to this behavior as context matching. However, this assumption may be weakened in networks containing contextual anomalies: normal nodes still tend to share common features with their normal neighbors, whereas contextual anomalies have different features from their neighbors. In light of this divergence, we introduce a context matching task that is specifically designed to detect contextual anomalies by evaluating how well a node matches its local context.

Subgraph Encoder. Our approach initiates by encoding the subgraph $\mathcal{G}_i$ corresponding to a target node v_i. In our work, we employ a one-layer Graph Convolutional Network (GCN) [5] as the subgraph encoder due to its effectiveness in representing graph data. The formulation of the GCN can be expressed as follows:

$$\mathbf{H}_i = \phi \left(\widetilde{\mathbf{D}}_i^{-\frac{1}{2}} \widetilde{\mathbf{A}}_i \widetilde{\mathbf{D}}_i^{-\frac{1}{2}} \mathbf{X}_i \mathbf{W} \right), \tag{2}$$

where $\widetilde{\mathbf{A}}_i = \mathbf{A}_i + \mathbf{I}$ denotes the adjacency matrix with self-loop of $\mathcal{G}_i$, $\widetilde{\mathbf{D}}_i$ is the degree matrix of $\widetilde{\mathbf{A}}_i$ and $\mathbf{W} \in \mathbb{R}^{F \times D}$ is a learnable weight matrix. ϕ is a non-linear activation function, such as ELU and ReLU. $\mathbf{H}_i$ is the learned node representation matrix. To prevent information leakage of the target node during the message aggregation process of GCN, we remove edges between the target node and its neighbor nodes, which we refer to as *center anonymization*.

Adaptive Center Generator. Drawing inspiration from GAT [16], we propose an attention-based center generator that can adaptively leverage contextual information of the target node to generate an ideal center node representation. Specifically, we first perform *self-attention* mechanism $att : \mathbb{R}^D \times \mathbb{R}^D \to \mathbb{R}$ on the center node to compute relationship coefficients and subsequently normalize them using the softmax function. This process is encapsulated in the following equation:

$$a_j = \frac{\exp \left(\text{LeakyReLU} \left(\mathbf{a}^{\mathrm{T}} [\mathbf{W}_\beta \mathbf{H}_{i,1} \| \mathbf{W}_\beta \mathbf{H}_{i,j}] \right) \right)}{\sum_{t=2}^{K} \exp \left(\text{LeakyReLU} \left(\mathbf{a}^{\mathrm{T}} [\mathbf{W}_\beta \mathbf{H}_{i,1} \| \mathbf{W}_\beta \mathbf{H}_{i,t}] \right) \right)}, \tag{3}$$

where a_j represents the attention coefficient between the center node and the j-th node in the subgraph $\mathcal{G}_i$. LeakyReLU is the activation function and the symbol $\parallel$ denotes concatenation operation. $\mathbf{H}_{i,j}$ is the embedding of the j-th node in $\mathcal{G}_i$, $\mathbf{H}_{i,1}$ denotes the embedding of target (center) node, $\mathbf{W}_\beta \in \mathbb{R}^{D \times D}$ is the projection matrix, and $\mathbf{a} \in \mathbb{R}^{2D}$ is a shared attention vector. The ideal center representation $\boldsymbol{g}_i$ is then generated by computing a weighted sum of the node embeddings:

$$\boldsymbol{g}_i = Generator\,(\mathbf{H}_i) = \sum_{j=2}^{K} a_j \mathbf{H}_{i,j}. \tag{4}$$

Discriminator. To evaluate the degree of context matching of target nodes, we introduce a discriminator component. This discriminator calculates the context matching score by evaluating the correlation between the original attributes of a target node v_i and the corresponding ideal center representation $\boldsymbol{g}_i$. For a positive instance pair $(v_i, \mathcal{G}_i)$, the context matching score of v_i can be calculated as follows:

$$s_i = Disc\,(\boldsymbol{g}_i, \mathbf{x}_i) = \sigma\left(\mathrm{ReLU}\,(\boldsymbol{g}_i \mathbf{W}_q + \mathbf{b}_q)^{\mathrm{T}}\,\mathrm{ReLU}\,(\boldsymbol{x}_i \mathbf{W}_k + \mathbf{b}_k)\right), \tag{5}$$

where $\sigma(x) = 1/(1 + \exp(-x))$ is the sigmoid function, $\mathbf{W}_q$, $\mathbf{W}_k$ are learnable projection matrices, $\mathbf{b}_q$, $\mathbf{b}_k$ are biases, and $\mathbf{x}_i$ is the attribute vector of v_i.

Contrastive Loss. As illustrated in Fig. 1, we construct the contextual anomaly subgraph $\mathcal{G}'_i$ by replacing the center of another subgraph with the target node v_i. Similarly, for a negative instance pair $(v_i, \mathcal{G}'_i)$, the corresponding context matching score s'_i is calculated using Eq. (5). In general, positive instance pairs exhibit a close alignment between the target node and the generated center representation, while negative pairs show a noticeable mismatch. To capture and amplify this difference, we utilize the binary cross-entropy (BCE) loss [17] as the optimization objective for the context matching task:

$$\mathcal{L}_{context} = -\frac{1}{2N}\sum_{i=1}^{N}\left(\log\,(s_i) + \log\left(1 - s'_i\right)\right), \tag{6}$$

After training, s_i tends to be close to 1 (high context matching), while s'_i tends to be close to 0 (low context matching).

3.3   Link Prediction Contrastive Learning

Structural anomalies typically refer to nodes that exhibit numerous irrational connections with other nodes, such as telecommunications fraudsters or members of money laundering groups. To enhance the identification of such anomalies, we introduce a link prediction contrastive task designed to identify anomalies by evaluating the rationality of the target node's local structure.

Link Prediction. We first predict the local structure of target nodes. Specifically, we infer the probability γ_{ij} of a connection between v_i and v_j by calculating the dot-product of two vectors and applying the sigmoid function σ:

$$\gamma_{ij} = P\,((j,i) \in E) = \sigma\left(\phi^{\mathrm{T}}\,(\mathbf{W}\mathbf{x}_i)\,\phi\,(\mathbf{W}\mathbf{x}_j)\right), \tag{7}$$

where the weight matrix $\mathbf{W}$ is shared with the GCN defined in Eq. (2). For each positive instance pair $(v_i, \mathcal{G}_i)$, we predict the probability of connections between the center node v_i and its immediate neighbors within the graph $\mathcal{G}_i$. The link prediction error, l_i, is then calculated as follows:

$$l_i = \frac{\sum_{j \in \widetilde{\mathcal{N}}_i} (1 - \gamma_{ij})}{|\widetilde{\mathcal{N}}_i|}, \tag{8}$$

where $\widetilde{\mathcal{N}}_i$ denotes the set of 1-hop neighbors of the target node in $\mathcal{G}_i$, and the edge weights are assumed to default to 1.

Contrastive Loss. To simulate anomalous connectivity patterns, we construct a structural anomaly subgraph $\widetilde{\mathcal{G}}_i$ by replacing the neighbors of v_i with other arbitrary nodes, as shown in Fig. 1. Similarly, we can calculate the link prediction error l'_i for the negative instance pair $(v_i, \widetilde{\mathcal{G}}_i)$. The objective of the link prediction contrastive task is to minimize the link prediction error for positive instance pairs and maximize the link prediction error for negative instance pairs. For optimization, we employ the triplet loss [14]:

$$\mathcal{L}_{link} = \frac{1}{N} \sum_{i=1}^{N} \left(\max \left(l_i - l'_i + \delta, 0 \right) \right), \tag{9}$$

where the magin value δ is a positive constant. After model training, nodes having illogical connections with multiple neighbors are assigned higher link prediction errors.

3.4 Model Training and Anomaly Score Inference

During the training phase, the model is trained by simultaneously optimizing the objectives for the context matching and link prediction tasks. The combined loss function can be expressed as:

$$\mathcal{L} = \alpha \mathcal{L}_{context} + (1 - \alpha) \mathcal{L}_{link}, \tag{10}$$

where α is a hyperparameter to balance the importance of the two objectives.

During the inference phase, we sample R subgraphs $(\{\mathcal{G}_i^{(1)}, \ldots, \mathcal{G}_i^{(R)}\})$ for each node v_i to capture sufficient contextual information for precise anomaly score calculations. In the context matching task, the contextual anomaly score of v_i is defined as the negative average context matching score:

$$score_{con}(v_i) = - \sum_{r=1}^{R} \frac{s_i^{(r)}}{R}, \tag{11}$$

where $s_i^{(r)}$ is the context matching score of the instance pair $(v_i, \mathcal{G}_i^{(r)})$, as determined by Eq. (5). Following Eq. (11), nodes that exhibit a severe mismatch with their context will be assigned higher anomaly scores.

In the link prediction task, the structural anomaly score of v_i is defined as the average link prediction error:

$$score_{str}(v_i) = \sum_{r=1}^{R} \frac{l_i^{(r)}}{R}, \tag{12}$$

where $l_i^{(r)}$ is the link prediction error of the instance pair $(v_i, \mathcal{G}_i^{(r)})$ calculated according to Eq. (8). Finally, we combine $score_{con}(v_i)$ and $score_{str}(v_i)$ into the final anomaly score:

$$score(v_i) = \alpha score_{con}(v_i) + (1 - \alpha)score_{str}(v_i), \tag{13}$$

where the parameter α defined in Eq. (10) serves as a trade-off term.

4 Experiments

4.1 Experimental Setup

Datasets. We conduct experiments on four widely used datasets [4,9,19,22]. Cora, Citeseer, and ACM are citation networks, where nodes represent papers and edges represent citation relationships between papers. Due to the lack of ground truth anomalies in these datasets, the anomalies are artificially injected by the perturbation scheme [2]. Specifically, we inject equal proportions of contextual and structural anomalies into the datasets following previous works [2,4,9,22]. Elliptic [19] is a bitcoin transaction network with real anomalies, where nodes are transaction entities and edges represent the flow of bitcoins. We aim to detect illicit entities. The statistical features of these datasets are shown in Table 1.

Table 1. The statistics of the datasets.

Dataset	# nodes	# edges	# features	# anomalies
Cora	2708	5429	1433	150
Citeseer	3227	4732	3703	150
ACM	16484	71980	8339	600
Elliptic	46564	73248	93	4545

Baselines and Metrics. We compare our proposed framework with six popular baseline methods, namely Radar [6], ANOMALOUS [12], DOMINANT [2], CoLA [9], ANEMONE [9], and Sub-CR [22]. Among these baselines, the latter three are state-of-the-art contrastive learning-based anomaly detection methods. For more detailed descriptions of these methods, please refer to Sect. 1. The Area Under the Receiver Operating Characteristic Curve (AUC-ROC) metric is used to evaluate the performance of anomaly detection.

Table 2. Anomaly detection performance (AUC scores) comparison. OOM means the issue of out-of-memory is incurred.

Methods	Cora	Citeseer	ACM	Elliptic
Radar	0.7254	0.6717	0.6389	OOM
ANOMALOUS	0.7461	0.7102	0.6634	OOM
DOMINANT	0.8155	0.8251	0.7601	0.3293
CoLA	0.8779	0.8968	0.8237	0.7319
ANEMONE	0.9057	0.9189	0.8627	0.7616
Sub-CR	0.9132	0.9303	0.8059	0.7515
SCL	**0.9438**	**0.9527**	**0.8902**	**0.8238**

Implementation Details. In our proposed framework, SCL, we set the subgraph size K to 4 and the batch size to 300 for all datasets to balance efficiency and performance. We employ a one-layer GCN with 64-dimensional embeddings as the subgraph encoder. The random walk probability p is set to 0.5 to achieve a balanced consideration of both low-order neighbors, for the immediate locality, and high-order neighbors, for the extended context of the nodes. The constant δ is set to 0.3. The learning rate is set to 0.001 for Cora, Citeseer, and ACM, and set to 0.0001 for Elliptic. The balance parameter α is set to 0.3 for Cora and Citeseer, 0.7 for ACM, and 1 for Elliptic. The number of epochs for model training is set to 300. The model is optimized with the Adam optimizer during the training phase. In the anomaly inference phase, we sample 300 subgraphs for each node to calculate the final anomaly scores.

4.2 Result and Analysis

Table 2 shows the comparison results of SCL and baselines. Based on the results, we draw the following conclusions. (1) SCL demonstrates a robust performance advantage over the six baseline methods across all evaluated datasets. Specifically, SCL reaches AUC-ROC gains of 3.06%, 2.24%, 2.75%, and 6.22% on Cora, Citeseer, ACM, and Elliptic, respectively. Remarkably, SCL maintains its excellent performance on the real-world bitcoin transaction network Elliptic. (2) The shallow methods Radar and ANOMALOUS exhibit inherent limitations in handling high-dimensional and large-scale datasets, potentially leading to out-of-memory (OOM) problems. In addition, the results demonstrate that most deep learning-based methods outperform shallow methods. (3) Compared to the graph autoencoder-based model DOMINANT, contrastive learning-based models such as COLA and the proposed SCL achieve significant performance gains.

4.3 Ablation Study

In this section, we conduct ablation studies to evaluate the effectiveness of each contrastive learning module in SCL. We introduce two variants of SCL: SCL-C,

Table 3. The AUC scores of ablation study.

Dataset	Cora	Citeseer	ACM
SCL-C	0.9141	0.9307	0.8682
SCL-L	0.9364	0.9336	0.8023
SCL	**0.9438**	**0.9527**	**0.8902**

which excludes the link prediction module, and SCL-L, which lacks the context matching module. Table 3 presents the results of the ablation experiments on datasets Cora, Citeseer, and ACM, which contain various types of anomalies. Firstly, it can be observed that both contrastive learning modules contribute substantially to anomaly detection. It is worth noting that SCL-C demonstrates notable effectiveness, even surpassing the performance of most baseline methods. Overall, through the joint optimization of two modules, SCL achieves significant performance enhancements. Specifically, on the mentioned datasets, SCL obtains an average improvement of 2.46% in AUC-ROC when compared to SCL-C. Furthermore, SCL demonstrates an average improvement of 3.81% in AUC-ROC when compared to SCL-L. Independent contrastive tasks tend to identify specific anomaly types. In contrast, the SCL framework utilizes joint contrastive learning, fully integrating the advantages of both complementary tasks. This integration enhances the model's ability to capture various anomaly patterns, leading to an overall improvement in performance.

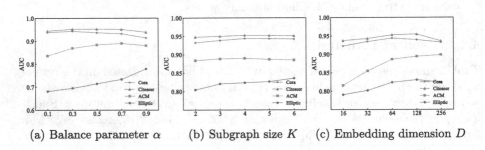

(a) Balance parameter α (b) Subgraph size K (c) Embedding dimension D

Fig. 2. The experimental results for parameter study.

4.4   Parameter Study

In this subsection, we investigate the impacts of three important parameters on the performance of the proposed framework: the balance parameter α, the subgraph size K, and the embedding dimension D. As shown in Fig. 2(a), the proposed framework performs better on Cora, Citeseer, and ACM when $0.3 \leq \alpha \leq 0.7$. This confirms that the SCL can effectively capture multiple anomaly patterns through joint learning of the context matching task and the link prediction task. On the Elliptic dataset, the model achieves better performance as the weight of the context matching task increases, indicating that the anomalies in the Elliptic dataset are more characteristic of contextual anomalies. Figure 2(b) and Fig. 2(c) illustrate the AUC scores under different subgraph sizes K and embedding dimensions D, respectively. Here, we identify $K = 4$ as an ideal choice that balances memory consumption and model performance. Additionally, a suitable embedding dimension is $D = 64$ for most datasets.

5 Related Works

Attributed networks, as a kind of graph-structured data, are widely used for modeling real-world scenarios. The identification of anomalous nodes in these networks is crucial, particularly in security-related fields such as fraud detection [18] and log anomaly detection [21]. In earlier studies, traditional machine learning methods were commonly employed for anomaly detection, including density estimation techniques, e.g., LOF [1], structural clustering methods, e.g., SCAN [20], residual analysis, e.g., Radar [6], and matrix decomposition techniques, e.g., ANOMALOUS [12]. Recently, deep learning (DL) techniques, particularly Graph Neural Networks (GNNs), have gained significant attention in anomaly detection on attributed networks. DL-based anomaly detection approaches can be broadly categorized as autoencoder-based models and contrastive learning-based models. Autoencoder-based models, such as DOMINANT [2], are trained to reconstruct attribute matrix or adjacency matrix and detect anomalies based on the reconstruction errors. Contrastive learning-based models aim to capture differences between different views to identify anomalies, e.g., CoLA [9], ANEMONE [4], and Sub-CR [22]. In contrast to autoencoder-based models that focus on reconstructing the entire network, contrastive learning-based models generate pairs of instances in batches for model training, resulting in improved scalability for large-scale datasets. However, existing contrastive learning-based models have not designed fine-grained contrastive learning tasks tailored to different anomaly patterns, resulting in suboptimal performance. In addition, how to leverage self-supervised information from the network structure to enhance the identification of anomalous nodes remains underexplored.

6 Conclusions

In this paper, we propose a novel multi-task contrastive learning framework for anomaly detection on attributed networks. The framework can effectively capture various anomaly patterns by jointly optimizing the well-designed context contrastive task and the link prediction contrastive task. Specifically, the context matching task aims to detect contextual anomalies by measuring the degree of matching between the target node and its contextual information. On the other hand, the link prediction task aims to identify structural anomalies by evaluating the rationality of the target node's connections. Experiments on four widely used datasets show that the proposed framework outperforms the baseline methods.

References

1. Breunig, M.M., Kriegel, H.P., Ng, R.T., Sander, J.: LOF: identifying density-based local outliers. In: SIGMOD, pp. 93–104 (2000)
2. Ding, K., Li, J., Bhanushali, R., Liu, H.: Deep anomaly detection on attributed networks. In: SDM, pp. 594–602 (2019)
3. Fei, G., Mukherjee, A., Liu, B., Hsu, M., Castellanos, M., Ghosh, R.: Exploiting burstiness in reviews for review spammer detection. In: ICWSM (2013)
4. Jin, M., Liu, Y., Zheng, Y., Chi, L., Li, Y.F., Pan, S.: ANEMONE: graph anomaly detection with multi-scale contrastive learning. In: CIKM, pp. 3122–3126 (2021)
5. Kipf, T.N., Welling, M.: Semi-supervised classification with graph convolutional networks. In: ICLR (2017)
6. Li, J., Dani, H., Hu, X., Liu, H.: Radar: residual analysis for anomaly detection in attributed networks. In: IJCAI, pp. 2152–2158 (2017)
7. Li, Y., Huang, X., Li, J., Du, M., Zou, N.: SpecAE: Spectral autoencoder for anomaly detection in attributed networks. In: CIKM, pp. 2233–2236 (2019)
8. Liu, K., et al.: Bond: benchmarking unsupervised outlier node detection on static attributed graphs. In: NeurIPS, pp. 27021–27035 (2022)
9. Liu, Y., Li, Z., Pan, S., Gong, C., Zhou, C., Karypis, G.: Anomaly detection on attributed networks via contrastive self-supervised learning. TNNLS **33**(6), 2378–2392 (2022)
10. Ma, X., et al.: A comprehensive survey on graph anomaly detection with deep learning. TKDE **35**(12), 12012–12038 (2023)
11. McPherson, M., Smith-Lovin, L., Cook, J.M.: Birds of a feather: homophily in social networks. Ann. Rev. Sociol. **27**(1), 415–444 (2001)
12. Peng, Z., Luo, M., Li, J., Liu, H., Zheng, Q., et al.: ANOMALOUS: a joint modeling approach for anomaly detection on attributed networks. In: IJCAI (2018)
13. Peng, Z., Luo, M., Li, J., Xue, L., Zheng, Q.: A deep multi-view framework for anomaly detection on attributed networks. TKDE **34**(6), 2539–2552 (2022)
14. Schroff, F., Kalenichenko, D., Philbin, J.: Facenet: a unified embedding for face recognition and clustering. In: CVPR (2015)
15. Tong, H., Faloutsos, C., Pan, J.Y.: Fast random walk with restart and its applications. In: ICDM, pp. 613–622 (2006)
16. Velickovic, P., Cucurull, G., Casanova, A., Romero, A., Liò, P., Bengio, Y.: Graph attention networks. In: ICLR (2018)
17. Veličković, P., Fedus, W., Hamilton, W.L., Liò, P., Bengio, Y., Hjelm, R.D.: Deep Graph Infomax. In: ICLR (2019)
18. Wang, J., Wen, R., Wu, C., Huang, Y., Xiong, J.: Fdgars: fraudster detection via graph convolutional networks in online app review system. In: WWW (2019)
19. Weber, M., et al.: Anti-money laundering in bitcoin: Experimenting with graph convolutional networks for financial forensics. arXiv preprint arXiv:1908.02591 (2019)
20. Xu, X., Yuruk, N., Feng, Z., Schweiger, T.A.: SCAN: a structural clustering algorithm for networks. In: KDD, pp. 824–833 (2007)
21. Zhang, C., et al.: DeepTraLog: trace-log combined microservice anomaly detection through graph-based deep learning. In: ICSE, pp. 623–634 (2022)
22. Zhang, J., Wang, S., Chen, S.: Reconstruction enhanced multi-view contrastive learning for anomaly detection on attributed networks. In: IJCAI (2022)

SATJiP: Spatial and Augmented Temporal Jigsaw Puzzles for Video Anomaly Detection

Liheng Shen$^{(\boxtimes)}$ ⓘ, Tetsu Matsukawa ⓘ, and Einoshin Suzuki ⓘ

ISEE, Kyushu University, Fukuoka 819-0395, Japan
shen.liheng.020@s.kyushu-u.ac.jp, {matsukawa,suzuki}@inf.kyushu-u.ac.jp

Abstract. Video Anomaly Detection (VAD) is a significant task, which refers to taking a video clip as input and outputting class labels, e.g., normal or abnormal, at the frame level. Wang et al. proposed a method called DSTJiP, which trains the model by solving Decoupled Spatial and Temporal Jigsaw Puzzles and achieves impressive VAD performance. However, the model sometimes fails to detect abnormal human actions where abnormal motions are accompanied by normal motions. The reason is that the model learns representations of little- and non-motion parts of training examples, resulting in being insensitive to abnormal motions. To circumvent this problem, we propose to solve Spatial and Augmented Temporal Jigsaw Puzzles (SATJiP) as an extension of DSTJiP. SATJiP encourages the model to focus on motions by a novel pretext task, enabling it to detect abnormal motions accompanied by normal motions. Experiments conducted on three standard VAD benchmarks demonstrate that SATJiP outperforms the state-of-the-art methods.

Keywords: video anomaly detection · self-supervised learning · spatio-temporal data

1 Introduction

Anomaly detection refers to a data mining problem that aims to find data that do not conform to expected behavior [5]. As a branch of this task, Video Anomaly Detection (VAD) aims to detect anomalies in a video clip, in which anomalies are unexpected events that involve spatio-temporal information [35,37]. VAD is significant in real-world applications, such as surveillance [16,34]; yet, difficult to conduct as a supervised classification task because gathering all kinds of diverse anomalous event data is almost infeasible [20,41]. Thus, VAD is commonly performed in a one-class learning paradigm, i.e., to train a one-class classifier which can learn the representations of regular events using only normal, unlabeled training examples [20,35], and use it to detect anomalies in the test data.

VAD is challenging since it requires modeling diverse examples which have complex spatio-temporal information, and detecting various anomalies at a fine-grained level, e.g., detecting running persons in an outdoor scenario among

© The Author(s), under exclusive license to Springer Nature Singapore Pte Ltd. 2024
D.-N. Yang et al. (Eds.): PAKDD 2024, LNAI 14645, pp. 27–40, 2024.
https://doi.org/10.1007/978-981-97-2242-6_3

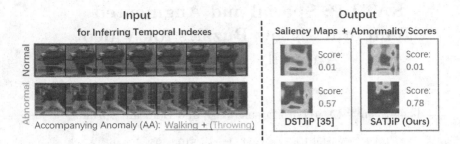

Fig. 1. Visualization examples. **Left:** Input examples for inferring temporal indexes, where the model's inference basis is randomly marked with different bounding boxes. **Right:** A saliency map of each model for each input is generated by Grad-CAM [28], where the high saliency region denoted by red color can be regarded as the model's inference basis. Different from DSTJiP [35], SATJiP fails to focus on the regions which contain regular motions in AA for temporal index inference. SATJiP outputs a higher abnormality score than DSTJiP for AA, indicating a better detection performance. (Color figure online)

crowded walking pedestrians as abnormal. In these anomalies, a category of abnormal human actions are relatively difficult to detect, e.g., throwing a bag while walking, where the leg and body motions are normal but the arm motions are abnormal. The abnormal motions in this example are localized, making it easy to deceive a model that is insensitive to motions, resulting in judging this example as not having a large deviation from the training examples. We denote abnormal human actions where abnormal motions are accompanied by normal motions as Accompanying Anomalies (AA), which commonly involve notable motions. Detecting AA could be highly significant, e.g., neglecting a thief who is stealing a bag while walking on the street, can lead to economic loss.

Previous methods can be divided into unsupervised generation-based methods and Self-Supervised Learning (SSL) based methods. Unsupervised generation-based methods often train Auto-Encoders (AEs) to recover frames at the pixel level [6,12,15,18,25]. However, pixel-level generation makes the model extract low-level features which do not help to detect anomalies, e.g., colors. SSL-based methods apply diverse self-supervised tasks, e.g., instance contrast tasks [38] or multiple tasks including binary classification, prediction, and distillation [11]. However, self-supervision signals in [11,38] fail to make the model learn high-level semantics, e.g., human actions, at a fine-grained level. In an SSL manner, Wang et al. [35] proposed to solve two multi-class classification tasks, i.e., Decoupled Spatial and Temporal Jigsaw Puzzles (DSTJiP). The Spatial and Temporal Jigsaw Puzzles (SJiP and TJiP) respectively shuffle patches spatially or temporally, and require a model to take these shuffled patches as input and infer the index of each patch. DSTJiP assumes that only indexes of the patches generated from normal examples can be accurately inferred while those generated from abnormal examples cannot. DSTJiP makes the model neglect task-irrelevant low-level features and extract highly discriminative representations [35], reaching impressive VAD performance.

However, by experiments, we found that DSTJiP [35] cannot accurately detect AA. The reason is that, by solving TJiP, the model learns spatio-temporal representations of normal examples concerning notable-, little-, and non-motion parts. However, little- and non-motion parts are irrelevant for judging abnormal motions. Affected by these parts, the model becomes relatively insensitive to abnormal motions, making the model easy to judge AA as similar to those learned training examples.

Inspired by the information bottleneck theory [32], we aim to decrease the information that does not help to detect abnormal motions in the latent space representations. To this end, we perform data augmentation to encourage the model to learn information relevant to detecting abnormal motions, i.e., representations of notable motion parts. Specifically, we propose a VAD method which solves Spatial and Augmented Temporal Jigsaw Puzzles (SATJiP). We design Masked Temporal Jigsaw Puzzles (MTJiP) as a pretext task, to augment the TJiP in DSTJiP [35]. For some of the patches in MTJiP, we generate masks to reserve notable motion parts while masking out parts with little or no motion, which encourages the model to focus on regions of notable motions. The model thus becomes sensitive to notable motions which deviate from those learned motions, improving the possibility of finding abnormal motions and detecting AA as abnormal. Figure 1 shows examples to support our motivation and proposal.

Our contributions are summarized as follows:

- We extend DSTJiP [35] to SATJiP by proposing an extra pretext task, i.e., MTJiP. SATJiP makes the model find abnormal motions in accompanying anomalies, boosting the detection performance.
- We conduct comprehensive experiments on three standard VAD benchmarks. The experimental results show that SATJiP achieves better performance than the state-of-the-art methods.

2   Related Works

Most of VAD methods rely on unsupervised generation, i.e., training Auto-Encoders (AEs) to reconstruct a frame or frames [6,22], or predict the missing frame at the pixel-level [17,18,41]. They assume that a large error in reconstruction or prediction indicates an anomaly. However, large reconstruction errors for abnormal events and small prediction errors for normal events cannot be guaranteed. To solve this problem, hybrid solutions of both reconstruction and prediction [31,40], and methods of learning the prototypes of normal examples [12,20,25] are proposed. However, pixel-level generation is not robust to the deviation of low-level features, e.g., colors, limiting the model's performance.

SSL-based VAD methods, e.g., solving an instance contrast task [38] or multiple binary classification tasks [11], are robust to the deviation in low-level features. However, models in [11,38] fail to learn high-level semantics, e.g., human actions, at a fine-grained level, resulting in sub-optimal VAD performance. This failure can be attributed to the ease of tasks in [11,38], which do not compel the model to learn a detailed progression of actions. Wang et al. proposed DSTJiP [35], where each patch for index inference is associated with multiple class labels, providing more fine-grained self-supervision signals. However, the model is relatively insensitive to notable motions and sometimes fails to detect AA.

Inspired by TJiP in [35], we propose MTJiP as a pretext task, to counter this problem in DSTJiP. The main difference between TJiP and MTJiP is that the latter applies a specific masking strategy. Specifically, using optical-flow-based masks to highlight notable motion parts in patches, and randomly mixing masked and unmasked patches. Generation-based methods [9,26,33] often mask out the parts that need to be focused on, while ours masks out the redundant parts. The reason is that generation-based models learn to recover the masked parts while a jigsaw puzzle solver tends to neglect those parts since they do not help to infer indexes of patches.

3   Problem Formulation: Frame-Level VAD

We define a test dataset of VAD as $D^{\text{test}} = \{V_k | k = 1, ..., K\}$, where V_k represents to the k-th video clip and K is the number of the video clips. We further denote V_k as $V_k = \{I_t^k | t = 1, ..., T^k\}$, where I_t^k represents the t-th frame and T^k is the number of frames. I_t^k and its temporally adjacent frames are stacked to construct a mini-clip v_t^k with length l ($l \ll T^k$). We define the label of I_t^k as $Y(I_t^k) \in \{0, 1\}$, where 0 refers to normal while 1 refers to abnormal. Generally, the goal of frame-level VAD is to train a model F^{vad} with D^{train} whose structure is the same as D^{test} to achieve $F^{\text{vad}}(v_t^k) = Y(I_t^k)$.

4   Proposal: SATJiP

Since SATJiP is an extension of DSTJiP [35], we first introduce DSTJiP as a preliminary, and then explain our main contribution, i.e., proposing an extra pretext task, MTJiP, to train the model. Figure 2 shows an overview of SATJiP in the training phase, where SJiP, TJiP, and MTJiP are constructed as the model's input. The model solves jigsaw puzzles and outputs the inferred probability matrices of spatial and temporal indexes. In the test phase, same as DSTJiP [35], the model solves only SJiP and TJiP, and detects the corresponding examples of inaccurate index inference as abnormal.

4.1   Preliminary

Input. DSTJiP [35] applies a YOLOv3 detector [27] to extract all objects frame by frame. For each object detected in frame i, DSTJiP constructs an Object-Centric Spatio-Temporal (OCST) cube $\mathbf{c}$ by simply stacking patches cropped from its temporally adjacent frames $\{i - \frac{l-1}{2}..., i - 1, i, i + 1, ..., i + \frac{l-1}{2}\}$ using

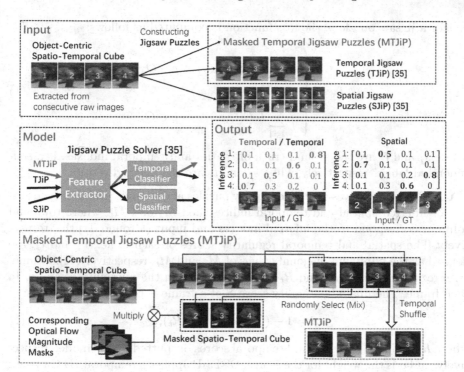

Fig. 2. Overview of SATJiP in the training phase. In the input, **MTJiP denoted by red color is our original part** and its details are shown in the bottom. (Color figure online)

the same bounding box, where l is the length of the cube. The OCST cube is denoted as $\mathbf{c} = c_{1:l}$, where c_t is a patch at temporal index t within the cube. DSTJiP rescales all the extracted patches into a fixed spatial size $(H \times W)$. $c_t^{h,w}$ denotes the pixel value at a spatial position (h, w) in the patch c_t.

DSTJiP decomposes c_t into $n \times n$ equal-sized mini-patches and shuffles them with one of the all possible random permutations P^s to construct Spatial Jigsaw Puzzles (SJiP). The shuffled mini-patches in each patch share the same permutation. It also shuffles each patch in $\mathbf{c}$ in the temporal direction with one of the all random permutations P^t to construct TJiP.

Self-supervised Model Training. DSTJiP constructs both SJiP and TJiP to train the model to infer the index of each spatially shuffled mini-patch or each temporally shuffled patch in a cube. A mini-batch consists of two sets: Q_s and Q_t, denoting the sets of SJiP and TJiP, respectively. DSTJiP optimizes the model using the Cross-Entropy (CE) loss though Mean Square Error (MSE) loss can be also used, as discussed in [35].

For a jigsaw puzzle p, the loss function is computed as follows:

$$L_p = \begin{cases} \dfrac{1}{n^2} \displaystyle\sum_{j=1}^{n^2} CE(s_j, \hat{s}_j), & p \in Q_s, \\[2em] \dfrac{1}{l} \displaystyle\sum_{i=1}^{l} CE(t_i, \hat{t}_i), & p \in Q_t, \end{cases} \tag{1}$$

where s_j and t_i are the ground truth of spatial and temporal indexes, i.e., self-generated labels, respectively, while $\hat{s}_j$ and $\hat{t}_i$ are the inferred results.

VAD. The model solves unshuffled jigsaw puzzles and outputs probability matrices for inferred spatial and temporal indices, i.e., M_s and M_t, where rows and columns correspond to the inferred indexes and input (Ground Truth), respectively. The spatial and temporal regularity scores of each $\mathbf{c}$, i.e., r_s and r_t, are defined as the minimum diagonal values of M_s and M_t, respectively. The frame-level regularity scores R_s and R_t are generated from the minimum r_s and r_t in each frame. The final abnormality score S of a frame is computed as follows:

$$S = 1 - ((1 - \sigma)R_s + \sigma R_t), \tag{2}$$

where R_s and R_t are spatial and temporal scores outputted by the model, respectively, and σ is the weight for the scores. DSTJiP applies a temporal 1D Gaussian filter to smooth the scores and perform max-min normalization.

4.2   Masked Temporal Jigsaw Puzzles (MTJiP)

Selecting. Abnormal motions only lie in notable motions, but notable motions are possibly normal or abnormal. Notable motions are thus crucial for distinguishing AA and normal human actions. To this end, only those $\mathbf{c}$ involving notable motions should be selected to construct MTJiP. To select them, inspired by [29], SATJiP evaluates dynamic degree $g(\cdot)$ of each $\mathbf{c}$ based on the average absolute frame difference between the first and last patches as $g(\mathbf{c}) = \frac{1}{HW} \sum_{h=1}^{H} \sum_{w=1}^{W} \|c_1^{h,w} - c_l^{h,w}\|$ and selects $\mathbf{c}$ whose $g(\mathbf{c})$ is greater than a threshold θ. D denotes a set of the selected $\mathbf{c}$.

Masking. To encourage the model to focus on regions of notable motions during temporal index inference, SATJiP highlights the notable motion parts by masking little- and non-motion parts in patches, based on optical flow magnitude $\mathbf{f} = F(\mathbf{c}) = f_{2:l}$, which indicates the intensity of pixel changes. For the optical flow extractor $F(\cdot)$, SATJiP uses FlowNet 2.0 [14], which is also applied in other previous VAD methods [20,41]. Then SATJiP obtains a masked cube $\mathbf{m} = m_{2:l}$, where $m_t^{h,w} = \mathbb{I}(G(f_t)^{h,w} > \gamma)c_t^{h,w}$. $\mathbb{I}(\cdot)$ is an indicator function. $G(\cdot)$ is a 2D Gaussian filter to remove the noise and γ is a masking threshold. If $G(f_t)^{h,w}$ is lower than γ, this pixel is considered with little or no motion, and thus masked.

Mixing and Shuffling. A random mixture of $\mathbf{c}$ and $\mathbf{m}$ encourages the model to extend its attention on highlighted parts in the masked patches to the related

---

Algorithm 1: Construction of MTJiP in mini-batches

   input : set of cubes C, frame length l, thresholds θ and γ, unmasked patch
          ratio δ

   output: set of MTJiP Z

1   $Z \leftarrow \varnothing$, $P^t \leftarrow$ all permutations $[P_1^t, P_2^t, ..., P_{(l)!}^t]$;
2   $D \leftarrow \{\mathbf{c} | g(\mathbf{c}) > \theta, \mathbf{c} \in C\}$; // select cubes that have notable motions
3   **for** $\mathbf{c}$ **in** D **do**
4      $\mathbf{f} \leftarrow G(F(\mathbf{c}))$; // generate and apply masks
5   **for** h, w **in** H, W **do**
6   **if** $f_t^{h,w} > \gamma$ **then** $m_t^{h,w} \leftarrow c_t^{h,w}$;
7   **else** $m_t^{h,w} \leftarrow 0$;

8   **for** $t \leftarrow 1$ **to** l **do**
9          $r \leftarrow U_{\text{float}}[0, 1]$; // randomly mix patches
10   **if** $t = 1$ **then** $a_t \leftarrow c_t$;
11   **else if** $r > \delta$ **then** $a_t \leftarrow m_t$;
12   **else** $a_t \leftarrow c_t$;

13      $n \leftarrow U_{\text{int}}[1, (l)!]$;
14      $\mathbf{z} \leftarrow$ TemporallyShuffle$(\mathbf{a}, P_n^t)$;
15      $Z \leftarrow Z \cup \{\mathbf{z}\}$;

---

regions in the unmasked patches, making the model learn to focus on notable motion parts in unmasked patches. Let $\mathbf{a} = a_{1:l}$ be an OCST cube, where each a_t is randomly picked from c_t or m_t with a ratio $\delta : 1 - \delta$, and $a_1 = c_1$ is set for aligning other patches. SATJiP shuffles the mixed cubes in the temporal direction to construct MTJiP $\mathbf{z} = z_{1:l}$. By solving MTJiP, the model focuses on learning highlighted notable motion parts among all patches, which increases the sensitivity to abnormal motions and the possibility of accurately detecting AA. Algorithm 1 explains the construction of MTJiP in mini-batches, where each set of MTJiP, denoted as Z, is added to train the model. The training relies on the same loss function for TJiP in Eq. (1).

5 Experiments

5.1 Datasets and Evaluation Metric

We employ three public benchmark VAD datasets, i.e. **UCSD Ped2 (Ped2)** [24], **CUHK Avenue (Avenue)** [21], and **ShanghaiTech Campus (STC)** [23]. They are widely applied in VAD [11,18,20,35,41]. Their anomalies involve abnormal objects, actions, and their combinations, and are rich in diversity.

In frame-level VAD, the abnormality score for a normal frame should be close to 0 (negative), while that of an abnormal frame should be 1 (positive). We use the Area Under the ROC Curve (AUROC) as the evaluation metric for frame-level VAD. Specifically, we concatenate the scores of all frames in the dataset and compute AUROC. This evaluation approach is widely adopted in VAD [12, 18,20,35,38,41]. A higher AUROC value indicates better performance. We use the official codes and models of baseline methods to conduct experiments. The results in Table 2 are exceptions, where we directly copy the baseline methods' AUROC results from [35], and other papers.

5.2   Implementation Details

For common parameters in DSTJiP [35] and ours, we keep the same values as [35] except for the training epoch and batch size. Since the training loss does not decrease sufficiently, we increase the training epoch to 70 on Ped2, and 120 on Avenue and STC, while decreasing the batch size to 64. We set θ, δ, and σ, to 0.5, 0.5, and 0.6, respectively, for all the datasets. We set γ to 0.1 for Ped2 and Avenue. Due to the illumination and capturing distance, the estimated optical flow of some examples in STC has low magnitudes. Since a too-high value of γ makes some of the patches masked too much, we set γ to 0.05 for STC.

Table 1. Verification of detecting Accompanying Anomalies (AA). AUROC (%). **Bold** figures: best score. <u>Underlined</u> figures: 2nd and 3rd best score.

Method	Avenue			STC		
	NoAA	Original	OnlyAA	NoAA	Original	OnlyAA
MNAD-P [25]	87.1	87.2	90.9	67.0	67.7	<u>83.3</u>
HF²-VAD [20]	<u>89.3</u>	<u>89.5</u>	**95.0**	<u>74.3</u>	<u>74.4</u>	74.5
STEAL-Net [2]	87.0	87.1	<u>92.1</u>	73.5	73.7	78.4
DSTJiP [35]	<u>92.5</u>	<u>92.1</u>	80.0	**84.3**	**84.2**	<u>82.7</u>
SATJiP (Ours)	**93.2**	**93.2**	<u>94.0</u>	<u>84.0</u>	<u>84.1</u>	**86.6**

5.3   Comparison in Detecting Accompanying Anomalies (AA)

In Table 1, we analyze the model's performance in detecting AA based on corresponding frame-level abnormality scores. We set three cases, i.e., deleting scores of abnormal inputs involving AA (NoAA), keeping all scores (Original), and deleting scores of other abnormal inputs except those involving AA (OnlyAA). The results in "OnlyAA" directly show the performance in detecting AA, while those in "NoAA" and "Original" serve as references. If a method possesses an advantage in detecting AA, its AUROC values should increase from the first case (NoAA) to the last case (OnlyAA). The methods listed on the top are unsupervised generation-based while those listed on the bottom are SSL-based.

In "OnlyAA", SATJiP performs the 2nd and 1st best in Avenue and STC, respectively, demonstrating its better performance in detecting AA than other baseline methods. Comparing all the results in "NoAA" and "Original", the differences are minor. The reason is that frames of AA only occupy 0.7% and 1.6% of the total frames, 2.8% and 3.9% of the abnormal frames in the test sets of Avenue and STC, respectively. SATJiP shows its advantage in detecting AA, which is a disadvantage in DSTJiP [35]. Unsupervised generation-based methods also prove their strength in detecting AA, but fail in other examples, showing sub-optimal overall performance.

5.4   Comparison in Detecting Diverse Video Anomalies

Table 2 shows the performance comparison of state-of-the-art methods on Ped2, Avenue, and STC datasets. Here, we follow the official settings of these datasets.

Table 2. AUROC (%) comparison of the state-of-the-art VAD methods. **Bold** figures: best score. Underlined figures: 2nd and 3rd best score.

Type	Method	Ref	Ped2	Avenue	STC
Unsupervised Generation-based	IntegratAE	PRL [31]	96.3	85.1	73.0
	ClusterAE	ECCV'20 [6]	96.5	86.0	73.3
	MNAD-Pred	CVPR'20 [25]	97.0	88.5	70.5
	VEC	MM'20 [41]	97.3	90.2	74.8
	BMAN	TIP'20 [17]	96.6	90.0	76.2
	LNRA	BMVC'21 [1]	96.5	84.7	76.0
	CT-D2GAN	MM'21 [10]	97.2	85.9	77.7
	STEAL-Net	ICCVW'21 [2]	98.4	87.1	73.7
	HF^2-VAD	ICCV'21 [20]	99.3	91.1	76.2
	PGDLE	ICDM'21 [16]	97.6	87.8	74.5
	AMMC-Net	AAAI'21 [4]	96.6	86.6	73.7
	Multipath-Pred	TNNLS'22 [36]	96.3	88.3	76.6
	OGMR-Net	ICME'22 [42]	97.4	92.6	74.9
	STM-AE	ICME'22 [19]	98.1	89.8	73.8
	BDPN	AAAI'22 [7]	98.3	90.3	78.1
	STATE	ICDM'22 [37]	-	90.3	77.8
	AMSRC	ICASSP'23 [13]	99.3	**93.8**	76.3
	BFI-VAD	WACV'23 [8]	98.9	89.7	75.0
	LERF-VAD	AAAI'23 [30]	**99.4**	91.5	78.6
	USTN-DSC	CVPR'23 [39]	98.1	89.9	73.8
SSL-based	CAC	MM'20 [38]	-	87.0	79.3
	SS-MTL	CVPR'21 [11]	97.5	91.5	82.4
	DSTJiP	ECCV'22 [35]	99.0	92.2	**84.3**
	SSMTL++v1	CVIU'23 [3]	-	93.7	82.9
	SSMTL++v2	CVIU'23 [3]	-	91.6	83.8
	SATJiP	Ours	**99.4**	93.2	84.1

For a VAD model, a robust performance in detecting diverse anomalies is essential since anomalies are various and unpredictable in the real world. Considering their performance among all the datasets, Unsupervised Generation-based methods are less competitive than SSL-based methods in detecting diverse video anomalies. Especially in STC, generation-based methods fall behind by a notable margin, since the scenarios and illuminations in STC are more diverse than those in Avenue and Ped2, resulting in richer task-irrelevant low-level features. Despite the very few AA in these datasets, SATJiP still reaches the 1st, 3rd, and 2nd best on Ped2, Avenue, and STC, respectively. This fact verifies that SATJiP has a robust ability to detect diverse video anomalies.

5.5 Ablation Study

Table 3(a) shows a comparison of spatial and temporal scores, which correspond to the model's spatial and temporal view, respectively. The AUROC values of "R_t (Ours)" prove that the performance of SATJiP's temporal view improves notably compared with that of DSTJiP [35]. This fact verifies the effect of training with MTJiP.

Table 3. Ablation Study of SATJiP based on AUROC (%). **Bold** figures: best score. Underlined figures: 2nd best score.

(a) Analysis of Spatio-Temporal View.

Score	Ped2	Avenue	STC
R_s [35]	86.7	88.9	81.1
R_s (Ours)	93.1	86.6	81.7
R_t [35]	98.9	86.6	80.0
R_t (Ours)	99.2	92.8	81.9
R_s+R_t [35]	99.0	92.1	**84.2**
R_s+R_t (Ours)	**99.4**	**93.2**	84.1

(b) Analysis of the components in MTJiP.

S+T (base)	+T	+masking	+mixing	+noshuffle	Ped2	Avenue	STC
✓					98.4	91.3	83.6
✓	✓				98.8	92.1	83.1
✓	✓	✓			**99.4**	91.9	82.7
✓	✓	✓	✓		**99.4**	**93.2**	**84.1**
✓	✓	✓	✓	✓	97.9	92.1	81.3

Table 3(b) aims to verify the effect of each component in MTJiP. We train DSTJiP [35] with the same parameter setting as ours, denote it as "S+T", and set it as the baseline. MTJiP can be seen as TJiP with masking and mixing, and thus we list these components, namely "+T", "+masking", and "+mixing". Since shuffling is a common manipulation for all jigsaw puzzles both in

DSTJiP [35] and ours, we only set "+noshuffle" for MTJiP as an example, aiming to show the effect of shuffling.

Due to the overfitting problem, repeatedly solving TJiP, i.e., "+T", brings almost no benefit. Solving fully masked TJiP in extra, i.e., "+T+masking", makes the model learn to handle masked and unmasked patches, respectively, resulting in an undesirable impact of MTJiP on the model's VAD performance. Without shuffling patches, i.e., "+T+masking+mixing+noshuffle", the model takes shortcuts, i.e., learning unexpected representations, resulting in a deteriorated performance of the model's temporal view. Based on our observation, the AUROC value of the temporal score in each dataset decreases by at least 5% compared with the corresponding value shown in Table 3(a). The results show that "+T+masking+mixing" performs the best, verifying the effectiveness of all components in MTJiP.

5.6 VAD Examples

Figure 3 shows a comparison of frame-level VAD in Video No. 12 in Avenue [21], where several frames and their corresponding scores are selected for interpretation. To avoid false detection, the abnormality scores of abnormal frames should be higher than those of normal frames, i.e., colored circles should be higher than corresponding colored dotted lines in the figure. DSTJiP [35] and HF^2-VAD [20] output a lower abnormality score for the abnormal frame #544 and #840 than that for the normal frame (#1005), respectively, while SATJiP avoids such a problem. SATJiP performs the best since it reaches the highest AUROC values.

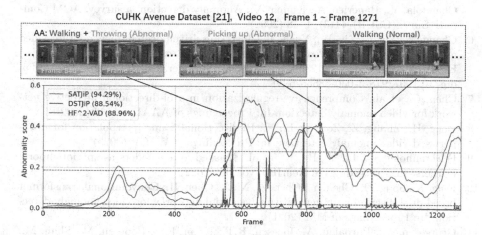

Fig. 3. VAD performance comparison of DSTJiP [35], HF^2-VAD [20], and SATJiP (ours). A larger value of a frame in score curves indicates a larger possibility of this frame being abnormal. Each method's AUROC (%) value for this video clip is given. Colored horizontal dotted lines indicate scores for the normal example, while colored circles indicate scores for the abnormal example. (Color figure online)

6   Conclusion

We have proposed an SSL-based VAD method, i.e., SATJiP, which is an extension of DSTJiP [35]. In SATJiP, our well-designed pretext task, i.e., MTJiP, highlights notable motion parts in some patches, making the model focus on learning notable motions and become more sensitive to abnormal motions in AA compared to [35]. Experiments on three standard public VAD benchmarks verify the superiority of SATJiP. Specifically, SATJiP shows a robust detection ability for diverse video anomalies in different datasets and displays its strength in detecting AA. One of our future works is to extend MTJiP to a generic pretext task, aiming to benefit other methods.

Acknowledgment. This work was partially supported by JST, the establishment of university fellowships towards the creation of science technology innovation, Grant Number JPMJFS2132.

References

1. Astrid, M., Zaheer, M.Z., Lee, J.Y., Lee, S.I.: Learning not to reconstruct anomalies. In: Proceedings of BMVC (2021)
2. Astrid, M., Zaheer, M.Z., Lee, S.I.: Synthetic temporal anomaly guided end-to-end video anomaly detection. In: Proceedings of ICCVW (2021)
3. Barbalau, A., et al.: SSMTL++: revisiting self-supervised multi-task learning for video anomaly detection. Comput. Vis. Image Underst. **229**, 103656 (2023)
4. Cai, R., Zhang, H., Liu, W., Gao, S., Hao, Z.: Appearance-motion memory consistency network for video anomaly detection. In: Proceedings of AAAI (2021)
5. Chandola, V., Banerjee, A., Kumar, V.: Anomaly detection: a survey. ACM Computi. Surv. (CSUR) **41**(3), 1–58 (2009)
6. Chang, Y., Tu, Z., Xie, W., Yuan, J.: Clustering driven deep autoencoder for video anomaly detection. In: Vedaldi, A., Bischof, H., Brox, T., Frahm, J.-M. (eds.) ECCV 2020. LNCS, vol. 12360, pp. 329–345. Springer, Cham (2020). https://doi.org/10.1007/978-3-030-58555-6_20
7. Chen, C., et al.: Comprehensive regularization in a bi-directional predictive network for video anomaly detection. In: Proceedings of AAAI, vol. 36 (2022)
8. Deng, H., Zhang, Z., Zou, S., Li, X.: Bi-directional frame interpolation for unsupervised video anomaly detection. In: Proceedings of WACV (2023)
9. Feichtenhofer, C., Li, Y., He, K., et al.: Masked autoencoders as spatiotemporal learners. In: Proceedings of NeurIPS, vol. 35 (2022)
10. Feng, X., Song, D., Chen, Y., Chen, Z., Ni, J., Chen, H.: Convolutional transformer based dual discriminator generative adversarial networks for video anomaly detection. In: Proceedings of MM (2021)
11. Georgescu, M., Barbalau, A., Ionescu, R.T., Khan, F.S., Popescu, M., Shah, M.: Anomaly detection in video via self-supervised and multi-task learning. In: Proceedings of CVPR (2021)
12. Gong, D., et al.: Memorizing normality to detect anomaly: memory-augmented deep autoencoder for unsupervised anomaly detection. In: Proceedings of ICCV (2019)

13. Huang, X., Zhao, C., Wu, Z.: A video anomaly detection framework based on appearance-motion semantics representation consistency. In: Proceedings of ICASSP (2023)
14. Ilg, E., Mayer, N., Saikia, T., Keuper, M., Dosovitskiy, A., Brox, T.: Flownet 2.0: evolution of optical flow estimation with deep networks. In: Proceedings of CVPR (2017)
15. Ionescu, R.T., Khan, F.S., Georgescu, M.I., Shao, L.: Object-centric auto-encoders and dummy anomalies for abnormal event detection in video. In: Proceedings of CVPR (2019)
16. Lai, Y., Han, Y., Wang, Y.: Anomaly detection with prototype-guided discriminative latent embeddings. In: Proceedings of ICDM (2021)
17. Lee, S., Kim, H.G., Ro, Y.M.: BMAN: bidirectional multi-scale aggregation networks for abnormal event detection. IEEE Trans. Image Process. **29**, 2395–2408 (2020)
18. Liu, W., Luo, W., Lian, D., Gao, S.: Future frame prediction for anomaly detection–a new baseline. In: Proceedings of CVPR (2018)
19. Liu, Y., Liu, J., Zhao, M., Yang, D., Zhu, X., Song, L.: Learning appearance-motion normality for video anomaly detection. In: Proceedings of ICME (2022)
20. Liu, Z., Nie, Y., Long, C., Zhang, Q., Li, G.: A hybrid video anomaly detection framework via memory-augmented flow reconstruction and flow-guided frame prediction. In: Proceedings of ICCV (2021)
21. Lu, C., Shi, J., Jia, J.: Abnormal event detection at 150 FPS in MATLAB. In: Proceedings of ICCV (2013)
22. Luo, W., Liu, W., Gao, S.: Remembering history with convolutional LSTM for anomaly detection. In: Proceedings of ICME (2017)
23. Luo, W., Liu, W., Gao, S.: A revisit of sparse coding based anomaly detection in stacked RNN framework. In: Proceedings of ICCV (2017)
24. Mahadevan, V., Li, W., Bhalodia, V., Vasconcelos, N.: Anomaly detection in crowded scenes. In: Proceedings of CVPR (2010)
25. Park, H., Noh, J., Ham, B.: Learning memory-guided normality for anomaly detection. In: Proceedings of CVPR (2020)
26. Pathak, D., Krahenbuhl, P., Donahue, J., Darrell, T., Efros, A.A.: Context encoders: feature learning by inpainting. In: Proceedings of CVPR (2016)
27. Redmon, J., Farhadi, A.: YOLOv3: An incremental improvement. CoRR abs/ arXiV: 1804.02767 (2018)
28. Selvaraju, R.R., Cogswell, M., Das, A., Vedantam, R., Parikh, D., Batra, D.: Grad-CAM: Visual explanations from deep networks via gradient-based localization. In: Proceedings of CVPR (2017)
29. Shen, L., Matsukawa, T., Suzuki, E.: Detecting video anomalous events with an enhanced abnormality score. In: Proceedings of PRICAI, vol. 13629 (2022)
30. Sun, C., Shi, C., Jia, Y., Wu, Y.: Learning event-relevant factors for video anomaly detection. In: Proceedings of AAAI, vol. 37 (2023)
31. Tang, Y., Zhao, L., Zhang, S., Gong, C., Li, G., Yang, J.: Integrating prediction and reconstruction for anomaly detection. Pattern Recogn. Lett. **129**, 123–130 (2020)
32. Tishby, N., Zaslavsky, N.: Deep learning and the information bottleneck principle. In: Proceedings of ITW, pp. 1–5 (2015)
33. Tong, Z., Song, Y., Wang, J., Wang, L.: VideoMAE: masked autoencoders are data-efficient learners for self-supervised video pre-training. In: Procedings of NeurIPS, vol. 35 (2022)
34. Vu, H., Nguyen, T.D., Travers, A., Venkatesh, S., Phung, D.: Energy-based localized anomaly detection in video surveillance. In: Proceedings of PAKDD (2017)

35. Wang, G., Wang, Y., Qin, J., Zhang, D., Bao, X., Huang, D.: Video anomaly detection by solving decoupled spatio-temporal jigsaw puzzles. In: Proceedings of ECCV (2022). https://doi.org/10.1007/978-3-031-20080-9_29
36. Wang, X., Wang, X., et al.: Robust unsupervised video anomaly detection by multipath frame prediction. IEEE Trans. Neural Netw, Learn. Syst. **33**(6), 2301–2312 (2022)
37. Wang, Y., Qin, C., Bai, Y., Xu, Y., Ma, X., Fu, Y.: Making reconstruction-based method great again for video anomaly detection. In: Proceedings of ICDM (2022)
38. Wang, Z., Zou, Y., Zhang, Z.: Cluster attention contrast for video anomaly detection. In: Proceedings of MM (2020)
39. Yang, Z., Liu, J., Wu, Z., Wu, P., Liu, X.: Video event restoration based on keyframes for video anomaly detection. In: Proceedings of CVPR (2023)
40. Ye, M., Peng, X., Gan, W., Wu, W., Qiao, Y.: AnoPCN: video anomaly detection via deep predictive coding network. In: Proceedings of MM (2019)
41. Yu, G., et al.: Cloze test helps: effective video anomaly detection via learning to complete video events. In: Proceedings of MM (2020)
42. Zhou, W., Li, Y., Zhao, C.: Object-guided and motion-refined attention network for video anomaly detection. In: Proceedings of ICME (2022)

STL-ConvTransformer: Series Decomposition and Convolution-Infused Transformer Architecture in Multivariate Time Series Anomaly Detection

Yu-Xiang Wu [ID] and Bi-Ru Dai[✉] [ID]

National Taiwan University of Science and Technology, Taipei 106335, Taiwan
m11115020@mail.ntust.edu.tw, brdai@csie.ntust.edu.tw

Abstract. In rapidly evolving industrial IT systems, the integration of sensor networks has become the cornerstone of operational workflows. These networks diligently collect data in the form of time series, where the format intertwines closely with temporal dependencies, crucial for anomaly detection models. Hence, the extraction of information in the time domain is advantageous for anomaly detection. To address this, we adopt a method of time series decomposition to delve into seasonality, trend, and residual components. Additionally, we design a novel algorithm that combines Transformer architecture with convolutional layers, focusing on subtle local dependencies within time series data. Extensive validation on three different real-world datasets highlights the robustness of our approach, demonstrating its proficiency in anomaly detection in time series materials. This underscores the advantage of combining convolutional strategies with Transformer architecture in capturing complex patterns and anomalies.

Keywords: time series · anomaly detection · series decomposition

1 Introduction

Time series anomaly detection is a key research area with broad applications in both academia and industry. It involves identifying data deviations from expected patterns. The swift advancement of the Internet of Things (IoT) and 5G has led to an upsurge in smart devices in factories, increasing the need for sensor monitoring. This surge makes manual anomaly detection impractical, highlighting the need for effective techniques for management and preventative maintenance. The last decade has seen a growth in sensor complexity and data volume, presenting new challenges. Research has focused on three approaches: traditional statistical, supervised, and unsupervised learning. Traditional methods use mean and variance for anomaly detection. Supervised learning treats it as a classification task, using machine learning or neural networks for feature extraction. However, anomalies are scarce, and expert labeling is costly and time-consuming, making unsupervised learning, which uses ample normal data for training, the most explored

© The Author(s), under exclusive license to Springer Nature Singapore Pte Ltd. 2024
D.-N. Yang et al. (Eds.): PAKDD 2024, LNAI 14645, pp. 41–52, 2024.
https://doi.org/10.1007/978-981-97-2242-6_4

approach. It avoids the scarcity of anomalies and applies various methods for multi-dimensional time series, with deep learning often surpassing traditional methods due to data complexity. Time series anomaly detection aims to capture temporal dependencies, a task for which conventional deep learning methods like RNNs [1] and GRUs [2] are not always effective. The transformer model with self-attention has proven successful in learning sequential representations [3, 10–12], handling multiple signal channels. Each token processed can represent single or multiple channels, leading to diverse transformer architectures [3, 10, 11].

However, the task of time series anomaly detection in long-term environments remains challenging. Long-term time series exhibit highly complex temporal patterns, making it difficult to extract temporal dependence. Many studies have begun to explore the characteristics of time series decomposition, such as traditional moving average decomposition (MA), Seasonality and Trend decomposition using Loess (STL) decomposition [4], and in recent years many studies have also begun to combine decomposition methods with deep learning [3, 13–15] applied to time series related tasks such as time series prediction and anomaly detection. The above studies all indicate that temporal decomposition has a good effect on deep learning. For example, a decomposition method to clarify complex temporal patterns was proposed in [3]. The time series is initially decomposed into trend and seasonality components and Fourier transform is used to extract frequency domain features. However, time series data not only contain trend and seasonality, but also residual elements [4], which may have a significant impact on the time series. In [15], Seasonality and Trend decomposition using Loess (STL) was employed to decompose time series data into seasonality and trend components. This is different from the MA decomposition. Using this decomposition can reduce the impact of anomalies during the decomposition process.

In summary, given that time series decomposition can extract features from time series data and the transformer architecture is capable of retaining long-term temporal dependencies while capturing relationships between different features, we propose a framework design. In this design, we adopt Seasonality and Trend decomposition using Loess (STL) to disentangle time series data, thereby extracting seasonal, trend, and residual features. Moreover, we enhance the transformer architecture by modifying the feed-forward network (FFN) to use one-dimensional convolution. This design improvement allows for more effective capture of local relationships within the data. Our contributions are summarized as follows:

1. This paper adopts the STL decomposition method and modifies the feed-forward network (FFN) of the traditional transformer to use a one-dimensional convolutional network. This design allows the model to better capture the local dependencies in the decomposed time series data.
2. Experimental results on multiple real datasets demonstrate that our proposed method achieves state-of-the-art performance in time series anomaly detection.

2   Related Work

2.1   Prediction-Based Models

Prediction-based models use sophisticated deep learning components to capture spatial-temporal relationships within data and assess anomalies based on prediction errors. These models benefit from well-designed modules that enhance prediction accuracy, ultimately improving their ability to detect abnormal instances. For instance, [7] introduce a Graph Neural Network (GNN)-based approach to facilitate information aggregation among sensors. Similarly, [8] combines Variational AutoEncoder (VAE) for feature extraction and system representation with Graph Neural Network (GNN) and stochastic graph relational learning to capture inter-sensor dependencies.

2.2   Reconstruction-Based Models

The reconstruction-based models attempt to detect the anomalies by the reconstruction error. The OmniAnomaly model [5], as introduced by Su et al. (2019), extends the LSTM-VAE model by incorporating a normalizing flow and leveraging reconstruction probabilities for anomaly detection. In their work, Su et al. (2019) employ a stochastic Recurrent Neural Network (RNN) to discover robust representations for multivariate time series data. The architecture of an autoencoder that is inspired by Generative Adversarial Networks (GANs) and utilizes adversarial-style learning is detailed in [6]. This setup features a singular encoder paired with dual decoders, all working in unison to reconstruct the input from time series data.

2.3   Transformers for Time Series Analysis

Recently, Transformers [9] have shown great power in sequential data processing, such as natural language processing, audio processing and computer vision. For time series analysis, benefiting from the advantage of the self-attention mechanism, Transformers are used to discover reliable long-range temporal dependencies. The Anomaly Transformer [10], utilizing a novel Anomaly-Attention mechanism and the concept of association discrepancy, effectively addresses the challenging problem of unsupervised time series anomaly detection. A deep transformer-based anomaly detection [11] and diagnosis model that efficiently addresses the challenges of pinpointing anomalies in multivariate time series data by leveraging attention-based sequence encoders, focus score-based self-conditioning, adversarial training, and model-agnostic meta-learning (MAML). Decomposes time series into seasonality and trend components and utilizes transformer [3] to effectively capture complex temporal dependencies.

3 Methodology

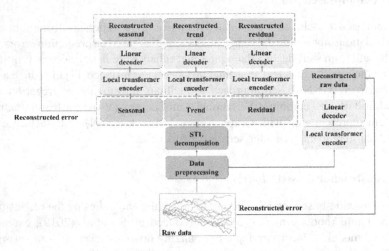

Fig. 1. The STL-transformer Architecture

3.1 Problem Formulation

We consider a multivariate time series as a sequence of data points of size T, each with a timestamp. It can be defined as $X = \{x_0, x_1, x_2 \ldots x_{T-1}\}$, where T is the length of the time series, x_t is an observation at timestamp t, and $0 \leq t < T$. Define the number of sensors as m, $x_t \in R^m$ and $X \in R^{T \times m}$.

Anomaly Detection: Given an original time series, X, for training, our objective is to identify irregularities in a new, unseen time series, X_{test}, which has a length of $\widehat{T}$ and shares the same characteristics as the training series. The unsupervised time series anomaly detection problem is to predict an output $Y = \{y_0, y_1, y_2 \ldots y_{\widehat{T}-1}\}$ where $y_t \in \{0, 1\}$, $0 \leq t < \widehat{T}$ to denote whether X_{test} is an anomaly.

3.2 Overall Architecture

The STL-ConvTransformer framework is shown in Fig. 1, integrates STL sequence decomposition with the encoder of transformer. We have revamped the traditional feed-forward network (FFN) in the transformer by incorporating a one-dimensional convolution, enhancing the focus of the model on local dependencies. In our preprocessing, feature selection is applied to eliminate irrelevant columns, optimizing the number of input fields and thus reducing computational time. This framework splits time series data into two parts: the undecomposed sequence maintaining original feature relationships, and the decomposed sequence revealing complex characteristics. Subsequently, our proposed local transformer extracts finer-grained features, and after a linear decoder, the reconstructed data is used to calculate the reconstructed error. We will provide a detailed explanation of the methodology.

3.3 Data Preprocessing

In this section, to enhance the capability of the model to establish an accurate representation, we undertake several preprocessing steps on the data. Multivariate time series inherently consists of data spanning multiple dimensions. However, not all these dimensions necessarily contribute beneficially to the predictive performance of the model. Our initial approach involves employing the Pearson correlation coefficient [12] to filter out specific features. The Pearson coefficient quantifies the linear relationship between features and the label. Given that our task label is binary, oscillating between 0 and 1, any feature exhibiting only 0 and 1 values, resulting in a Pearson coefficient of 0, is deemed non-contributory to our primary objective and is consequently disregarded. After feature selection, in order to prevent numerical instability and improve the performance of the model, we normalize the data.

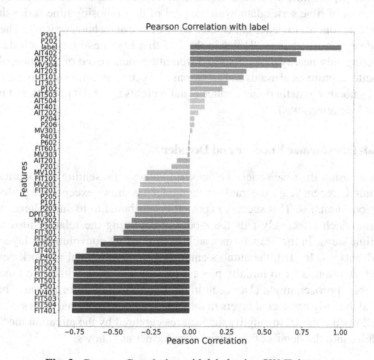

Fig. 2. Pearson Correlation with label using SWaT dataset

In Fig. 2, the SWaT [18] dataset, which consists of sensor data collected from underground water pipelines, as an example. Upon analysis, we found that the sensors P202 and P301 exhibited a Pearson coefficient value of 0. Hence, we deduced that these features have little relevance to our anomaly detection and thus are excluded in the following steps.

3.4 Decomposition Block

There is a variety of techniques available for decomposing time series, ranging from basic approaches like moving averages to more classical strategies, including additive and multiplicative decompositions. Other notable methods are the X11 decomposition [20], seat decomposition [20], and the STL decomposition [4]. However, in our study, we have chosen to use the STL decomposition [4] for its simplicity in implementation and its proven efficacy in real-world applications. In the approach presented by [3], the series data is decomposed into two components: trend and seasonality, represented as $X = S + P$. Here, S represents the seasonality of the series, while P denotes the trend of the series. In traditional statistical methods, the trend is extracted using the Moving Average (MA) approach, leading to the formula Hence, $S = X - MA(X)$. However, as mentioned in [4], a time series can further be decomposed into residuals represented as $X = S + P + R$. Here, R represents the residuals of the series. Relying solely on Moving Average (MA) for trend extraction might be influenced by outliers in the data. Thus, we adopt the STL (Seasonal Decomposition of Time Series by Loess) method introduced in [4] for time series decomposition. STL decomposition is a method used for the analysis of time series data, with the goal of decomposing time series data into three main components to better understand its structure and characteristics. The reason why this design is better than [3] lies in the fact that Loess estimates each data point by considering only nearby data points and weighting them based on distance, allowing Loess to better capture local trends and variations in the data. In contrast, Moving Average is a global smoothing method that applies equal weights to all data points and may not handle local variations well.

3.5 Local-Transformer Encoder and Decoder

After decomposing the time series, we obtained values representing seasonality, trend, and residuals. In recent years, the transformer model has shown exceptional performance across various domains. This success is primarily attributed to its multi-head attention mechanism, which effectively aids the model in capturing the relationships between different time steps. In the Transformer architecture, using convolutional layers offers distinct advantages. In [16], the authors employed a convolutional network before the multi-head attention layer to initially process local information within the time series. However, this approach might also result in the loss of certain information. Therefore, we replaced the fully connected layers in the FFN with convolutional layers. This modification allows the model to amplify the features captured by the attention mechanism. We will delve into the details of the Local-transformer as follows.

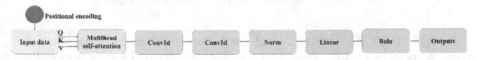

Fig. 3. Local-transformer encoder

As shown in Fig. 1, suppose we have N encoder layers in Local-transformer block. The overall equations for l^{th} encoder layer are summarized as $X^l = Encoder\,(X^{l-1})$. We will present the details of the Local-transformer encoder as shown in Fig. 3 as follows:

Initialization

Query: $Q = W_q \cdot X$, Key: $K = W_k \cdot X$, Value: $V = W_v \cdot X$ where, X is the input and W_q, W_k, W_v are weight matrices. Attention score:

$$Attention(Q, K) = Softmax\left(\frac{Q \times K}{\sqrt{d_k}}\right), \tag{1}$$

where d_k is the dimension of each head. Attention output:

$$O = Attention(Q, K) \cdot V. \tag{2}$$

The multi-head output is the concatenation of outputs from all heads followed by a linear transformation:

$$MultiHead(Q, K, V) = W_o \cdot Concat(head_1 \cdots head_h), \tag{3}$$

where h is the number of heads and W_o is the output weight matrix. For each encoder layer:

$$X\prime = LayerNorm(X + MultiHead(Q, K, V)), \tag{4}$$

$$Y = Activation\left(Conv1\left(X'\right)\right), \tag{5}$$

$$Z = LayerNorm(Conv2(Y)). \tag{6}$$

For each encoder layer, the input X is combined with the multi-head attention output, and then passed through layer normalization to produce $X\prime$, $X\prime$ is then passed through a 1D convolutional layer named *Conv1*, followed by an activation function to generate Y. Finally, Y undergoes another 1D convolutional layer named *Conv2* and is then subjected to layer normalization to produce Z. We have replaced the original fully connected layers with *Conv1* and *Conv2*, which are 1D convolution layers. This allows the model to focus more on capturing local dependencies after attention weights are applied.

Finally, our model reconstructs the time series using the embeddings X^N (N is the number of encoder layers), i.e.,

$$\widehat{X} = Decoder\left(X^N\right) \tag{7}$$

where $\widehat{X}$ is the restructured time series, $Decoder(\cdot)$ is a simple linear model.

3.6 Loss Function and Anomaly Score

In this study, the reconstructed error is then computed relative to the original decomposed values. At the same time, the raw data is utilized in conjunction with the transformer-extracted features. Our loss function is formalized as:

$$L_{raw}(X) = \left|\widehat{X} - X\right|^2 \tag{8}$$

$$L_{decompose}(S) = \left|\hat{S} - S\right|^2 \tag{9}$$

$$L_{decompose}(P) = \left|\hat{P} - P\right|^2 \tag{10}$$

$$L_{decompose}(R) = \left|\hat{R} - R\right|^2 \tag{11}$$

$$L_{Final} = L_{raw}(X) + L_{decompose}(S) + L_{decompose}(P) + L_{decompose}(R) \tag{12}$$

We employ the Mean Squared Error (MSE) for the computation of the loss value. Here, X represents the original data and $\hat{X}$ represents the reconstruction of the original data, S represents the seasonal data and $\hat{S}$ denotes the reconstruction of the seasonal data, P represents the trend data and $\hat{P}$ signifies the reconstruction of the trend data, R represents the residual data and $\hat{R}$ corresponds to the reconstruction of the residual data. Consequently, this design allows for the preservation of features from the original data while also capturing characteristics from the decomposed data. After the training phase is completed, during the testing phase, we apply the test data to the trained model to produce a reconstructed dataset, and Mean Squared Error (MSE) is utilized to compute the error between the original and reconstructed data as an anomaly score. Subsequently, a threshold is set, we establish our threshold by using a loss average, and by selecting the configuration that performs best on the test dataset; values exceeding this threshold are considered anomalies, while those below it is deemed normal.

$$\text{AnomalyScore}(X_{test}) = \sum_{i=0}^{t-1} \left|\widehat{x_{test_i}} - x_{test_i}\right|^2. \tag{13}$$

The AnomalyScore is also calculated using the MSE (Mean Squared Error). We feed the test data X_{test} into the trained model. Here, x_{test_i} represents the data at the ith time point, and $\widehat{x_{test_i}}$ stands for the reconstructed data.

4 Experiments

In this section, we will introduce these public datasets we have utilized and provide an overview of the current state-of-the-art models, followed by a presentation of our results.

Table 1. Statistics of datasets used in experiments.

Datasets	#Features	#Train	#Test	Anomalies
WADI	127	1209601	172801	5.99%
SWaT	51	495000	449919	11.97%
PSM	26	132481	87841	27.76%

Datasets
Here is a description of the three experiment datasets, and Table 1 summarizes the
statistics of the three datasets.

(1) WADI (Water Distribution) [17] is a sophisticated water distribution system equipped
with 123 sensors and actuators, designed to operate continuously and mon-itor its
vast network of pipelines. Over a period of 16 days, it was subjected to a comprehensive evaluation, including 14 days of regular functioning and 2 days of simulated attack scenarios. These attack scenarios were based on an advanced attack
model, specifically crafted to test the resilience of cyber-physical systems (CPS),
re-sulting in a series of 15 distinct attacks executed during the trial period.
(2) SWaT (Secure Water Treatment) [18] is a monitored critical infrastructure system,
tested over 11 days with 51 sensors. It operated normally for 7 days and faced 41
attacks over 4 days, based on an advanced CPS attack model. Data from sensors,
actuaries, and network traffic were labeled to identify abnormal behaviors.
(3) Pooled Server Metrics (PSM) [19], collected from eBay's server nodes, contains 26
anonymized features over 21 weeks—13 for training and 8 for testing. It includes
labeled anomalies in the testing phase, identified by experts, providing insights into
server metrics like CPU usage and memory.

Table 2. Performance comparison of the baseline methods (as %, red: best, blue: second best).
The '-' in the table on the WADI dataset indicates that the values are not provided in DATN [3].

Dataset	WADI			SWaT			PSM		
Method	P	R	F1	P	R	F1	P	R	F1
Omianomaly	98.25	64.97	78.22	81.42	84.3	82.83	88.39	74.46	80.83
USAD	99.47	13.18	23.28	98.51	66.18	79.17	92.1	57.6	70.9
GRELEN	77.3	61.3	68.2	95.6	83.5	89.1	94.2	92.1	93.1
GDN	97.5	40	57	99.35	68.12	81	43.4	76	55.2
DATN	-	-	-	95.38	85.32	90.12	97.39	87.49	92.25
TranAD	35.29	82.96	49.51	97.6	69.97	81.51	88.15	84.91	89.72
anomaly-transformer	63.8	99	77.95	86.38	92.14	88.17	94.75	88.59	90.56
STL ConvTransformer	84.68	86.26	85.46	97.32	82.77	89.46	94.71	98.26	96.45

Baselines

We compare our model with seven state-of-the-art baselines:

(1) OmniAnomaly [5]: The model, a stochastic recurrent neural network that combines the Gated Recurrent Unit (GRU) and Variational Autoencoder (VAE), is designed to capture the normal patterns of multivariate time series.
(2) USAD [6]: The model uses adversarially trained autoencoders for efficient unsupervised anomaly detection.
(3) GRELEN [8]: The model utilizes VAE (Variational AutoEncoder) as the overarching framework for feature extraction and system representation. Additionally, Graph Neural Networks (GNN) and stochastic graph relation learning strategies are employed to capture dependencies between sensors.
(4) GDN [7]: The model integrates a structure learning approach with graph neural networks, further leveraging attention weights to offer explain ability for the identified anomalies.
(5) DATN [3]: The model decomposes the time series into seasonality and trend components, and restructures it as a fundamental inner block of a deep model.
(6) AnomalyTran [10]: The model is designed with a unique Anomaly-Attention mechanism to assess the association discrepancy. Through the implementation of a minimax strategy, it amplifies the contrast between normal and abnormal patterns within the association discrepancy.
(7) TranAD [11]: The model uses an adversarial training program to amplify the reconstruction error because the simple transformer-based networks often miss small abnormal deviations.

Table 3. Ablation study of different variants of STL-ConvTransformer. The best result is in bold.

Components			F1 Measure (as %)					
			WADI		SWaT		PSM	
S	T	R	w/	w/o	w/	w/o	w/	w/o
✓	✓	✓	85.46	75.22	89.46	79.5	**96.45**	93.93
✓	✓		**86.76**	73.74	87.32	85.84	96.42	**94.26**
✓		✓	81.51	75.61	**90**	79.71	96.01	91.66
	✓	✓	84.75	67.31	87.67	78.78	96.21	94.1
✓			86.51	79.2	86.95	**88.53**	95.1	91.23
	✓		82.6	**80.42**	85.99	86.26	96.25	89.68
		✓	83.87	60.93	87.08	87.97	95.17	88.43

Results

We adopted precision (P), recall (R), and F1 as our evaluation metrics. To comprehensively evaluate the three datasets, we replicated some results from the baselines, as shown in Table 2. In our experimental results, we achieved the best performance on the PSM dataset in terms of the F1 metric, and secured the second-best performance on the SWaT dataset. These outcomes validate that decomposing the time series into seasonality, trend, and residuals can enhance the ability of the model to learn robustness over longer time frames. In our ablation study, we also observed the significance of the decomposed sequences to the model, which we will detail further below.

Ablation Study
For a better understanding of the advantages brought by each component of our method, we conducted an ablation study on three datasets, as shown in Table 3. We averaged the results over two runs for each dataset configuration. The components of each time series decomposition, including seasonality (S), trend (T), and residual (R), were combined separately with the complete time series data. In Table 3, the term 'w/' indicates that the model incorporates the undecomposed sequence $L_{raw}(X)$, while 'w/o' signifies its exclusion.

1. Firstly, when the undecomposed sequence branch is incorporated (w/), the performance is better than when it is not used (w/o).
2. In the case of the WADI dataset, excluding the R component yields better F1 scores, suggesting that for this dataset, the R component may not be important and may even hinder performance and that for the SWaT dataset, the omission of the trend component improves F1 score. Regarding the PSM dataset, the F1 scores of the models all decreased slightly when specific components were removed, indicating that each component has an impact on the PSM dataset.

In summary, STL-ConvTransformer improves the performance of anomaly detection through temporal decomposition and novel Local-transformer, especially in terms of F1 score. Its core strength lies in its temporal decomposition capabilities, paired with the innovative Local-transformer, which unlocks new potentials for detailed local feature extraction. This method does not just excel in isolation; it thrives when amalgamated with various decomposition elements, such as seasonality, trend, and residual components. Such integration magnifies the acuity of model in capturing nuanced local patterns, thereby dramatically elevating the overall anomaly detection performance.

5   Conclusion

In this study, we innovatively propose a method that integrates STL decomposition with the transformer architecture to handle time series data. Furthermore, we have revamped the traditional transformer architecture by replacing its fully connected feed-forward network (FFN) with one-dimensional convolution, enhancing the ability of the model to capture local features within time series. Our approach has been extensively tested on real-world time series datasets and compared with other state-of-the-art anomaly detection algorithms. The experimental results have clearly demonstrated that the decomposed data are more focused on local features, thereby significantly improving the overall performance of the model.

References

1. Hoshi, Y., Park, D., Kemp, C.C.: A multimodal anomaly detector for robot-assisted feeding using an LSTM-based variational autoencoder. IEEE Robot. Autom. Lett. **3**, 1544–1551 (2018)
2. Chung, J., Gulcehre, C., Cho, K.H., Bengio, Y.: Empirical evaluation of gated recurrent neural networks on sequence modeling. arXiv preprint arXiv:1412.3555 (2014)
3. Wu, B., et al.: Decompose auto-transformer time series anomaly detection for network management. Electronics **12**(2), 354 (2023)
4. Cleveland, R.B., et al.: STL: a seasonality -trend decomposition. J. Off. Stat **6**(1), 3–73 (1990)
5. Su, Y., et al.: Robust anomaly detection for multivariate time series through stochastic recurrent neural network. In: Proceedings of the 25th ACM SIGKDD International Conference on Knowledge Discovery and Data Mining (2019)
6. Audibert, J., et al.: USAD: unsupervised anomaly detection on multivariate time series. In: Proceedings of the 26th ACM SIGKDD International Conference on Knowledge Discovery and Data Mining (2020)
7. Deng, A., Hooi, B.: Graph neural network-based anomaly detection in multivariate time series. In: Proceedings of the AAAI Conference on Artificial Intelligence, vol. 35, no. 5 (2021)
8. Zhang, W., Zhang, C., Tsung, F.: Grelen: multivariate time series anomaly detection from the perspective of graph relational learning. In: Proceedings of the Thirty-First International Joint Conference on Artificial Intelligence, IJCAI-22, vol. 7 (2022)
9. Vaswani, A., et al.: Attention is all you need. In: Advances in Neural Information Processing Systems, vol. 30 (2017)
10. Xu, J., et al.: Anomaly transformer: time series anomaly detection with association discrepancy. arXiv preprint arXiv:2110.02642 (2021)
11. Tuli, S., Casale, G., Jennings, N.R.: Tranad: deep transformer networks for anomaly detection in multivariate time series data. arXiv preprint arXiv:2201.07284 (2022)
12. Cohen, I., et al.: Pearson correlation coefficient. In: Noise Reduction in Speech Processing, pp. 1–4 (2009)
13. Zhang, C., et al.: TFAD: a decomposition time series anomaly detection architecture with time-frequency analysis. In: Proceedings of the 31st ACM International Conference on Information and Knowledge Management (2022)
14. He, X., et al.: OneShotSTL: one-shot seasonality -trend decomposition for online time series anomaly detection and forecasting. arXiv preprint arXiv:2304.01506 (2023)
15. Ouyang, Z., Jabloun, M., Ravier, P.: STLformer: exploit STL decomposition and rank correlation for time series forecasting. In: 31th European Signal Processing Conference (EUSIPCO) (2023)
16. Li, S., et al.: Enhancing the locality and breaking the memory bottleneck of transformer on time series forecasting. In: Advances in Neural Information Processing Systems, vol. 32 (2019)
17. Ahmed, C.M., Palleti, V.R., Mathur, A.P.: WADI: a water distribution testbed for research in the design of secure cyber physical systems. In: Proceedings of the 3rd International Workshop on Cyber-Physical Systems for Smart Water Networks (2017)
18. Mathur, A.P., Tippenhauer, N.O.: SWaT: a water treatment testbed for research and training on ICS security. In: 2016 International Workshop on Cyber-Physical Systems for Smart Water Networks (CySWater). IEEE (2016)
19. Abdulaal, A., Liu, Z., Lancewicki, T.: Practical approach to asynchronous multivariate time series anomaly detection and localization. In: Proceedings of the 27th ACM SIGKDD Conference on Knowledge Discovery and Data Mining (2021)
20. Dagum, E.B., Bianconcini, S.: Seasonal Adjustment Methods and Real Time Trend-Cycle Estimation. Springer, Heidelberg (2016). https://doi.org/10.1007/978-3-319-31822-6

TOPOMA: Time-Series Orthogonal Projection Operator with Moving Average for Interpretable and Training-Free Anomaly Detection

Shanfeng Hu[1(✉)] and Ying Huang[2]

[1] Department of Computer and Information Sciences, Northumbria University,
Newcastle-upon-Tyne NE1 8ST, UK
`shanfeng2.hu@northumbria.ac.uk`
[2] Institute of Virtual Reality and Intelligent Systems, Hangzhou Normal University,
Hangzhou 311121, China
`yw52@hznu.edu.cn`

Abstract. We present *TOPOMA*, a time-series orthogonal projection operator with moving average that can identify anomalous points for multivariate time-series, without requiring any labels nor training. Despite intensive research the problem has received, it remains challenging due to 1) scarcity of labels, 2) occurrence of non-stationarity in online streaming, and 3) trust issues posed by the black-box nature of deep learning models. We tackle these issues by avoiding training a complex model on historical data as in previous work, rather we track a moving average estimate of variable subspaces that can compute the deviation of each time step via orthogonal projection onto the subspace. Further, we propose to replace the popular yet less principled global thresholding function of anomaly scores used in previous work with an adaptive one that can bound the occurrence of anomalous events to a given small probability. Our algorithm is shown to compare favourably with deep learning methods while being transparent to interpret.

Keywords: anomaly detection · multivariate time series · orthogonal projection operator · moving average · subspace

1 Introduction

Industrial systems are getting increasingly complex. For safe running, sensors are adopted to monitor critical machinery, and capturing abnormal states online that deviate from baselines is essential for timely intervention [20].

One line of work on multivariate time-series anomaly detection phrases the problem as supervised learning, in which the sensor state of each moment is classified into normal or abnormal based on hand-crafted features [12] or learned ones [18]. Despite their high performance, these methods have certain flaws. First, they require anomaly labels for training, which can be hard to collect.

© The Author(s), under exclusive license to Springer Nature Singapore Pte Ltd. 2024
D.-N. Yang et al. (Eds.): PAKDD 2024, LNAI 14645, pp. 53–65, 2024.
https://doi.org/10.1007/978-981-97-2242-6_5

Second, they need to be trained on historical data prior to deployment, making them inapplicable for system start periods. And third, they transfer patterns learned from the past to future, which can be invalid for data generated from online systems that undergo unknown changes.

Devoid of the label issue, a body of unsupervised learning methods have also been proposed to mine anomalous patterns. They aim to compute an anomaly score for each time step based on probability density estimation [9], distance from similar past patterns [3], or prediction error [28]. Due to training on historical data, these methods can still suffer from distributional shifts of online time-series streams as for supervised learning.

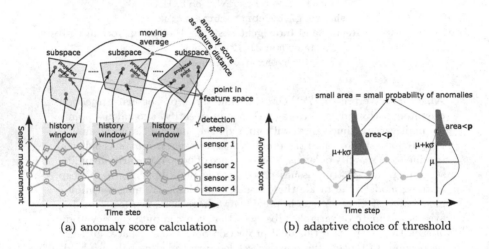

(a) anomaly score calculation   (b) adaptive choice of threshold

Fig. 1. Illustration of our anomaly detection algorithm. (a): A moving average of orthogonal projection operators associated with each sensor measurement subspace is estimated to compute anomaly scores. (b): The detection threshold is dynamically chosen at each step to bound the occurrence of anomalies to a small fixed probability value based on online estimated distributional parameters.

Deep learning has revolutionized anomaly detection [8]. Their strive for performance, however, comes with the pitfall that the resulting models are often highly parameterized and the decisions made are hard to explain to people [4]. This is in opposition to engineering in which the working principles of each component need to be sufficiently characterized, especially in safety-critical scenarios. Anomaly detectors built on deep learning do not satisfy this criterion.

To tackle those difficulties, we propose a new anomaly detector that is training-free (hence requiring no labels) and naturally handles online streaming data from system start, while admitting a simple algebraic form for fast computation and easy interpretation. One key of our algorithm (Fig. 1a) is that we construct an orthogonal projection operator associated with the subspace spanned by a window of variable measurements preceding each time step, which allows us to measure the distance from each step and its projection onto the subspace to derive the anomaly score. On top of this, we leverage the linearity of these operators to maintain an online moving average of them to incorporate both short and long-range time dependencies, thereby adapting to changing system dynamics. Achieving these goals, still, we have avoided introducing any parameters in the algorithm that need to be determined through training on historical records - supervised or unsupervised. As a result, interpretability is retained compared with black-box deep learning [4].

The other core of our algorithm (Fig. 1b) is the adaptive choice of anomaly score thresholds using probabilistic reasoning. We argue that setting a global threshold to decide if each computed score is high enough to warrant an alert, as widely adopted in previous work, lacks a principled foundation. It has long been overlooked that anomalous events are by nature small probability events according to an unknown distribution of scores, which means that we need to choose a global small probability that can in turn determine an appropriate local threshold at each moment. This motivates us to estimate the distribution of anomaly scores online and calculate each threshold so that the likelihood of any score falling above it is bounded by a given small value using Chebyshev's Inequality. This technique is applicable to all score-based methods.

The proposed contributions of this paper include the following:

- A new online algorithm that uses the moving average of time-series orthogonal projection operators for anomaly score calculation, which is interpretable and training-free by design.
- A principled technique that adaptively selects an anomaly score threshold for each step to enforce the small probability nature of anomaly events.

We evaluate the proposed algorithm on two open datasets [2,19] for comparison with recent deep learning methods. On top of its training-free and interpretable nature, it performs favourably against recent work.

2   Related Work

Time-series anomaly detection is an extensively studied topic [7]. Recent methods can be broadly categorized into supervised or unsupervised, depending on whether they require ground-truth labels to train. While supervised methods [10,12,18,27] are effective at discriminative modelling, they can be limited by the availability of scarce anomaly labels and suffer from distributional shifts on online streaming time-series due to training on historical data. Unsupervised methods, in contrast, approach the problem by discovering patterns that significantly deviate from estimated baselines. Some methods in this category use parametric [23] or non-parametric [5,9] density estimation to locate data points that are far from high-density regions. Methods [3,6] that use distances to identify abnormal/salient points are also implicitly related to non-parametric density estimation. Deep learning has also been used for anomaly detection, with auto-encoders [28], convolutional neural networks [24], recurrent neural networks

[21], generative adversarial networks [15,16,26], variational auto-encoding networks [17,29], as well as graph neural networks [8]. These methods are often highly parameterized, which can be hard to interpret. Our method, in contrast to existing work, does not require any anomaly labels nor training while being inherently interpretable thanks to its clear algebraic structure.

3   *TOPOMA*: Our Proposed Anomaly Detector

3.1   Problem Formulation

Suppose a collection of N sensors, $i \in \{1, 2, \cdots, N\}$, each of which records a variable of interest, $x_t^i \in \mathbb{R}$, that is real-valued and may change over time for $t \geq 1$. We denote $x_{t_1:t_2}^i \in \mathbb{R}^{1 \times (t_2 - t_1 + 1)}$ as the row vector of measurements by the i-th sensor from step t_1 to t_2. Similarly, we denote $x_t^{1:N} \in \mathbb{R}^{N \times 1}$ as the column vector of measurements by all sensors at step t. Combining both, we can define $x_{t_1:t_2}^{1:N} \in \mathbb{R}^{N \times (t_2 - t_1 + 1)}$ as the matrix of measurements by all sensors from t_1 to t_2. Anomaly detection is constructing a function $f : x_{1:t}^{1:N} \to y_t \in \{0, 1\}$, where $y_t = 0$ means that the system state at time step t is normal and $y_t = 1$ otherwise. Approaching the problem without requiring labels, we represent f as the composition of two functions: $f = h \circ s$, where $s : x_{1:t}^{1:N} \to s_t \in \mathbb{R}^{\geq 0}$ is a scoring function that measures how much the system state deviates from previous baselines, and $h : s_t \to y_t$ is the decision-making function with a threshold Δ_t: $h(s_t) = 1$ if $s_t > \Delta_t$ and 0 otherwise.

3.2   Moving Average of Orthogonal Projection Operators

Here, we propose an interpretable and training-free method to calculate the anomaly score s_t for each time step t by computing the distance from the sensor measurements $x_t^{1:N}$ to the subspace spanned by the vectors of a few previous steps. The key is representing the subspace using its associated orthogonal projection operator. We then use moving average to fuse both short and long-range time dependencies, enabling online anomaly detection.

Following previous work [8], we assume that the historical behaviours of all sensors up to time step t are encapsulated by the preceding M steps as $x_{t-M:t-1}^{1:N} \in \mathbb{R}^{N \times M}$ for $t \geq (M + 1)$. To motivate our operator, we start with the idea of representing each sensor's historical data, $x_{t-M:t-1}^i$, as a linear combination of all sensors' measurements $\{x_{t-M:t-1}^i\}_{1 \leq i \leq N}$, solving for a matrix $\mathcal{O} \in \mathbb{R}^{N \times N}$ such that $x_{t-M:t-1}^{1:N} = \mathcal{O} x_{t-M:t-1}^{1:N}$. This way, each coefficient $\mathcal{O}_{ij}$ in the matrix captures the relationship between the pair of i-th and j-th sensor's historical behaviours. The solution to this problem is not uniquely determined when $M < N$ but it can be reformulated as

$$\mathcal{O} = \arg \min \|O\|_F^2 \text{ s.t. } x_{t-M:t-1}^{1:N} = O x_{t-M:t-1}^{1:N} \tag{1}$$

in which the objective is the Frobenius norm, $\|O\|_F^2 = \sum_{i=1}^{N} \sum_{j=1}^{N} O_{ij}^2$. Solving this leads to the optimal solution as

$$\mathcal{O} = x_{t-M:t-1}^{1:N} [(x_{t-M:t-1}^{1:N})^T x_{t-M:t-1}^{1:N}]^{-1} (x_{t-M:t-1}^{1:N})^T \tag{2}$$

Algorithm 1. *TOPOMA*: multivariate time-series anomaly detector

---

Input: $\boldsymbol{x}_{1:L}^{1:N} \in \mathbb{R}^{N \times L}$, a time series with N sensors, L is set to infinity for online detection for streaming data.

Meta Parameters: $1 \le M$, history size; $0 < p < 1$, probability; $\alpha = 0.95$, $\alpha_\mu = 0.999$, $\alpha_v = 0.999$, moving average weight; norm=L2, for anomaly score calculation.

Output: $\boldsymbol{y}_{1:L} \in \{0, 1\}^{1 \times L}$, anomaly detection result.

1: Let $t = 1$
2: **while** $t \le L$ **do**
3:    Update the operator $\mathcal{O}_{t-1}$ using Eq. (4)
4:    Compute the anomaly score s_t using Eq. (5)
5:    Update the statistic $\hat{\mu}_{t-1}$ using Eq. (6)
6:    Update the statistic $\hat{v}_{t-1}$ using Eq. (7)
7:    Estimate the threshold Δ_t using Eq. (8)
8:    **if** $s_t > \Delta_t$ **then**
9:       Let $y_t = 1$
10:    **else**
11:       Let $y_t = 0$
12:    **end if**
13:    $y_t \leftarrow y_t$
14:    $t \leftarrow t + 1$
15: **end while**
16: **return** $\boldsymbol{y}_{1:L}$

---

where T is transpose and $[\cdot]^{-1}$ takes the inverse of a square matrix. In practice, we add $\epsilon \boldsymbol{I}_{M \times M}$ to prevent inverting a potentially singular matrix, where ϵ is a small constant and $\boldsymbol{I}_{M \times M}$ is the identity matrix.

It is easy to verify from (2) that $\boldsymbol{x}_{t-M:t-1}^{1:N} = \mathcal{O}\boldsymbol{x}_{t-M:t-1}^{1:N}$, indicating that it is a projection operator onto the subspace spanned by the column vectors of the sensor history matrix. Further, the property that $\mathcal{O} = \mathcal{O}^T$ and $\mathcal{O} = \mathcal{O}^2$ makes $\mathcal{O}$ the subspace's unique orthogonal projection operator [11]. Given any time step within the history window, the corresponding sensor measurements are perfectly represented under the operator, $\boldsymbol{x}_k^{1:N} = \mathcal{O}\boldsymbol{x}_k^{1:N}$ for $t - M \le k \le t - 1$, which is consistent with our intuition that a record should not deviate from the baseline when it is part of the evidences for deriving the baseline. Nevertheless, most anomaly detectors do not explicitly impose this constraint.

Now, we can compute the anomaly score s_t for sensors at time step t as

$$s_t = \|\mathcal{O}\boldsymbol{x}_t^{1:N} - \boldsymbol{x}_t^{1:N}\|_{\text{norm}} \tag{3}$$

where we use the operator derived from immediate history (as in Eq. (2)) to project the current record onto the subspace of system baselines and measure how far the record is from its projection. If the deviation is large, then the system may have undergone an abrupt change that may signal an anomaly; otherwise, the system behaves consistently with the preceding history. The choice of a suitable norm in (3) depends on applications.

Having derived Eq. (3) for a single step, we now extend the calculation to a full time-series. The key idea is that we leverage the linearity of the operator to maintain a moving average of all previous operators up to each time step. Slightly abusing the notation, we define $\mathcal{O}_{t-1}$ as the operator used for anomaly score calculation at time step t, which is computed recursively as

$$\mathcal{O}_{t-1} = \begin{cases} \alpha \mathcal{O}_{t-2} + (1-\alpha)\mathcal{O} & \text{if } t \geq M+1 \\ \alpha \mathcal{O}_{t-2} + (1-\alpha)\mathcal{O}' & \text{if } 1 < t < M+1 \\ \boldsymbol{I}_{N \times N} & \text{otherwise} \end{cases} \quad (4)$$

where $\boldsymbol{I}_{N \times N}$ is the identity matrix, $\mathcal{O}$ is computed in Eq. (2) using the M-steps sensor history $\boldsymbol{x}_{t-M:t-1}^{1:N}$, and $\mathcal{O}'$ is computed the way as for $\mathcal{O}$ but only with the history $\boldsymbol{x}_{1:t-1}^{1:N}$ that has fewer than M steps. The coefficient $\alpha \in (0,1)$ controls the degree of weight decay, with a larger value incorporating contributions from more distant pasts.

Plugging Eq. (4) into (3), we obtain the anomaly score formula for a full time-series

$$s_t = \|\mathcal{O}_{t-1}\boldsymbol{x}_t^{1:N} - \boldsymbol{x}_t^{1:N}\|_{\text{norm}} \quad (5)$$

in which it is clear that $\mathcal{O}_{t-1}\boldsymbol{x}_t^{1:N} = (1-\alpha)\mathcal{O}_{t-2}\boldsymbol{x}_t^{1:N} + \alpha\mathcal{O}\boldsymbol{x}_t^{1:N}$ accounts for both old and new evidences for determining whether the current sensor statuses are normal or not. The evolution rules in (4) are valid because our operators are linear. For non-linear models like neural networks, such rules do not generally hold. While fine-tuning could be explored to evolve a model [13], it is non-trivial to control how much each segment of the past can contribute to the current decision-making. Our algorithm is free of this issue.

3.3   Adaptive Choice of Anomaly Score Thresholds

Having derived the anomaly score function s, we now move on to the decision-making function h. In contrast to previous work, our idea is choosing a suitable threshold Δ_t at each time step so that the chance of the score s_t falling above Δ_t is bounded by a small probability $p \in (0,1)$, thereby enforcing the small probability nature of anomaly events. Since the distribution of anomaly scores is unknown and can change through time, we propose to rely on an empirical estimate of moments of the distribution to bound the probability.

The intuition behind this is that the sequence of anomaly scores $\{s_k\}_{k=1}^{t-1}$ up to time step t are expected to fluctuate, and only when the latest score s_t is unreasonably high can it be considered as a true anomaly. To formalize this, we assume that s_t is a random variable with finite expectation μ and finite non-zero variance v. The Chebyshev's Inequality [22] states that for any real number $c > 0$, $\Pr(|s_t - \mu| \geq c\sqrt{v}) \leq \frac{1}{c^2}$. The key here is that by assuming a symmetric form of the unknown distribution, we can bound the right-hand side of the inequality as $2p$, giving us $c = \frac{1}{\sqrt{2p}}$ and the associated threshold $\Delta_t = \mu + \sqrt{\frac{v}{2p}}$. Now, we can derive an estimation of the threshold as follows:

$$\hat{\mu}_{t-1} = \begin{cases} \alpha_\mu \hat{\mu}_{t-2} + (1 - \alpha_\mu)s_{t-1} & \text{if } t \geq 3 \\ s_{t-1} & \text{if } t = 2 \\ 0 & \text{otherwise} \end{cases} \tag{6}$$

$$\hat{v}_{t-1} = \begin{cases} \alpha_v \hat{v}_{t-2} + (1 - \alpha_v)(s_{t-1} - \hat{\mu}_{t-1})(s_{t-1} - \hat{\mu}_{t-2}) & \text{if } t \geq 3 \\ 0 & \text{otherwise} \end{cases} \tag{7}$$

$$\Delta_t = \hat{\mu}_{t-1} + \sqrt{\frac{\hat{v}_{t-1}}{2p}} \tag{8}$$

where we use the Welford's method [25] in Eq. (6) and (7) for online update of the estimations of μ and v, while $\alpha_\mu \in (0, 1)$ and $\alpha_v \in (0, 1)$ are two parameters to control the rate of weight decay in moving average.

It is easy to see from Eq. (8) and Fig. 2 that the threshold becomes higher if the probability p is set to be lower, and that the online updated mean and variance of previous anomaly scores adapt the threshold through time. This is more intuitive and principled than setting a global threshold as in previous work.

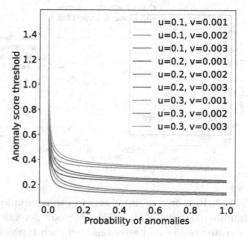

Fig. 2. Using probability to derive anomaly threshold.

3.4 Complexity Analysis

The dominant part of our algorithm (1) is computing the operator for each time step. The time complexity for processing a time-series with L steps is $\Theta(LN^2M)$ with $N \geq M$, where N is the number of sensors and M is the size of history window. The space complexity of storing the average operator is $\Theta(N^2)$. Both are quite low compared with previous learning or non-learning-based methods.

4   Results and Discussion

4.1   Synthetic Data

Here, we create a time-series (Fig. 3) with anomalies to test our algorithm ($M = 3$, $p = 0.1$, other parameters as default). First, our algorithm captures the onset of each attack accurately by producing a strong increase of anomaly score, missing none despite the variety of these anomalies. The response corresponds to the duration of each attack well, as indicated by the fall of anomaly scores. Second, our algorithm handles the severity of each anomaly proportionably, which generates the three highest spikes when the readings of all sensors experience an abrupt disturbance: spiking (#2), zeroing (#4), and dropping (#7). Smaller perturbations (e.g., #5 and #8), in contrast, elicit lower scores.

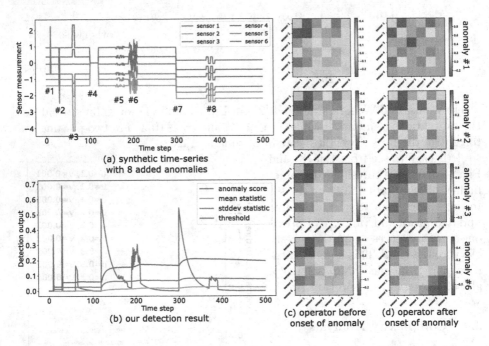

Fig. 3. Response of our algorithm to anomalies. (a): a time-series with various attacks (numbered from #1 to #8). (b): our detection output. (c) and (d): visualization of operator before and after onset of each type of anomaly.

Third, comparing operators before and after each anomaly event, it can be seen that changes arise due to underlying sensor disturbances, with more masses moved to the diagonal elements where the corresponding sensors are most significantly disrupted (#1: sensor 3, #2: sensor 2, #3: sensor 2, #6: sensor 6). By decomposing Eq. (2) we obtain $\mathcal{O}_{ij} = \frac{(x_{t_1:t_2}^i - \bar{x}_{t_1:t_2})C^{-1}(x_{t_1:t_2}^j - \bar{x}_{t_1:t_2})^T}{N} + \frac{1}{N}$ in which $\bar{x}_{t_1:t_2} = \frac{1}{N}\sum_{k=1}^{N} x_{t_1:t_2}^k$ is the average reading from time step t_1 to t_2 across all sensors and $C = \frac{1}{N}\sum_{k=1}^{N}(x_{t_1:t_2}^k - \bar{x}_{t_1:t_2})^T(x_{t_1:t_2}^k - \bar{x}_{t_1:t_2})$ is the covariance of all sensors. This shows that non-diagonal elements tend to become smaller compared with diagonal ones when certain sensors behave abnormally, reducing the pairwise correlations among them.

And fourth, the adaptive thresholds our algorithm compute (using $p = 0.1$) for each time step work well expect for anomaly #5 and #8, due to their relative mildness and growing variances of anomaly scores. This can greatly simplify the task of choosing thresholds manually and globally in applications.

4.2   Real-World Data

Now, we apply our algorithm to real-world application scenarios and compare it with recent learning-based methods. Following the recent work of [8], we choose two datasets for the study: Secure Water Treatment (SWAT) [19] and Water Distribution (WADI) [2]. The former was generated from a water treatment environment by Singapore's Public Utility Board and the latter arose from a realistic water treatment, storage, and distribution network consisting of multiple water distribution pipelines. The test sets of both datasets comprise simulated cyber-physical attacks to the systems and hence contain ground-truth anomaly labels for evaluation. We downsample both test sets every 10 time steps by taking the median values and the most common label within each period of 10 steps as the ground-truth, as done in [8]. We also apply standardization using the mean and standard deviation of each sensor readings from the corresponding training sets. We set the meta parameters of our algorithm as follows: $M = 5$ (same as in [8] for the history window), $p = 0.1$, $\alpha = \alpha_\mu = \alpha_v = 0.9999$ (higher than default to combat severe sensor noises in both datasets), norm= L_∞. Our algorithm does not require any training. Once set, it runs in online mode quickly.

Table 1. Comparison of our algorithm with machine/deep learning methods for multivariate time series anomaly detection on the SWAT [19] and WADI [2] test datasets respectively. Evaluation metrics include: Precision, Recall, and F1-Measure, as calculated with regards to true anomalies.

Method	SWAT			WADI			Feature		
	Prec.	Rec.	F1	Prec.	Rec.	F1	Parameterized	Training	Online
PCA [23]	24.92	21.63	0.23	39.53	5.63	0.10	light	light	no
KNN [3]	7.83	7.83	0.08	7.76	7.75	0.08	non-parametric	lazy	no
FB [14]	10.17	10.17	0.10	8.60	8.60	0.09	light	light	no
AE [1]	72.63	52.63	0.61	34.35	34.35	0.34	heavy	deep	no
DAGMM [30]	27.46	69.52	0.39	54.44	26.99	0.36	heavy	deep	no
LSTM-VAE [21]	96.24	59.91	0.74	87.79	14.45	0.25	heavy	deep	no
MAD-GAN [15]	98.97	63.74	0.77	41.44	33.92	0.37	heavy	deep	no
GDN [8]	99.35	68.12	0.81	97.50	40.19	0.57	heavy	deep	no
TOPOMA (ours)	24.57	**73.36**	0.37	19.66	**46.53**	0.28	no param.	not required	yes

We reproduce Table 1 from [8] and compare our method against several machine and deep learning-based methods: Princioal Component Analysis (PCA) [23], K Nearest Neighbors (KNN) [3], Feature Bagging (FB) [14], Autoencoders (AE) [1], Deep Autoencoding Gaussian Mixture Model (DAGMM) [30], LSTM-VAE [21], MAD-GAN [15], and (Graph Deviation Network) GDN [8]. It can be seen that our algorithm, despite being the only non-learning-based method in the table, achieves a good balance of detection recall and precision on both datasets. Our scores for recall are the highest, meaning that TOPOMA is

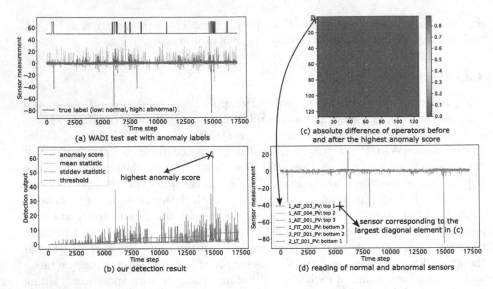

Fig. 4. WADI test results. (a): subsampled WADI test set with 127 sensors and ground-truth anomaly labels [2]. (b): our detection output. (c): absolute difference of operators right before and after the time step where our algorithm responds the strongest to the underlying signals. (d): largest changes in the diagonal elements of operator indicate abnormal sensors.

able to capture a larger number of ground-truth attacks compared with the competitors. On the SWAT test set, it surpasses the three light learning algorithms (PCA, KNN, FB) while being closely behind DAGMM - a heavy deep learning model - for F1 score evaluation. On WADI, it outperforms PCA, KNN, FB as well as LSTM-VAE which is also a highly parameterized deep neural network.

We note that the detection precision of our algorithm is not yet on par with these deep learning solutions. As shown in Fig. 4 (a, b), while our algorithm responds well to visually recognizable sensor disturbances on the WADI test set and rests quickly when situation abates, a large proportion of such events are not true anomalies as indicated by the ground-truth. This is likely caused by the fact that our algorithm works directly on raw features as opposed to learned ones in other methods, which can have much lower signal-to-noise ratios and are therefore harder to process. This may be improved via feature transformation/learning in future work. Nevertheless, our algorithm successfully localizes sensors that behave abnormally by detecting significant changes to the diagonal elements of operators right before and after a strong response, as demonstrated in Fig. 4 (c, d). This is consistent with the interpretation we make of Fig. 3 (c) on a controlled synthetic time-series.

5   Conclusion

We have presented *TOPOMA*, a new algorithm for multivariate time-series anomaly detection. It does not require any labels nor training for ready use on batch or online streaming data. The key of our algorithm is encoding system baselines using an orthogonal projection operator that evolves across time. We have also developed a new technique to adapt detection thresholds using probability bounding.

Future Work. Our algorithm currently works on raw features, which can be improved by transforming raw data into a better represented space to improve detection accuracy. The Eq. (2) can be differentiable and hence permits feature learning. Estimating a more accurate distribution form of anomaly scores (e.g., using higher order moments to remove our current assumption of symmetry) to bound the probability of anomalies tighter is another future work.

References

1. Aggarwal, C.C.: Data Mining. Springer, Cham (2015). https://doi.org/10.1007/978-3-319-14142-8
2. Ahmed, C.M., Palleti, V.R., Mathur, A.P.: WADI: a water distribution testbed for research in the design of secure cyber physical systems. In: Proceedings of the 3rd International Workshop on Cyber-Physical Systems for Smart Water Networks, pp. 25–28 (2017)
3. Angiulli, F., Pizzuti, C.: Fast outlier detection in high dimensional spaces. In: Elomaa, T., Mannila, H., Toivonen, H. (eds.) PKDD 2002. LNCS, vol. 2431, pp. 15–27. Springer, Heidelberg (2002). https://doi.org/10.1007/3-540-45681-3_2
4. Asatiani, A., et al.: Challenges of explaining the behavior of black-box AI systems. MIS Q. Exec. **19**(4), 259–278 (2020)
5. Aydin, I., Karaköse, M., Akin, E.: A robust anomaly detection in pantograph-catenary system based on mean-shift tracking and foreground detection. In: 2013 IEEE International Conference on Systems, Man, and Cybernetics, pp. 4444–4449. IEEE (2013)
6. Boniol, P., Linardi, M., Roncallo, F., Palpanas, T.: Automated anomaly detection in large sequences. In: 2020 IEEE 36th International Conference on Data Engineering (ICDE), pp. 1834–1837. IEEE (2020)
7. Cook, A.A., Mısırlı, G., Fan, Z.: Anomaly detection for IOT time-series data: a survey. IEEE Internet Things J. **7**(7), 6481–6494 (2019)
8. Deng, A., Hooi, B.: Graph neural network-based anomaly detection in multivariate time series. In: Proceedings of the AAAI Conference on Artificial Intelligence, vol. 35, pp. 4027–4035 (2021)
9. Desforges, M., Jacob, P., Cooper, J.: Applications of probability density estimation to the detection of abnormal conditions in engineering. Proc. Inst. Mech. Eng. C J. Mech. Eng. Sci. **212**(8), 687–703 (1998)
10. Görnitz, N., Kloft, M., Rieck, K., Brefeld, U.: Toward supervised anomaly detection. J. Artif. Intell. Res. **46**, 235–262 (2013)
11. Hogben, L.: Handbook of linear algebra. CRC press (2006)

12. Joshi, M.V., Agarwal, R.C., Kumar, V.: Predicting rare classes: can boosting make any weak learner strong? In: Proceedings of the Eighth ACM SIGKDD International Conference on Knowledge Discovery and Data Mining, pp. 297–306 (2002)
13. Käding, C., Rodner, E., Freytag, A., Denzler, J.: Fine-tuning deep neural networks in continuous learning scenarios. In: Chen, C.-S., Lu, J., Ma, K.-K. (eds.) ACCV 2016. LNCS, vol. 10118, pp. 588–605. Springer, Cham (2017). https://doi.org/10.1007/978-3-319-54526-4_43
14. Lazarevic, A., Kumar, V.: Feature bagging for outlier detection. In: Proceedings of the Eleventh ACM SIGKDD International Conference on Knowledge Discovery in Data Mining, pp. 157–166 (2005)
15. Li, D., Chen, D., Jin, B., Shi, L., Goh, J., Ng, S.-K.: MAD-GAN: multivariate anomaly detection for time series data with generative adversarial networks. In: Tetko, I.V., Kůrková, V., Karpov, P., Theis, F. (eds.) ICANN 2019. LNCS, vol. 11730, pp. 703–716. Springer, Cham (2019). https://doi.org/10.1007/978-3-030-30490-4_56
16. Liang, H., et al.: Robust unsupervised anomaly detection via multi-time scale DCGANS with forgetting mechanism for industrial multivariate time series. Neurocomputing **423**, 444–462 (2021)
17. Liu, Y., Lin, Y., Xiao, Q., Hu, G., Wang, J.: Self-adversarial variational autoencoder with spectral residual for time series anomaly detection. Neurocomputing **458**, 349–363 (2021)
18. Ma, D., Yuan, Y., Wang, Q.: Hyperspectral anomaly detection via discriminative feature learning with multiple-dictionary sparse representation. Remote Sens. **10**(5), 745 (2018)
19. Mathur, A.P., Tippenhauer, N.O.: SWaT: a water treatment testbed for research and training on ICS security. In: 2016 International Workshop on Cyber-physical systems for Smart Water Networks (CySWater), pp. 31–36. IEEE (2016)
20. Namuduri, S., Narayanan, B.N., Davuluru, V.S.P., Burton, L., Bhansali, S.: Deep learning methods for sensor based predictive maintenance and future perspectives for electrochemical sensors. J. Electrochem. Soc. **167**(3), 037552 (2020)
21. Park, D., Hoshi, Y., Kemp, C.C.: A multimodal anomaly detector for robot-assisted feeding using an LSTM-based variational autoencoder. IEEE Robot. Autom. Lett. **3**(3), 1544–1551 (2018)
22. Saw, J.G., Yang, M.C., Mo, T.C.: Chebyshev inequality with estimated mean and variance. Am. Stat. **38**(2), 130–132 (1984)
23. Shyu, M.L., Chen, S.C., Sarinnapakorn, K., Chang, L.: A novel anomaly detection scheme based on principal component classifier. MIAMI UNIV CORAL GABLES FL DEPT OF ELECTRICAL AND COMPUTER ENGINEERING, Tech. rep. (2003)
24. Tang, Z., Chen, Z., Bao, Y., Li, H.: Convolutional neural network-based data anomaly detection method using multiple information for structural health monitoring. Struct. Control. Health Monit. **26**(1), e2296 (2019)
25. Welford, B.: Note on a method for calculating corrected sums of squares and products. Technometrics **4**(3), 419–420 (1962)
26. Xia, X., et al.: GAN-based anomaly detection: a review. Neurocomputing (2022)
27. Zavrtanik, V., Kristan, M., Skočaj, D.: Draem-a discriminatively trained reconstruction embedding for surface anomaly detection. In: Proceedings of the IEEE/CVF International Conference on Computer Vision, pp. 8330–8339 (2021)

28. Zhou, C., Paffenroth, R.C.: Anomaly detection with robust deep autoencoders. In: Proceedings of the 23rd ACM SIGKDD International Conference on Knowledge Discovery and Data Mining, pp. 665–674 (2017)
29. Zhou, Y., Liang, X., Zhang, W., Zhang, L., Song, X.: VAE-based deep SVDD for anomaly detection. Neurocomputing **453**, 131–140 (2021)
30. Zong, B., et al.: Deep autoencoding gaussian mixture model for unsupervised anomaly detection. In: International Conference on Learning Representations (2018)

Latent Space Correlation-Aware Autoencoder for Anomaly Detection in Skewed Data

Padmaksha Roy[✉][ID], Himanshu Singhal[ID], Timothy J O'Shea[ID], and Ming Jin[ID]

Virginia Tech, Blacksburg, VA, USA
{padmaksha,himanshusinghal,oshea,jinming}@vt.edu

Abstract. Unsupervised learning-based anomaly detection using autoencoders has gained importance since anomalies behave differently than normal data when reconstructed from a well-regularized latent space. Existing research shows that retaining valuable properties of input data in latent space helps in the better reconstruction of unseen data. Moreover, real-world sensor data is skewed and non-Gaussian in nature rendering mean-based estimators unreliable for such cases. Reconstruction-based anomaly detection methods rely on Euclidean distance as the reconstruction error which does not consider useful correlation information in the latent space. In this work, we address some of the limitations of the Euclidean distance when used as a reconstruction error to detect anomalies (especially near anomalies) that have a similar distribution as the normal data in the feature space. We propose a latent dimension regularized autoencoder that leverages a robust form of the Mahalanobis distance (MD) to measure the latent space correlation to effectively detect near as well as far anomalies. We showcase that incorporating the correlation information in the form of robust MD in the latent space is quite helpful in separating both near and far anomalies in the reconstructed space.

Keywords: Anomaly Detection · Unsupervised learning · Latent Space Regularization

1 Introduction

Autoencoder has an encoding network that provides a mapping from the input domain to a latent dimension and the decoder tries to reconstruct the original data from the latent dimension. Autoencoders have been successfully used in the context of unsupervised learning to effectively learn latent representations in low-dimensional space when density estimation is difficult in the high-dimensional

Supplementary Information The online version contains supplementary material available at https://doi.org/10.1007/978-981-97-2242-6_6.

© The Author(s), under exclusive license to Springer Nature Singapore Pte Ltd. 2024
D.-N. Yang et al. (Eds.): PAKDD 2024, LNAI 14645, pp. 66–77, 2024.
https://doi.org/10.1007/978-981-97-2242-6_6

space [1, 14]. In the context of anomaly detection, it happens most of the time that the anomaly samples are rare and tedious to label making unsupervised learning the holy grail in this field. Training auto-encoders using a reconstruction error corresponds to maximizing the lower bound of mutual information between input and learned representation. Therefore, regularization methods are of great significance for preserving useful information from input space in latent space. In kernel-based autoencoders, the pairwise similarities in the data are encoded as a prior kernel matrix and the auto-encoder tries to reconstruct it from the learned latent dimension representations. This helps in learning data representations with pre-defined pairwise relationships encoded in the prior kernel matrix [15]. Although the majority of the reconstruction-based methods assume that the outlier class cannot be effectively compressed and reconstructed, empirical results suggest that reconstruction-based approaches alone fail to capture particular anomalies that lie near the latent dimension manifold of the inlier class [8]. Previously, authors have highlighted the challenging scenario where anomalies with high reconstruction error and residing far away from the latent dimension manifold can be easily detected but those having low reconstruction error and lying close to the manifold with inlier samples are highly unlikely to be detected. It is difficult to find a novelty score as a threshold for such cases. In this context, we try to highlight the importance of treating the far and near anomalies separately based on the feature correlation and the amount of skewness present in the data. In our case, we derive a robust form of the well-known MD distance that provides a reliable estimate of location and scale especially when the data is skewed and correlated in nature. The median and median absolute deviation(MAD) as estimators of location and scale have been studied in the robust statistics community in order to understand outliers [9, 12]. They proved to be efficient estimators when the data follow a skewed or non-Gaussian distribution. In our latent space model, we define a strategy to deal with the near anomalies - the anomalies that lie close to the normal data in the feature space. Although autoencoders have been around for a while now, the robust autoencoder equipped with the robust MD, that we propose, can detect both near and far anomalies accurately in real-time sensor datasets which are mostly skewed and correlated in nature.

In short, our contributions can be summarized below:

- We formulate a problem of detecting both near and far anomalies in real-time sensor datasets that are skewed(non-Gaussian) and have correlated features by leveraging a robust form of the well-known Mahalanobis distance that captures useful correlation information in the latent space. Our method can accurately classify those near anomalies that have a similar distribution as the normal data in the feature space while also correctly classifying the distant anomalies.
- Our method jointly optimizes the autoencoder reconstruction loss and the latent space regularization loss using the robust MD distance to balance the detection and generative performance of the model.

– Evaluation results showcase significant improvement both in terms of classification metrics and reconstruction error when experimented with different
cybersecurity, and medical datasets and compared with state-of-the-art baselines.

2   Related Work

Unsupervised anomaly detection in latent space has been an interesting area
of research for a long time. Anomaly detection with reconstruction error relies
on the fact that anomalies cannot be effectively compressed and reconstructed
from the latent subspace of an autoencoder when the latent space is trained
on the normal data. These methods include PCA, Deep AE [18], Robust
deep AE [10,11,14,18], high dimensional anomaly detection [22]. Widely known
approaches employ a two-step process where dimensionality reduction is followed by a density estimation technique in the latent space. However, during
compression, there can be a loss of useful information from the high dimensional space in the latent space. In this context, the authors [1] proposed a
joint optimization of the parameters of the deep autoencoder and a mixture
model for density estimation to detect anomalies in latent space. The model is
trained on normal data points and then assigns low probabilities to new data
points that are far from the learned normal distribution. Others propose to analyze the reconstruction error of latent dimension autoencoders and demonstrate
promising results [10,14]. Zhou et al. proposed robust autoencoders [14] that
can distinguish anomalies from random noise along with discovering important
non-linear features for detecting anomalies. The authors [19] propose to train
neural networks using unsupervised representation learning to predict data distances in a randomly projected space and then the model is optimized to learn
the class structures embedded in the projected space. However, reconstruction-
based methods are limited by the fact that they do not consider any latent
dimension correlation information. The authors [15] have developed kernelized
autoencoders that aim to optimize a joint objective of minimizing reconstruction
error from latent space and misalignment error between prior and latent space
codes which helps in detecting some kind of rare anomalies. MD-based outlier
detection methods have been explored in the past for multivariate outlier detection [2,8,17,20]. Lee et al. [13] proposed an out-of-distribution (OOD) detection
model using Mahalanobis distance on the features learned by a deep classification
model to detect OOD samples. The authors [24] also emphasized the problem of
overlapping of normal and anomaly class distributions in the latent space and
developed a score based on how well the point fits the learned distribution of
normal data. In deep probabilistic generative modeling, the anomalies or the
OOD samples are detected by setting a threshold on the likelihood and selecting
an efficient OOD score is often difficult. The authors of the paper [7] proposed an
efficient OOD score metric to detect OOD data for VAEs. The problem of near
anomaly detection and OOD has been emphasized in [17,23]. In [3], the authors
propose detecting outliers in an augmented or enlarged latent space as anomalies tend to lie in sparse regions. In other applications [6], the authors propose

a two-stage approach where a stacked denoising autoencoder is used to extract diverse features and followed by a KNN classifier to detect anomalies in the trained latent space. A similar kind of approach is also followed here [4], where a deep neural net is used to extract the features, and a KNN classifier is leveraged in the encoded space. In addition to these, there are also other industrial applications where feature selection techniques are combined with unsupervised learning algorithms to improve anomaly detection performance [5]. Our approach, on the other hand, relies on the distribution of the reconstruction error of the normal data from a regularized latent space which considers the latent dimension feature correlation in the form of robust MD distance as an additional measure to distinguish between near and far anomalies.[1]

3   Problem Formulation

With autoencoders, appropriately modeling the distribution of the normal data is important to separate the normal and the anomaly samples in the reconstructed space. We aim to detect both near and far anomalies successfully by taking advantage of another distance metric in the latent space. Our methodology combines a robust form of the Mahalanobis Distance(MD) to measure useful feature correlation in the latent space in addition to the reconstruction loss of an AE to achieve better performance in detecting both near and far anomalies.

3.1   Robust Hybrid Error with MD in Latent Space

The robust MD loss using MAD and median as estimators of scale and location is derived based on the principles stated in the theorems below. We request the readers refer to the supplementary material section for detailed proof of the theorems.

Theorem 1. *The sample median is $2/n$ times more efficient than the sample mean at exponential distribution. The result is consistent with the fact that the skewed distribution has a heavy right tail, which can cause the mean to be affected by outliers and skewness. Here, n denotes the sample size.*

Theorem 2. *Similarly, the relative efficiency of the sample median absolute deviation (MAD) to sample variance is $\approx \frac{2}{(4-\pi)}$ at exponential distribution.*

The robust distance is calculated by measuring the Mahalanobis distance (D_M) between the encoded samples (Z) and the median of the features of the encoded samples in the latent space. The purpose of the robust MD is to effectively capture useful feature correlation information in the latent dimension feature space. The robust form of the Mahalanobis distance (D_M) is calculated based on how many standard deviations an encoded sample z_i is from the median of the encoded data features in the latent space. The median is calculated individually

---

[1] https://github.com/padmaksha18/DRMDIT-AE/.

for each latent dimension feature variable. In the encoded space (Z), the robust form of the MD can be estimated as

$$\hat{D}_M(Z) = \sqrt{(Z - median)^T \hat{R}^{-1} (Z - median)},$$

where $\hat{R}$ is the estimated feature-based correlation matrix of encoded data in the latent space and the robust correlation coefficient(ρ) between two latent features i and j is calculated as

$$\rho_{z_i,z_j} = \frac{\mathbf{E}[(Z_i - median_i)(Z_j - median_j)]}{MAD_{z_i}, MAD_{z_j}},$$

where the MAD of a latent space feature i is given by

$$MAD_{z_i} = median|Z_i - median(Z_i)|.$$

Here, we aim to develop a method for estimating the feature correlation in the latent space when the data is highly skewed and features are correlated. The distance metric leverages the robust correlation estimator in the latent dimension feature space with the median and MAD as location and scale estimators respectively. On the other hand, a mean-based covariance coefficient is highly biased when the data is skewed and non-Gaussian in nature and is also unable to capture the feature correlation in the latent space without effectively estimating the standard deviation in the features.

3.2 Objective Function

The deep latent space correlation-aware autoencoder enabled with the robust MD distance is trained by minimizing the following loss function

$$\mathcal{L}(\theta, \phi) = \min_{\theta,\phi} \alpha_1 D_M(Z_{\theta,\phi}) + \alpha_2 \mathcal{L}_e(X, D_{\theta,\phi}(E_{\theta,\phi}(X))), \tag{1}$$

where D_M is the robust MD loss, X and Z are the input and latent dimensions, $\mathcal{L}_e$ is the reconstruction loss, θ and ϕ are the encoder and decoder parameters, α_1, α_2 are the regularization parameters that determine the weightage assigned to each part of the objective. The joint objective tries to establish a pareto optimal solution between the two objectives. We assign different weightage on robust MD loss in the latent space and the reconstruction loss and try to balance the classification and generative performance of the model.

4 Experiments

Our model is a deep-stacked autoencoder model with several hidden layers and we constrain our encoder (E) and decoder (D) network to have the same architecture, that is, $W_E = W_D^T$. We experiment with different numbers of stacked layers for the encoder and the decoder model. During training, we use only the normal

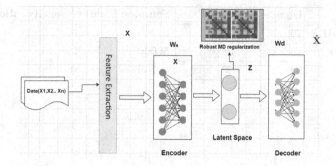

Fig. 1. Deep Latent Space Correlation-Aware Autoencoder(DLSCA-AE).

data as input to our encoder model so that the model learns a latent dimension feature representation of the normal data. The model performance is dependent on tuning two hyperparameters – the weightage to the robust MD regularization parameter α_1 which balances the latent dimension correlation using the robust MD and the reconstruction weightage α_2. These two hyper-parameters must be tuned appropriately to balance the classification and generative performance of the model. It is difficult to detect the near anomalies using density-based methods or nearest neighbor measures as they lie very close to the normal data in the feature space. In our experiment design, we separate the anomalies into two separate classes - the near and far ones. Here, we define the near anomalies as those that are less skewed and are similar to the normal data in the feature space representation whereas the far anomalies are the ones that have highly skewed features. We refer the reader to the supplementary section to have a clear visualization of the far and near anomalies in the feature space.

4.1 Datasets

- **CSE-CIC-IDS2018** This is a publicly available cybersecurity dataset that is made available by the Canadian Cybersecurity Institute (CIC). We consider the data from two different days which consists of different kinds of attacks. The anomaly data is labeled as '1'.
- **NSL-KDD** This is also a publicly available benchmark cybersecurity dataset made available by CIC. It has a total of 43 different features of internet traffic flow. We use 21 correlated features to develop our model.
- **Arrythmia** It is a multi-class classification dataset and the aim is to distinguish between the presence and absence of cardiac arrhythmia and to classify it in one of the normal or anomaly groups.

4.2 Baseline Methods

In order to compare the performance of our model, we consider standard baselines like kernel-based autoencoders (DKAE), density estimation models

Datasets	Dimensions	Corr Feats	Samples(Train)	Anomaly ratio(Test)
CSE-CIC-IDS	79	29	50000	0.5
NSL-KDD	43	20	50000	0.5
Arrythmia	34	20	50000	0.5

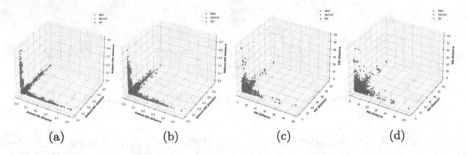

(a)   (b)   (c)   (d)

Fig. 2. (a), (b) shows the reconstructed space when the robust MD is used as a regularizer and (c), (d) corresponds to the reconstructed space with standard MD with mean and covariance as regularizer.

(DAGMM), VAE and, AE, AE-KNN, mean-based MD Autoencoder, and some statistical models.

- **DKAE** [15] The Deep Kernelized Autoencoder [15] has a kernel alignment loss that is calculated as the normalized Frobenius distance between the latent dimension code matrix and the prior kernel matrix and a reconstruction loss.
- **DAGMM** [1] The Deep Autoencoding Gaussian Mixture Model [1] is an unsupervised anomaly detection model that optimizes the parameters of the deep autoencoder and the mixture model simultaneously using an estimation network to facilitate the learning of a Gaussian Mixture Model (GMM).
- **VAE** VAE leverages a probabilistic encoder-decoder network and the reconstruction probability is used for detecting anomalies. It also tries to regularise the organization of the latent space by making the distributions returned by the encoder close to a standard Gaussian distribution.
- **MD-based Autoencoder** [8] This model also leverages MD in the latent space with mean as the estimator of location and the sample covariance.
- **PCA** PCA performs a linear transformation to convert a set of correlated variables into a set of linearly unrelated correlated variables of a smaller dimension. Its primary aim is to perform dimensionality reduction and it achieves it by identifying the principal components along which the data points vary the most.
- **OCSVM** One class SVM is a popular anomaly detection algorithm that finds the decision boundary of the normal data based on SVM with kernel approximation and separates the data from the origin in the transformed high-dimensional predicted space.
- **Stacked AE** Autoencoders are encoder-decoder neural networks that learn a compressed latent space representation of the normal data. The representa-

tion of the anomaly data deviates significantly from the learned latent space and thus has a higher reconstruction error.

- **Isolation Forest** Isolation Forest builds a random ensemble of trees by randomly selecting features and splitting them based on a chosen split value. Anomalies are expected to have a smaller depth, i.e., distance between the root and the nodes as they have lesser and distinct and can be separated easily.
- **Stacked AE-KNN** [4] Autoencoders are used to perform dimensionality reduction upon the data and then the K-nearest neighbors algorithm is applied on the latent space representation obtained from the autoencoder.

4.3   Ablation Study

During training, we use only the normal data as input to our encoder model so that the model learns a latent dimension feature representation of the normal data. We consider the standard classification metrics such as accuracy, precision, and recall to demonstrate the detection performance of the model. We kept the training distribution comprising of 50k normal data samples that are used as training data and the test data consists of an equal proportion of normal and anomaly OOD samples. The near anomalies while getting reconstructed from the latent space which is regularized with the robust MD distance of the normal data are projected to a different subspace in the reconstructed space which efficiently helps them to be separated from the normal data in the reconstructed space. The reconstructed space as depicted in Fig. 2, shows that the robust MD regularized AE can separate both near and far anomalies from the normal data compared to an MD regularized AE with mean and sample covariance estimator. The threshold for anomaly detection depends on the distribution of the reconstruction distance of the normal data. We choose the threshold for anomaly detection to be two standard deviations from the mean of the reconstructed distance of the normal data. In Fig. 2, we try to detect the near and far anomalies using a threshold that is chosen based on the reconstruction range of the normal data. Our proposed autoencoder DLSCA-AE showcases an improvement of 5%-15% in MSE and 5%-8% in MAE while reconstructing unseen data compared to the DKAE model which uses simple Euclidean distance as reconstruction error. Better MSE and MAE of reconstruction do not just imply good learning of the input representations in the latent space but also the performance of the encoder in back-mapping to the higher dimensional space. Due to the good generative performance of the model, it can be also leveraged to generate synthetic data in applications where data is not easily available.

5   Hyperparameter Sensitivity

While training, we put more weightage on the MD-based regularization (α_1) and less weightage on the reconstruction error (α_2). We train our model up to

Table 1. Here, we compare the performance of the proposed model(DLSCA-AE) and baseline models while detecting near and far anomalies.

Model	Type	CSE-CIC-IDS-1 Dataset				NSL-KDD			
		Accuracy	Precision	Recall	AUC	Accuracy	Precision	Recall	AUC
DLSCA-AE	Near	**91.5**	**90.7**	93.4	**94.5**	92.4	**94.6**	**93.1**	**96.0**
	Far	98.9	90.1	92.4	98.9	91.2	95.2	95.4	97.3
DKAE	Near	50	75.2	50.8	71.4	82.0	81.6	92.2	95.6
	Far	79.8	85.2	79.8	74.5	74.5	74.7	74.7	74.5
DAGMM	Near	67.6	68.1	67.9	67.9	**95.4**	86.9	89.3	88.8
	Far	66.6	75.6	66.8	65.1	94	86.8	89.1	91.9
MD-AE	Near	57.3	57.2	52.2	53.4	52.3	53.9	52.5	62.6
	Far	52.8	50.6	51.6	71.6	51.3	53.6	53.5	69.5
PCA	Near	52.2	51.1	96.2	18.6	78.5	73.7	88.4	72.8
	Far	99.8	99.6	99.9	99.9	93.9	99.5	88.2	93.8
OCSVM	Near	63.7	83.7	34.0	60.1	97.1	97.1	97.0	99.5
	Far	97.7	**98.9**	96.4	99.6	98.4	99.7	96.8	99.7
Stacked AE	Near	50.0	0.0	0.0	2.9	83.1	82.3	84.2	87.4
	Far	93.9	97.0	90.6	98.3	96.8	94.8	99.0	99.5
VAE	Near	49.9	0.0	0.0	1.9	82.8	81.2	85.2	85.6
	Far	**99.9**	99.8	99.9	99.9	95.7	92.3	99.6	99.1
Isolation Forest	Near	84.8	79.4	**93.8**	80.5	90.1	86.7	94.6	94.9
	Far	94.0	96.2	91.6	97.6	95.8	93.5	98.4	98.6
Stacked AE-KNN	Near	82.7	90.7	72.8	82.4	84.7	83.3	86.2	88.0
	Far	97.4	96.1	**98.8**	**99.1**	**98.3**	**98.5**	**98.0**	**99.8**
Model	Type	CSE-CIC-IDS-2				Arrythmia			
		Accuracy	Precision	Recall	AUC	Accuracy	Precision	Recall	AUC
DLSCA-AE	Near	**91.4**	91.7	93.4	**94.5**	**92.4**	94.6	**93.1**	**96.0**
	Far	98.9	90.1	92.4	98.9	91.2	95.2	95.4	97.3
DKAE	Near	50	75.2	50.8	71.4	82.0	81.6	92.2	95.6
	Far	79.8	85.2	79.8	74.5	74.5	74.7	74.7	74.5
DAGMM	Near	67.6	68.1	67.9	67.9	81.64	**94.72**	66.46	78.11
	Far	66.6	75.6	66.8	65.1	99.1	99.2	99.2	99.2
MD-AE	Near	57.3	57.2	52.2	53.4	52.3	53.9	52.5	62.6
	Far	52.8	50.6	51.6	71.6	51.3	53.6	53.5	69.5
PCA	Near	51.8	50.9	96.5	18.2	87.3	95.9	77.5	83.0
	Far	99.8	99.6	99.9	99.9	99.7	99.8	**99.9**	99.8
OCSVM	Near	56.5	80.2	17.2	56.6	81.6	90.4	70.1	80.8
	Far	98.4	**99.0**	97.9	99.7	99.6	99.7	96.8	**99.7**
Stacked AE	Near	49.9	0.0	0.0	3.3	66.7	93.8	34.7	50.9
	Far	66.6	66.6	99.9	74.2	99.2	99.3	99.2	99.4
VAE	Near	49.9	0.0	0.0	2.2	71.9	90.9	47.7	64.7
	Far	**99.7**	99.5	**99.9**	**99.9**	**99.7**	**99.8**	99.6	99.7
Isolation Forest	Near	84.9	79.1	**94.9**	79.9	91.5	90.3	92.7	97.3
	Far	93.7	97.5	89.6	97.4	99.9	99.8	99.9	99.9
Stacked AE-KNN	Near	80.5	**92.8**	66.1	77.2	88.2	88.1	77.3	99.7
	Far	96.0	95.9	96.0	98.8	83.7	88.5	77.4	79.8

Table 2. It shows the comparison of reconstruction MSE and MAE with the baselines

Model	Type	CSE-CIC-IDS		NSL-KDD	
		MSE	MAE	MSE	MAE
DLSCA-AE	Near	**0.173 ± 0.008**	**0.229 ± 0.006**	**0.935 ± 0.003**	**0.722 ± 0.003**
	Far	1.413 ± 0.020	0.760 ± 0.004	**0.935 ± 0.040**	0.890 ± 0.009
DKAE	Near	0.334 ± 0.005	0.396 ± 0.005	1.686 ± 0.008	0.890 ± 0.002
	Far	1.870 ± 0.020	0.950 ± 0.030	1.673 ± 0.050	0.913 ± 0.005
DAGMM	Near	0.258 ± 0.006	0.390 ± 0.003	1.790 ± 0.005	0.998 ± 0.005
	Far	1.454 ± 0.030	0.7925 ± 0.010	1.6006 ± 0.040	0.903 ± 0.040
AE	Near	0.183 ± 0.006	0.304 ± 0.003	1.302 ± 0.009	0.815 ± 0.005
	Far	0.775 ± 0.030	0.562 ± 0.010	1.083 ± 0.040	**0.743 ± 0.008**
VAE	Near	0.202 ± 0.009	0.383 ± 0.003	1.280 ± 0.006	0.832 ± 0.003
	Far	**1.411 ± 0.020**	**0.673 ± 0.030**	1.051 ± 0.050	0.775 ± 0.005

150-200 epochs. Batch size is another important training hyper-parameter that directly relates to the number of samples used to estimate the correlation matrix in the latent space. We observe that the initial training is more stable when we use a higher number of samples to estimate the feature correlation in the latent space. After experimenting with different weightage for both the reconstruction loss (α_2) and the MD regularizer (α_1), we find that the best results in terms of classification metrics are obtained when α_2 are in the range of {0.1 to 0.2 } and α_1 is chosen between {0.8 to 0.9}. Otherwise, we suggest keeping the reconstruction weightage significantly less than the robust MD regularizer to determine a good threshold for detecting both near and far anomalies. The best results are reported in Table 1. We showcase the effectiveness of the latent space regularization in improving the generative performance of the model with both near and far anomalies in Table 2.

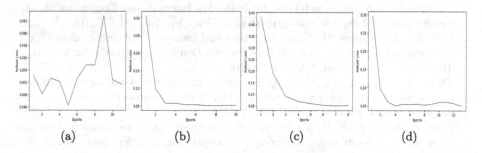

| (a) | (b) | (c) | (d) |

Fig. 3. (a), (b), (c), (d) shows the validation error during training with sample sizes 50, 100, 200, and 300 respectively. We see that the training is more stable when the batch size is higher during each epoch of training.

6   Conclusion

In this paper, we propose a correlation-aware regularized autoencoder that leverages a robust form of MD to capture correlation information in the latent space which helps to detect certain kinds of anomalies more efficiently. The MAD- and median-based robust correlation estimators are useful indicators of specific kinds of anomalies especially when the data is skewed and features are correlated. We find significant improvement in detecting the near anomalies while having equally good performance in detecting the far anomalies with standard classification metrics on standard datasets. We also showcase the improvement of the generative performance of the model while reconstructing data. In the future, we would be interested in understanding the effectiveness of bounding the robust correlation between related data domains to achieve out-of-distribution(OOD) generalization.

Acknowledgment. We would like to thank Virginia Tech National Security Institute (VTNSI) for supporting our work and Dr. Lamine Mili (ECE, Virginia Tech) for the introductory course on Robust Statistics.

References

1. Zong, B., et al.: Deep autoencoding Gaussian mixture model for unsupervised anomaly detection. In: International Conference on Learning Representations (2018)
2. Ren, J., Fort, S., Liu, J., Roy, A. G., Padhy, S., Lakshminarayanan, B.: A simple fix to Mahalanobis distance for improving near-OOD detection. arXiv preprint arXiv:2106.09022 (2021)
3. Angiulli, F., Fassetti, F., Ferragina, L.: Latent out: an unsupervised deep anomaly detection approach exploiting latent space distribution. Mach. Learn. **112**, 4323–4349 (2022). https://doi.org/10.1007/s10994-022-06153-4
4. Guo, J., Liu, G., Zuo, Y., Wu, J.: An anomaly detection framework based on autoencoder and nearest neighbor. In: 2018 15th International Conference on Service Systems and Service Management (ICSSSM), pp. 1-6. IEEE (2018)
5. Rashid, A.B., Ahmed, M., Sikos, L.F., Haskell-Dowland, P.: Anomaly detection in cybersecurity datasets via cooperative co-evolution-based feature selection. ACM Trans. Manage. Inf. Syst. **13**(3), 1–39 (2022)
6. Zhang, Z., Jiang, T., Li, S., Yang, Y.: Automated feature learning for nonlinear process monitoring-an approach using stacked denoising autoencoder and k-nearest neighbor rule. J. Process Control **64**, 49–61 (2018)
7. Xiao, Z., Yan, Q., Amit, Y.: Likelihood regret: an out-of-distribution detection score for variational auto-encoder. Adv. Neural. Inf. Process. Syst. **33**, 20685–20696 (2020)
8. Denouden, T., Salay, R., Czarnecki, K., and Abdelzad, V. Phan, B., Vernekar, S.: Improving reconstruction autoencoder out-of-distribution detection with mahalanobis distance (2018)
9. Hampel, Frank R: Robust statistics: a brief introduction and overview. Seminar für Statistik, Eidgenössische Technische Hochschule,vol 04 (2001)

10. Zhai, S., Cheng, Y., Lu, W., Zhang, Z.: Deep structured energy based models for anomaly detection. In: International Conference on Machine Learning, pp. 1100–1109 (2016)

11. Yang, X., Huang, K., Goulermas, J.Y., Zhang, R.: Joint learning of unsupervised dimensionality reduction and gaussian mixture model. Springer **45**, 791–806 (2017)

12. Huber, P.J., 2004. Robust statistics (Vol. 523). John Wiley and Sons

13. Lee, K., Lee, K., Lee, H., Shin, J.: A simple unified framework for detecting out-of-distribution samples and adversarial attacks. In: Advances in Neural Information Processing Systems, vol. 31 (2018)

14. Zhou, C., Paffenroth, R.C.: Anomaly detection with robust deep autoencoders. In: Proceedings of the 23rd ACM SIGKDD International Conference on Knowledge Discovery and Data Mining, (pp. 665-674) (2017)

15. Kampffmeyer, M., Løkse, S., Bianchi, F.M., Jenssen, R., Livi, L.: Deep kernelized autoencoders. In: Sharma, P., Bianchi, F.M. (eds.) SCIA 2017. LNCS, vol. 10269, pp. 419–430. Springer, Cham (2017). https://doi.org/10.1007/978-3-319-59126-1_35

16. Fan, H., Zhang, F., Wang, R., Xi, L., Li, Z.: Correlation-aware deep generative model for unsupervised anomaly detection. In: Lauw, H.W., et al. (eds.) PAKDD 2020. LNCS (LNAI), vol. 12085, pp. 688–700. Springer, Cham (2020). https://doi.org/10.1007/978-3-030-47436-2_52

17. Fort, S., Ren, J., Lakshminarayanan, B.: Exploring the limits of out-of-distribution detection. Adv. Neural. Inf. Process. Syst. **34**, 7068–7081 (2021)

18. Pang, G., Shen, C., Cao, L., Van Den Hengel, A.: Deep learning for anomaly detection: a review. ACM Comput. Surv. **54**(2), 1–38 (2021)

19. Wang, H., Pang, G., Shen, C., Ma, C.: Unsupervised representation learning by predicting random distances. arXiv preprint arXiv:1912.12186 (2019)

20. Ghorbani, H.: Mahalanobis distance and its application for detecting multivariate outliers. Facta. Univ. Ser. Math. Inform. **34**(3), 583–95 (2019)

21. Laforgue, P., Clémençon, S., d'Alché-Buc, F.: Autoencoding any data through kernel autoencoders. In: The 22nd International Conference on Artificial Intelligence and Statistics, (pp. 1061-1069). PMLR (2019)

22. Erhan, L., et al.: Smart anomaly detection in sensor systems: a multi-perspective review. Inf. Fusion **67**, 64–79 (2021)

23. Koner, R., Sinhamahapatra, P., Roscher, K., Günnemann, S., Tresp, V.: OOD-former: Out-of-distribution detection transformer. arXiv preprint arXiv:2107.08976 (2021)

24. Ando, S., Ayaka, Y.: Anomaly detection via few-shot learning on normality. In: Machine Learning and Knowledge Discovery in Databases: European Conference, ECML PKDD 2022, Grenoble, France, September 19-23, 2022, Proceedings, Part I, pp. 275-290. Cham: Springer International Publishing, 2023 https://doi.org/10.1007/978-3-031-26387-3_17

SeeM: A Shared Latent Variable Model for Unsupervised Multi-view Anomaly Detection

Phuong Nguyen[ID] and Tuan M. V. Le[✉][ID]

New Mexico State University, Las Cruces, NM 88003, USA
{ntphuong,tuanle}@nmsu.edu

Abstract. There have been multiple attempts to tackle the problem of identifying abnormal instances that have inconsistent behaviors in multi-view data (i.e., multi-view anomalies) but the problem still remains a challenge. In this paper, we propose an unsupervised approach with probabilistic latent variable models to detect multi-view anomalies in multi-view data. In our proposed model, we assume that views of an instance are generated from a shared latent variable that uniformly represents that instance. Since the latent variable is shared across views, an abnormal instance that exhibits inconsistencies across different views would have a lower likelihood. This is because, using a single latent variable, the model could not explain well all views that are inconsistent. Therefore, the likelihood of instances based on the proposed shared latent variable model can be used to detect multi-view anomalies. We derive a variational inference algorithm for learning the model parameters that scales well to large datasets. We compare our proposed method with several state-of-the-art methods for multi-view anomaly detection on several datasets. The results show that our method outperforms the existing methods in detecting multi-view anomalies.

Keywords: multi-view anomaly detection · latent variable models

1 Introduction

In some real-world applications, an object may be perceived from multiple sources or perspectives. For example, a webpage can be represented by its content (e.g., words) or by other webpages (URLs) linked to it; face images can be captured by different devices from multiple angles; a person's activity can be recorded by several sensors which give multiple views of that activity. Although single-view learning algorithms can be employed on multi-view data by concatenating multiple views into a single view, that approach could possibly ignore the statistical properties of each view and it could potentially cause the overfitting issue due to the increased data dimension [31]. Therefore, to improve the generalization performance, multi-view learning has emerged to tackle several challenges in learning with multi-view data such as multi-view classification [19,26,29], multi-view clustering [2,13,17], representation learning [5,21,27], and multi-view anomaly detection [11,20,24].

© The Author(s), under exclusive license to Springer Nature Singapore Pte Ltd. 2024
D.-N. Yang et al. (Eds.): PAKDD 2024, LNAI 14645, pp. 78–90, 2024.
https://doi.org/10.1007/978-981-97-2242-6_7

In this paper, we focus on the task of multi-view anomaly detection whose objective is to detect abnormal instances that have inconsistent behaviors across multiple views [8, 24]. For example, a research article that has content about a specific field but cites several articles from other fields can be considered as abnormal in a multi-view sense. Here, an article can be represented by the first view of content and the second view formed by cited articles. Another example is in movie recommendation where a movie of horror genre is also watched by users who likes romance or comedy genres. Figure 1 is an illustration showing the main difference between multi-view anomaly and single-view anomaly. In this figure, each instance has two views and is represented as a circle, a triangle, or a square depending on its class. Black circles and triangles are normal instances because they belong to the same cluster in both views. Focusing on the red triangle, it is a multi-view anomaly because in view one its neighbors are circle points but in view two it is close to triangle points. This is an inconsistent behavior based on its neighbors across views. In contrast, the green square is not a multi-view anomaly because it does not exhibit any cluster inconsistency. It is in fact an anomaly in each single view because it is relatively far from other points.

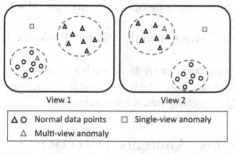

View 1   View 2

| △ ○ | Normal data points | □ | Single-view anomaly |
| △ | Multi-view anomaly | | |

Fig. 1. Illustration of multi-view anomaly in multi-view data.

Several methods have been proposed for multi-view anomaly detection. Clustering-based methods assume that an anomalous point would belong to different clusters in different views. The anomaly score is then computed by comparing the clustering structures in multiple views [7, 16, 30]. Therefore, the performance of these methods may degrade when data do not have clear clusters. To tackle this limitation, several methods have been proposed that rely on neighborhood structures [3, 20], latent space models [8, 9, 24, 25], and low-rank analysis [11, 12, 22]. Most of the these methods have a high model complexity partly due to the hierarchical structure of the Bayesian model with several priors or the high complexity of their inference or optimization algorithms. Moreover, some models are trying to detect different types of anomalies at the same time. Therefore, they may not have the best performance in solely detecting anomalies that behave inconsistently across views.

In this paper, we propose a simple but effective latent space model for multi-view anomaly detection that aims to solely detect samples exhibiting inconsistent behaviors across views. The proposed model is based on the assumption that views of a sample are generated from its representation in a latent space. We associate each instance with a shared latent variable that will be responsible for generating all views of the instance through view-specific Gaussian distributions. Compared to Bayesian-MVAD [24], a closely related method, our model has a lower model complexity where it has only one prior assigned to the latent variable, making the inference process more straightforward. Moreover, our method generalizes the process of data generation by using view projection functions that are parameterized by deep neural networks, allowing the model to express different relationships including linear and non-linear mappings between the latent space and the original space. In contrast, Bayesian-MVAD uses linear projection functions. It is not straightforward to generalize Bayesian-MVAD to non-linear cases because of the priors put on the weight parameters and the derivation of its inference algorithm and anomaly score. As shown in the experiments, by using non-linear mappings, our model can fit better to the data intrinsic structures and yield better performance in detecting multi-view anomalies. Our main contributions are as follows:

- We propose a **S**hared latent variable model for **M**ulti-view anomaly detection, called *SeeM*[1]. A shared latent variable is used for generating all views of an instance through view-specific Gaussian distributions. Although the model has low complexity where it has only one prior assigned to the latent variable, it still gains consistently good performance across several datasets. Moreover, our model generalizes the data generation process by parameterizing the view projection functions using deep neural networks, which will help increase the expressiveness of the model to non-linear relationships between the latent space and the original space.
- For inferring the model parameters, we derive a neural variational inference algorithm for our proposed model.
- We conduct extensive experiments to compare our method with state-of-the-art methods on several datasets. The results show that our method outperforms the existing methods in terms of detecting multi-view anomalies.

2 Proposed Model for Multi-view Anomaly Detection

2.1 The *SeeM* Model and Its Inference

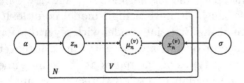

Fig. 2. The graphical model of *SeeM*.

Suppose that we are given a multi-view dataset, $\mathcal{X} = \{x_n\}_{n=1}^{N}$, where x_n is an instance, $x_n = \{x_n^{(v)}|x_n^{(v)} \in \mathbb{R}^{d_v}\}_{v=1}^{V}$. Here, N is the number of instances, V is the number of views, and d_v is the dimension of view v. To link multiple views of an instance, we assume that its views are generated from a single shared latent variable z_n that represents that instance, $z_n \in \mathbb{R}^{d_z}$. Each view v of an instance n is generated from its latent vector z_n through the corresponding view-specific Gaussian distribution centered at mean $\mu^{(v)}(z_n)$. By learning these shared latent variables, the model also learns the common neighborhood structures in different views in the latent

---

[1] https://github.com/thanhphuong163/SeeM.

space. For an abnormal instance that exhibits inconsistencies in its neighborhood structures across different views, the model may not find a single latent representation to explain well all of its views. Therefore, the likelihood of abnormal instances based on the proposed shared latent variable model should be low and can be used to detect multi-view anomalies as presented in Sect. 2.3.

We now present our model in details. For each instance x_n, its view v is generated from a Gaussian distribution centered at mean $\mu^{(v)}(z_n)$: $x_n^{(v)} \sim \mathcal{N}(\mu_\theta^{(v)}(z_n), \sigma^2 I)$, here $\mu^{(v)}(z_n)$ is a view project function that maps a latent vector z_n to the mean of the Gaussian. We parameterize the view project function using a multilayer perceptron (MLP) neural network to capture possibly nonlinear relationships between the latent space and each view: $\mu_\theta^{(v)}(z_n) = \text{MLP}(z_n)$, where θ collectively indicates the weight parameters of the neural network[2]. We also assume a Gaussian prior $\mathcal{N}(\mathbf{0}, \alpha^2 I)$ on the latent space. The generative process of *SeeM* to generate multi-view data is given as follows:

For each instance $n = 1 \ldots N$:
(a) Draw a latent vector:
$$z_n \sim \mathcal{N}(\mathbf{0}, \alpha^2 I)$$
(b) For each view $v = 1 \ldots V$, draw $x_n^{(v)}$:
$$x_n^{(v)} \sim \mathcal{N}(\mu_\theta^{(v)}(z_n), \sigma^2 I)$$

Figure 2 shows the corresponding graphical model of *SeeM*. Based on the proposed model, the marginal likelihood of the multi-view data is computed as:

$$p(\mathcal{X}|\alpha, \sigma) = \prod_{n=1}^{N} \int p(z_n|\alpha) \prod_{v=1}^{V} p(x_n^{(v)}|z_n, \sigma) dz_n \tag{1}$$

where $p(x_n^{(v)}|z_n, \sigma) = \mathcal{N}(\mu^{(v)}(z_n), \sigma^2 I)$, α is the standard deviation of the prior Gaussian distribution of the latent vectors, and σ is the standard deviation of the Gaussian distribution generating $x_n^{(v)}$[3]. Compared to previous latent variable models, our proposed model shares the same concept of using a shared latent variable model for multi-view anomaly detection. The main difference is that we parameterize view-specific mean projections by neural networks, which increases the expressiveness of the model to non-linear relationships between the latent space and original space. Moreover, our model is unsupervised and has less prior assumptions, which means less hyper-parameters to tune. For semi-supervised model such as [24], it uses linear mean projections. It is not straightforward to generalize that model to non-linear settings because several priors are assigned to its parameters and latent variables.

We derive an inference algorithm for our model on multi-view data using neural variational inference [10,18]. Given a multi-view dataset $\mathcal{X}$ containing

---

[2] Experimentally, non-linear neural networks work well for our problem in most of datasets. In the experiments, we use $\mu_\theta^{(v)}(z_n) = \text{Linear}(\text{ReLU}(\text{Linear}(z_n)))$.

[3] In our experiments, $\alpha = 1$ and $\sigma = 0.001$ work well for most of the datasets.

N instances with V views, we seek to maximize the following variational lower bound on the marginal log likelihood:

$$\mathcal{L}(\mathcal{X}|\alpha,\sigma;\theta,\phi) = \sum_{n=1}^{N} \left[\sum_{v=1}^{V} \mathbb{E}_{q_\phi(z_n|x_n)} \log p_\theta(x_n^{(v)}|z_n,\sigma) \right] - \text{KL}\left[q_\phi(z_n|x_n)\|p(z_n|\alpha)\right] \quad (2)$$

here $q_\phi(z_n|x_n)$ is the variational posterior distribution that is used to approximate the intractable true posterior $p_\theta(z_n|x_n)$. We assume that $p_\theta(z_n|x_n)$ is approximately a Gaussian distribution with a diagonal covariance matrix. Therefore, we let $q_\phi(z_n|x_n)$ take the Gaussian form with a diagonal covariance matrix, $q_\phi(z_n|x_n) = \mathcal{N}(\varphi_n, \eta_n^2 \mathbf{I})$. Following neural variational inference, we parameterize it using a neural network. Namely, we use an encoder that takes a concatenation of an instance's views as its input and produces the mean φ_n and the standard deviation η_n of the variational posterior distribution. More specifically,

$$h_n = \text{MLP}(x_n^{(1)} \oplus \ldots \oplus x_n^{(V)}) \quad (3)$$

$$\varphi_n = \text{Linear}(h_n) \quad (4)$$

$$\eta_n = \text{Softplus}(\text{Linear}(h_n)) \quad (5)$$

where $\text{MLP}(x_n^{(1)} \oplus \ldots \oplus x_n^{(V)})$ contains three hidden fully connected layers with the hyperbolic tangent activation function. The differentiable Monte Carlo estimate of the expectation in Eq. 2 is then given as[4]:

$$\mathbb{E}_{q_\phi(z_n|x_n)} \log p_\theta(x_n^{(v)}|z_n,\sigma) = \frac{1}{L} \sum_{l=1}^{L} \log p_\theta(x_n^{(v)}|z_n^{(l)},\sigma) \quad (6)$$

$$z_n^{(l)} = \varphi_n + \eta_n \odot \epsilon^{(l)}, \epsilon^{(l)} \sim \mathcal{N}(\mathbf{0},\mathbf{I}) \quad (7)$$

The KL divergence between two Gaussian distributions in Eq. 2 can be computed in closed form. We then maximize the variational lower bound $\mathcal{L}(\mathcal{X}|\alpha,\sigma;\theta,\phi)$ using stochastic gradient ascent[5] in Eq. 2 to estimate the weight parameters of the networks and the latent vector z_n for each instance x_n. In summary, Algorithm 1 shows the main steps of our inference algorithm.

2.2 Complexity Analysis

We analyze the complexity of the inference algorithm as shown in Algorithm 1. The dominant part is the calculation of the KL-divergence and log likelihood in Eq. 2. Since we parameterize all distributions using neural networks, the main computations involve several matrix multiplications. We calculate the

---

[4] $L = 1$ works well in our experiments.
[5] We use Adam optimization algorithm.

computational complexity per epoch of both KL-divergence and log likelihood in Eq. 2 as follows: $T(N, V) = \mathcal{O}(V \cdot (N \cdot \max\{d, d_h, d_z\} \cdot \max\{d, d_h, d_z\}))$, where $d = \max\{d_v\}_{v=1}^V$ is the largest dimension of all input views, d_h is the largest size of all hidden unit layers of the neural networks. The part $(N \cdot \max\{d, d_h, d_z\} \cdot \max\{d, d_h, d_z\})$ is the upper bound of matrix multiplications of all components. Compared to [24], our algorithm is linear in the number of views, and instances while having less hyper-parameters and making less assumptions on the priors.

2.3 Anomaly Score

Since we want to detect multi-view anomalies that have inconsistencies across views, we calculate the anomaly score of an instance x_n based on the summation of negative log likelihoods of all views given its latent vector:

$$\text{score}(x_n) = -\sum_{v=1}^{V} \log p_\theta(x_n^{(v)} | z_n) \quad (8)$$

For a multi-view anomaly, a single shared latent variable cannot explain well all views because there are inconsistencies across views of that instance. Therefore, the negative log likelihoods in Eq. 8 should be high. The higher the score is, the more anomalous that instance is.

Algorithm 1. Inference algorithm

Input: Data $\mathcal{X} = \{x_n\}_{n=1}^N$
Output: Parameters ϕ, θ, z_n
1: **while** not converged **do**
2:   **for** $n = 1 \ldots N$ **do**
3:     Sample instance $x_n \in \mathcal{X}$
4:     Calculate hidden vectors h_n (Eq. 3)
5:     Estimate posterior parameters φ_n, η_n (Eqs. 4, 5)
6:     Sample $z_n \sim q_\phi(z_n | x_n)$ (Eq. 7)
7:     **for** $v = 1 \ldots V$ **do**
8:       Estimate:
9:       Prior param. $\mu_{x_n^{(v)}} = \mu^{(v)}(z_n)$
10:       $\mathbb{E}_{q_\phi(z_n|x_n)}\left[\log p_\theta(x_n^{(v)}|z_n)\right]$ (Eq. 6)
11:     **end for**
12:     Evaluate $\mathcal{L}(x_n|\alpha, \sigma; \phi, \theta)$ (Eq. 2)
13:     Estimate gradient $\Delta_\phi \mathcal{L}$ and $\Delta_\theta \mathcal{L}$
14:     Update ϕ and θ using ADAM optimizer
15:   **end for**
16: **end while**

3 Experiments

3.1 Datasets and Baselines

We evaluate the proposed model quantitatively on 13 datasets: Breast Cancer Wisconsin (bcw)[6], Glass Identification (glass)[7], Heart Disease (heart)[8], Connectionist Bench Sonar (sonar)[9], Dry Bean (drybean)[10], htru2[11], Wine Quality (winequality)[12], Electrical Grid Stability (electrical_grid)[13], Magic Gamma

---

[6] https://archive.ics.uci.edu/dataset/15/breast+cancer+wisconsin+original.
[7] https://archive.ics.uci.edu/ml/datasets/glass+identification.
[8] https://archive.ics.uci.edu/ml/datasets/heart+disease.
[9] http://archive.ics.uci.edu/dataset/151/connectionist+bench+sonar+mines+vs+rocks.
[10] https://archive.ics.uci.edu/dataset/602/dry+bean+dataset.
[11] https://archive.ics.uci.edu/dataset/372/htru2.
[12] https://archive.ics.uci.edu/dataset/186/wine+quality.
[13] https://archive.ics.uci.edu/dataset/471/electrical+grid+stability+simulated+data.

Telescope (magic)[14], svmguide2[15], svmguide4[16], vehicle[17], Japanese Vowels (vowels)[18]. Table 1 shows details of the datasets. For a fair comparison, we follow the strategy as in [11,24] to generate multi-view anomalies for evaluation. We divide each dataset into two views by randomly splitting features into two separate sets. The view dimensions column in Table 1 shows the number of features of each view when we split a dataset into views. We then swap one random view of an instance with the same view of another instance that has a different class label to obtain a multi-view anomaly. For each dataset, we follow that process to generate multi-view anomalies with ratios 5%, 10%, 15%, 20%, 25%, 30%, and for each anomaly ratio we randomly generate 10 sample datasets. For performance evaluation, we use AUC (Area under the ROC curve) which is a widely used metric for evaluating anomaly detection performance [11,24]. A higher AUC score indicates a better performance in multi-view anomaly detection.

Table 1. Dataset statistics.

Dataset	Size	View dimensions	#classes	Dataset	Size	View dimensions	#classes
bcw	683	[4, 5]	2	vowels	1456	[6, 6]	2
glass	214	[4, 5]	6	drybean	13611	[8, 8]	7
heart	303	[6, 7]	5	htru2	17898	[4, 4]	2
sonar	208	[30, 30]	2	winequality	6497	[5,6]	10
svmguide2	391	[10, 10]	3	electrical_grid	10000	[6, 7]	2
svmguide4	612	[5, 5]	6	magic	19020	[5, 5]	2
vehicle	846	[9, 9]	4				

We compare *SeeM*[19] with several state-of-the-art methods for single-view and multi-view anomaly detection. Single-view methods include **LOF** [1], **IForest** [15], **OCSVM**[20], **DROCC** [6][21], and **DeepIForest** [28][22]. These are anomaly detection methods for single-view data. To run these methods, we merge all views into one single view. Multi-view methods include **MLRA** [12][23], **LDSR** [11][24], **Bayesian-MVAD** [24][25], **NCMOD** [3][26], and **SRLSP** [22][27]. MLRA and LDSR employ cross-view low-rank analysis for multi-view anomaly detection. Bayesian-MVAD builds a hierarchical Bayesian model by assigning priors

---

[14] https://archive.ics.uci.edu/ml/datasets/magic+gamma+telescope.
[15] https://www.csie.ntu.edu.tw/~cjlin/libsvmtools/datasets/multiclass/svmguide2.
[16] https://www.csie.ntu.edu.tw/~cjlin/libsvmtools/datasets/multiclass/svmguide4.
[17] https://www.csie.ntu.edu.tw/~cjlin/libsvmtools/datasets/multiclass/vehicle.original.
[18] http://odds.cs.stonybrook.edu/japanese-vowels-data/.
[19] https://github.com/thanhphuong163/SeeM.
[20] We use the implementations from scikit-learn.
[21] https://github.com/microsoft/EdgeML.
[22] https://github.com/xuhongzuo/deep-iforest.
[23] http://sheng-li.org/Codes/SDM15_MLRA_Code.zip.
[24] https://github.com/kailigo/mvod.
[25] https://github.com/zwang-datascience/MVAD_Bayesian/.
[26] https://github.com/auguscl/NCMOD.
[27] https://github.com/wy54224/SRLSP.

to all parameters and latent variables. NCMOD encodes intrinsic information of each view into a latent space by using the autoencoder and uniforms the neighborhood structures across views. SRLSP extracts the view-specific similarity and uniforms this information by learning a common cross-view similarity consensus with graph fusion for anomaly detection. In the experiments, all hyper-parameters of the baseline methods are set as default. For *SeeM*, the latent dimension is set $\min\{d_v|v \in V\} - 1$ (same as [24]). Batch size is set to 32. Dimensions of all hidden layers are set to 50. The number of epochs is set to 300. We employ Adam optimizer with initial learning rate 10^{-3}. All datasets are normalized using RobustScaler before feeding into our model. All methods are trained in a totally unsupervised setting where the training set includes both normal and multi-view anomalous instances.

Table 2. Average AUCs of all methods at anomaly ratio 5%.

Single-view Methods					
Dataset	DeepIForest	LOF	IForest	OCSVM	DROCC
bcw	0.8871 ± 0.027	0.7008 ± 0.046	0.8044 ± 0.032	0.8133 ± 0.039	0.4876 ± 0.182
glass	0.6107 ± 0.095	0.6856 ± 0.086	0.5926 ± 0.091	0.5934 ± 0.113	0.5421 ± 0.043
heart	0.5556 ± 0.075	0.6029 ± 0.048	0.5822 ± 0.037	0.5523 ± 0.034	0.5481 ± 0.065
sonar	0.6210 ± 0.088	0.6787 ± 0.110	0.5462 ± 0.108	0.5740 ± 0.119	0.3845 ± 0.068
svmguide2	0.5365 ± 0.045	0.6131 ± 0.052	0.5388 ± 0.052	0.5540 ± 0.042	0.5319 ± 0.061
svmguide4	0.6071 ± 0.050	0.6755 ± 0.062	0.6186 m ± 0.028	0.5361 ± 0.024	0.5155 ± 0.060
vehicle	0.7947 ± 0.049	0.9078 ± 0.041	0.7349 ± 0.031	0.7410 ± 0.032	0.4449 ± 0.103
vowels	0.7682 ± 0.030	0.9412 ± 0.010	0.6744 ± 0.025	0.7049 ± 0.033	0.6540 ± 0.059
magic	0.6974 ± 0.044	0.8207 ± 0.042	0.6520 ± 0.020	0.6452 ± 0.026	0.7758 ± 0.082
htru2	0.9138 ± 0.012	0.7071 ± 0.016	0.8639 ± 0.008	0.8578 ± 0.006	0.8419 ± 0.024
winequality	0.6384 ± 0.014	0.8265 ± 0.022	0.5820 ± 0.009	0.5995 ± 0.012	0.7539 ± 0.044
electrical_grid	0.5814 ± 0.010	0.7554 ± 0.038	0.5555 ± 0.015	0.5620 ± 0.016	0.8057 ± 0.080
drybean	0.8559 ± 0.014	0.7883 ± 0.020	0.7299 ± 0.017	0.7558 ± 0.011	0.9533 ± 0.024
Average	0.6975	0.7464	0.6519	0.6530	0.6338

Multi-view Methods					
Dataset	SeeM	Bayesian-MVAD	LDSR	NCMOD	SRLSP	MLRA
bcw	**0.9273** ± 0.028	0.8224 ± 0.051	0.6377 ± 0.046	0.8025 ± 0.034	0.8618 ± 0.044	0.6669 ± 0.039
glass	**0.8033** ± 0.081	0.7744 ± 0.092	0.6122 ± 0.102	0.6958 ± 0.084	0.7500 ± 0.072	0.5634 ± 0.105
heart	**0.9084** ± 0.051	0.6214 ± 0.067	0.5825 ± 0.062	0.5962 ± 0.065	0.5430 ± 0.099	0.5111 ± 0.085
sonar	0.9911 ± 0.009	0.9372 ± 0.033	**0.9936** ± 0.008	0.7006 ± 0.108	0.8610 ± 0.059	0.7871 ± 0.044
svmguide2	**0.9826** ± 0.012	0.7261 ± 0.114	0.7196 ± 0.057	0.6032 ± 0.046	0.5811 ± 0.062	0.8313 ± 0.052
svmguide4	0.6173 ± 0.073	0.7011 ± 0.049	0.5427 ± 0.075	**0.7150** ± 0.069	0.6433 ± 0.084	0.5856 ± 0.071
vehicle	**0.9893** ± 0.011	0.9651 ± 0.022	0.9496 ± 0.033	0.9334 ± 0.032	0.9120 ± 0.042	0.7538 ± 0.059
vowels	**0.9891** ± 0.006	0.9169 ± 0.025	0.9503 ± 0.018	0.9543 ± 0.010	0.8489 ± 0.024	0.8369 ± 0.036
magic	**0.8318** ± 0.028	0.7466 ± 0.072	0.6506 ± 0.051	0.8184 ± 0.045	0.8150 ± 0.035	0.6840 ± 0.025
htru2	**0.9870** ± 0.005	0.9052 ± 0.026	0.9480 ± 0.010	0.9596 ± 0.027	0.9353 ± 0.008	0.7941 ± 0.021
winequality	0.8648 ± 0.014	0.8441 ± 0.030	0.6123 ± 0.034	**0.8921** ± 0.018	0.7811 ± 0.015	0.6037 ± 0.024
electrical_grid	0.8114 ± 0.036	**0.9680** ± 0.024	0.6568 ± 0.041	0.8464 ± 0.021	0.8697 ± 0.016	0.6344 ± 0.036
drybean	**0.9983** ± 0.001	0.9949 ± 0.003	0.8512 ± 0.122	0.9913 ± 0.006	0.9879 ± 0.004	0.6648 ± 0.114
Average	**0.9001**	0.8402	0.7467	0.8084	0.7992	0.6859

3.2   Multi-view Anomaly Detection Performance

Table 2 shows average AUCs across 10 samples with anomaly rate 5% on 13 datasets. In general, all multi-view methods (*SeeM*, Bayesian-MVAD, LDSR, NCMOD, SRLSP, MLRA) outperform the other single-view methods (LOF, IForest, OCSVM, DROCC, DeepIForest) because single-view methods are not designed for detecting multi-view anomalies. Our proposed model has the best AUCs on 9 out of 13 datasets. NCMOD outperforms on 2 datasets (svmguide4 and winequality). Bayesian-MVAD produces the best results on one dataset (electrical_grid), but it gets a second place on average AUC over all datasets. LDSR and *SeeM* obtain the best performance on the sonar dataset. Our experiments also show that our model is significantly better than the multi-view counterparts. It could be explained by the fact that our model employs neural networks for the projections from latent space to views, which could help the model capture better the non-linear relationships between the latent space and the views. This will be explored more in our experiments in the below section where we present the performance of *SeeM* with different linear and non-linear projections.

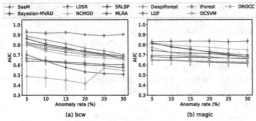

(a) bcw   (b) magic

Fig. 3. Average AUCs of all methods with different anomaly ratios.

In Fig. 3, we vary the anomaly ratios, from 5% to 30%, and show average AUCs of different methods across 10 samples. Due to limited space, we show results on two datasets. Overall, our method and the baselines have AUCs slightly decreasing when the anomaly ratio increases. However, the performance of our model are more stable than the others across different anomaly ratios. For larger anomaly ratios, our method outperforms the baselines by a bigger margin in terms of AUC averaged across samples and datasets.

3.3   Latent Dimension Analysis

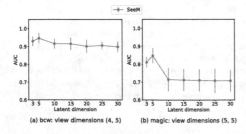

(a) bcw: view dimensions (4, 5)   (b) magic: view dimensions (5, 5)

Fig. 4. Average AUCs of *SeeM* with different latent dims. at anomaly ratio 5%.

In this section, we study the effectiveness of choosing the dimension for the latent variable z in our model on different datasets. Figure 4 shows the average AUCs of our model on two datasets bcw and magic on various latent dimensions including 3, 5, 10, 15, 20, 25, 30. From the figure, we can see that the performance is the best at the latent dimension around the views' dimensions. This supports our choice of latent dimension $\min \{d_v | v \in V\} - 1$ which generally gives the best performance in most of the datasets.

3.4 Non-linear Projections

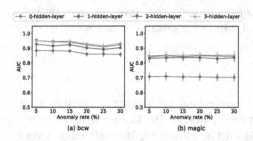

(a) bcw   (b) magic

Fig. 5. Performance of *SeeM* with linear and non-linear projections, varying anomaly ratios.

To study the effects of the projections (i.e., $\mu_\theta^{(v)}(z_n)$) in our model, we implement four variants of *SeeM* which have different structures in the projections from latent space to views. Namely, one variant where $\mu_\theta^{(v)}(z_n)$ has 0 hidden layer (i.e., it has only one linear transformation from latent space to views). This is a linear projection variant of *SeeM*. Three others have respectively 1, 2, and 3 hidden layers (with the ReLU activation function) before one linear transformation, which are considered as non-linear projections. The results of these four variants of *SeeM* are shown in Fig. 5. As we can see, the variants with non-linear projections totally outperform the one with the linear projection. It shows that the ability to capture the complex relationship between latent space and views of multi-view data helps improve the performance of multi-view anomaly detection. At some point, the performance will stop increasing even if we increase the number of hidden layers.

3.5 A Use Case with Real-World Multi-view Data

In this section, we show an application of multi-view anomaly detection by analyzing the inconsistency between content and citation of documents on CORA dataset[28]. The dataset contains 2708 documents. The content view of the dataset is a 0/1 matrix of 1433-word vocabulary. The citation view is a 2708 × 2708 0/1 matrix with value 1 indicates an undirected citation between two documents. Table 3 shows the top five documents with the highest anomaly scores on the left and the bottom five doc-

Table 3. Top 5 documents with highest anomaly scores (left) and bottom 5 documents with lowest anomaly scores (right) ranked by our model.

Doc. ID	score	Doc. ID	score
986	102.7246	2492	95.5906
2060	102.6642	2529	95.6560
2025	102.6371	392	95.8743
931	102.6366	1318	96.2232
1003	102.6159	569	96.4800

uments with the lowest anomaly scores on the right. Document 986 cites four other documents including documents 160, 1103, 2039 and 2067. These four documents have only zero or one common words with the document 986. In fact, document 986 has label *Probabilistic Methods* while the documents 160, 1103, 2039, and 2067 have labels *Theory*, *Rule Learning*, *Case Based*, and *Theory*

---

[28] https://lig-membres.imag.fr/grimal/data.html.

respectively. This is inconsistent because document 986 belongs to *Probabilistic Methods* but cites papers from other fields. Our method can detect such an inconsistency. Considering document 2492 which has the lowest anomaly score, this document cites document 1982 that has 15 common words in the vocabulary. They belong to the same category of *Neural Networks*. Therefore, it is consistent that document 2492 cites other documents that have similar content. That is why our method ranks document 2492 as a normal instance.

4 Related Work

There have been different approaches to detecting multi-view anomalies that exhibit inconsistencies across multiple views [8,11,14,16,20,22–24]. Clustering-based methods identify anomalies based on the clustering inconsistency across views [14,16]. In [14], data points are abnormal if they do not belong to any consensus cluster. Therefore, this method cannot find inconsistencies within a cluster. In [16], multi-view anomalies are detected by comparing the neighborhood structures in different views. The authors employed *affinity propagation* [4] as a clustering method in each view. However, these approaches would fail to find anomalies in data that do not have a clear clustering structure. To tackle this limitation, several methods have been proposed that rely on neighborhood structures [3,20], latent space models [8,9,24,25], and low-rank analysis [11,12,22].

Our method is closely related to latent space methods. Latent space models assume a normal instance has its views generated from a latent vector to solve the multi-view anomaly detection problem [8,9,24,25]. [8] assumes that an instance with views derived from more than one latent vector is abnormal. Anomaly score is calculated based on the probability that an instance has more than one latent vector. Wang et. al. [24] proposed a semi-supervised hierarchical Bayesian model by assigning priors to all parameters and latent variables. The anomalous score of an instance is formulated as its negative unscaled Student's t density. In both models, the views are projected from latent space via linear projections. However, in our model, we parameterize view-specific mean projections by neural networks. Moreover, our model is unsupervised and has less prior assumptions, which means less hyper-parameters to tune. For [24], it is not straightforward to generalize that model to non-linear settings because several priors are assigned to its parameters and latent variables.

5 Conclusion

We propose *SeeM*, a shared latent variable model for multi-view anomaly detection. The proposed method aims to learn latent representations of instances in multi-view data and uses projection functions that are parameterized by neural networks to generate multi-view data from the latent representations of instances. Extensive experiments demonstrate the effectiveness of our proposed model in detecting multi-view anomalies.

Acknowledgments. This research is sponsored by NSF #1757207 and NSF #1914635.

References

1. Breunig, M.M., Kriegel, H.P., Ng, R.T., Sander, J.: Lof: identifying density-based local outliers. In: ACM sigmod record. vol. 29, pp. 93–104. ACM (2000)
2. Chen, M.S., Huang, L., Wang, C.D., Huang, D.: Multi-view clustering in latent embedding space. In: Proceedings of the AAAI Conference on Artificial Intelligence, vol. 34, pp. 3513–3520 (2020)
3. Cheng, L., Wang, Y., Liu, X.: Neighborhood consensus networks for unsupervised multi-view outlier detection. In: Proceedings of the AAAI Conference on Artificial Intelligence, vol. 35, pp. 7099–7106 (2021)
4. Frey, B.J., Dueck, D.: Mixture modeling by affinity propagation. In: Advances in Neural Information Processing Systems **18** (2005)
5. Geng, Y., Han, Z., Zhang, C., Hu, Q.: Uncertainty-aware multi-view representation learning. In: Proceedings of the AAAI Conference on Artificial Intelligence. vol. 35, pp. 7545–7553 (2021)
6. Goyal, S., Raghunathan, A., Jain, M., Simhadri, H.V., Jain, P.: Drocc: deep robust one-class classification. ArXiv **abs/2002.12718** (2020)
7. Handong, Z., Yun, F.: Dual-regularized multi-view outlier detection. In: IJCAI, pp. 4077–4083 (2015)
8. Iwata, T., Yamada, M.: Multi-view anomaly detection via robust probabilistic latent variable models. In: Lee, D., Sugiyama, M., Luxburg, U., Guyon, I., Garnett, R. (eds.) Advances in Neural Information Processing Systems, vol. 29 (2016)
9. Ji, Y.X., et al.: Multi-view outlier detection in deep intact space. In: 2019 IEEE International Conference on Data Mining (ICDM), pp. 1132–1137. IEEE (2019)
10. Kingma, D., Welling, M.: Auto-encoding variational bayes (12 2014)
11. Li, K., Li, S., Ding, Z., Zhang, W., Fu, Y.: Latent discriminant subspace representations for multi-view outlier detection. In: Proceedings of the AAAI Conference on Artificial Intelligence **32**(1) (Apr 2018)
12. Li, S., Shao, M., Fu, Y.: Multi-view low-rank analysis for outlier detection. In: Proceedings of the SIAM International Conference on Data Mining (2015)
13. Lin, Y., Gou, Y., Liu, Z., Li, B., Lv, J., Peng, X.: Completer: incomplete multi-view clustering via contrastive prediction. In: Proceedings of the IEEE/CVF Conference on Computer Vision and Pattern Recognition, pp. 11174–11183 (2021)
14. Liu, A.Y., Lam, D.N.: Using consensus clustering for multi-view anomaly detection. In: 2012 IEEE Symposium on Security and Privacy Workshops, pp. 117–124 (2012)
15. Liu, F.T., Ting, K.M., Zhou, Z.H.: Isolation forest. In: 2008 Eighth IEEE International Conference on Data Mining, pp. 413–422 (2008)
16. Marcos Alvarez, A., Yamada, M., Kimura, A., Iwata, T.: Clustering-based anomaly detection in multi-view data. In: Proceedings of the 22nd ACM International Conference on Information and Knowledge Management, pp. 1545–1548 (2013)
17. Peng, X., Huang, Z., Lv, J., Zhu, H., Zhou, J.T.: Comic: Multi-view clustering without parameter selection. In: International Conference on Machine Learning, pp. 5092–5101. PMLR (2019)
18. Rezende, D.J., Mohamed, S., Wierstra, D.: Stochastic backpropagation and approximate inference in deep generative models. In: International Conference on Machine Learning, pp. 1278–1286. PMLR (2014)

19. Seeland, M., Mäder, P.: Multi-view classification with convolutional neural networks. PLoS ONE **16**(1), e0245230 (2021)
20. Sheng, X.R., Zhan, D.C., Lu, S., Jiang, Y.: Multi-view anomaly detection: neighborhood in locality matters. In: Proceedings of the AAAI Conference on Artificial Intelligence. vol. 33, pp. 4894–4901 (2019)
21. Wang, X., Peng, D., Hu, P., Sang, Y.: Adversarial correlated autoencoder for unsupervised multi-view representation learning. Knowledge-Based Systems (2019)
22. Wang, Y., Chen, C., Lai, J., Fu, L., Zhou, Y., Zheng, Z.: A self-representation method with local similarity preserving for fast multi-view outlier detection. ACM Trans. Knowl. Discov. Data **17**(1), 1–20 (2023)
23. Wang, Z., Fan, M., Muknahallipatna, S., Lan, C.: Inductive multi-view semi-supervised anomaly detection via probabilistic modeling. In: 2019 IEEE International Conference on Big Knowledge (ICBK), pp. 257–264. IEEE (2019)
24. Wang, Z., Lan, C.: Towards a hierarchical bayesian model of multi-view anomaly detection. In: Proceedings of the Twenty-Ninth International Joint Conference on Artificial Intelligence, IJCAI-20, pp. 2420–2426 (7 2020), main track
25. Wang, Z., et al.: Learning probabilistic latent structure for outlier detection from multi-view data. In: Pacific-Asia Conference on Knowledge Discovery and Data Mining, pp. 53–65 (2021)
26. Xie, X., Sun, S.: Multi-view support vector machines with the consensus and complementarity information. IEEE TKDE **32**(12), 2401–2413 (2019)
27. Xu, C., Guan, Z., Zhao, W., Niu, Y., Wang, Q., Wang, Z.: Deep multi-view concept learning. In: IJCAI, pp. 2898–2904. Stockholm (2018)
28. Xu, H., Pang, G., Wang, Y., Wang, Y.: Deep isolation forest for anomaly detection. IEEE Transactions on Knowledge and Data Engineering, pp. 1–14 (2023)
29. Xu, J., Li, W., Liu, X., Zhang, D., Liu, J., Han, J.: Deep embedded complementary and interactive information for multi-view classification. In: Proceedings of the AAAI Conference on Artificial Intelligence, vol. 34, pp. 6494–6501 (2020)
30. Zhao, H., Liu, H., Ding, Z., Fu, Y.: Consensus regularized multi-view outlier detection. IEEE Trans. Image Process. **27**(1), 236–248 (2017)
31. Zhao, J., Xie, X., Xu, X., Sun, S.: Multi-view learning overview: Recent progress and new challenges. Inform. Fusion **38**, 43–54 (2017)

Classification

QWalkVec: Node Embedding by Quantum Walk

Rei Sato$^{(\boxtimes)}$, Shuichiro Haruta , Kazuhiro Saito , and Mori Kurokawa

AI Division, KDDI Research, Inc., Fujimino, Ohara 2–1–15, Saitama 356-8502, Japan
{ei-satou,sh-haruta,ku-saitou,mo-kurokawa}@kddi.com

Abstract. In this paper, we propose QWalkVec, a quantum walk-based node embedding method. A quantum walk is a quantum version of a random walk that demonstrates a faster propagation than a random walk on a graph. We focus on the fact that the effect of the depth-first search process is dominant when a quantum walk with a superposition state is applied to graphs. Simply using a quantum walk with its superposition state leads to insufficient performance since balancing the depth-first and breadth-first search processes is essential in node classification tasks. To overcome this disadvantage, we formulate novel coin operators that determine the movement of a quantum walker to its neighboring nodes. They enable QWalkVec to integrate the depth-first search and breadth-first search processes by prioritizing node sampling. We evaluate the effectiveness of QWalkVec in node classification tasks conducted on four small-sized real datasets. As a result, we demonstrate that the performance of QWalkVec is superior to that of the existing methods on several datasets. Our code will be available at https://github.com/ReiSato18/QWalkVec.

Keywords: node classification · node embedding · quantum walk

1 Introduction

A graph consists of nodes that have labels and attribute information and a set of edges that connect the nodes. For example, the Worldwide Web [1] is a graph in which web pages are nodes, and the hyperlinks between pages are edges. In fields such as social networking services, a node classification problem, where graph structures and node features are used to predict user attributes [2], is important. Node classification requires the extraction of features from graph nodes. Node embedding methods have been studied using machine learning to represent the features of each node in a graph as a vector with a fixed length [3].

DeepWalk [3] and node2vec [4] are famous node embedding techniques that use random walks to sample the node sequences of graphs. DeepWalk consists of two parts: a random walk and Skip-gram [5]. The random walk samples nodes by randomly moving from each node to neighboring nodes, creating a node sequence with the same length as that of the walk. Skip-gram is a neural network that maximizes the co-occurrence probability of the sampled nodes and their surrounding nodes in a given sequence [6]. Node2vec uses biased random

© The Author(s), under exclusive license to Springer Nature Singapore Pte Ltd. 2024
D.-N. Yang et al. (Eds.): PAKDD 2024, LNAI 14645, pp. 93–104, 2024.
https://doi.org/10.1007/978-981-97-2242-6_8

walks, which efficiently search the neighborhoods of a given node through tunable parameters used for a breadth-first search (BFS) and a depth-first search (DFS). The BFS prioritizes sampling the neighboring nodes connected to a node, while the DFS aims to sample nodes as far away as possible from the current node. Node2vec enhances the effectiveness of the DFS while maintaining the benefits of the BFS. However, when the input graph is large, the walk length required to acquire sufficient sequence information via a random walk will also be large.

A quantum walk is the quantum version of a random walk and demonstrates faster propagation than a random walk on a graph [7–9]. Therefore, it has attracted attention as an alternative to random walks. A quantum walk describes the arbitrary direction states in terms of the probability amplitudes of quantum superposition. It exists stochastically at multiple positions at a given walk length. The diffusion rate in one-dimensional space is proportional to the walk length t for the variance of the position x of a random walk, i.e., $\sigma_x^2 \sim t$. In contrast, the variance of the position of a quantum walk is proportional to the square of the walk length, i.e., $\sigma_x^2 \sim t^2$ [10].

Balancing the BFS and DFS when using quantum walks for node feature representation is also essential in node classification tasks. The reason is that quantum walks are typically used as a superposition state. If we perform a quantum walk in this superposition state, we confirm that nodes with structural equivalence are preferentially sampled; that is, the DFS effect is dominant, as shown in a later section. In other words, this approach weakens the BFS effect, which is important for learning about neighboring nodes. Therefore, it is necessary to develop a flexible quantum walk that can balance BFS and DFS.

In this study, we propose QWalkVec, a quantum walk-based node embedding algorithm that can balance the BFS and DFS. QWalkVec starts with a superposition state where the DFS effects are optimized, and it incorporates parameters for maximizing the BFS effects with each walk length. Based on node2vec, QWalkVec integrates return and in-out parameters into a coin operator that determines the movement of the quantum walker to its neighboring nodes. QWalkVec creates a unique coin operator for each node, carries out the quantum walk, and generates features for each node. We use four datasets of labeled graphs and compare the node embedding accuracies of QWalkVec, DeepWalk, and node2vec. The contributions of this study are as follows.

– We formulate a new quantum walk with coin operators that incorporate the effects of the DFS and BFS processes. To the best of our knowledge, this is the first research that tries to formulate a quantum walk for node embeddings.
– Through extensive experiments, we demonstrate that QWalkVec outperforms DeepWalk and node2vec in several graphs.

The rest of this paper is organized as follows. In Sect. 2, we explain the essential preliminaries. Section 3 introduces the related works. In Sect. 4, we propose the QWalkVec algorithm for node classification. In Sect. 5, we evaluate the performances of QWalkVec by comparing it with DeepWalk and node2vec. Finally, we conclude the paper in Sect. 6.

2 Preliminaries

We first define the notations used throughout the paper. We also present the basic concepts of a quantum walk in this section.

2.1 Notations

$G(V, E, M)$ is a labeled undirected graph without edge weights. G consists of a set of N nodes $V = \{1, 2, .., i, .., N\}$ and a set of edges $E = \{e_{ij}\}$. M is the total number of labels. $\Phi^c \in \mathcal{R}^{N \times d}$ are feature representations of node2vec and DeepWalk, and $\Phi \in \mathcal{R}^{N \times t}$ are feature representations of QWalkVec. We write quantum states as $|i \rightarrow j\rangle$ to denote a particle at node i moving toward node j. k_i is a degree representing the number of links attached to node i. t is the walk length. v_0 is the source node. For every source node $v_0 \in V$, we define Φ^{v_0} as a feature representation of node v_0 generated through a sampling strategy.

2.2 Quantum Walks on Graphs

We define a quantum walk on a graph. The quantum state of a quantum walk with a length t is defined as

$$|\psi(t)\rangle = \sum_{i=1}^{N} \sum_{j=1}^{k_i} \psi_{ij}(t) |i\rangle \otimes |i \rightarrow j\rangle. \tag{1}$$

$|\psi(t)\rangle$ is defined in the Hilbert space $\mathcal{H} \equiv \mathcal{H}_N \otimes \mathcal{H}_k$, where $|i\rangle \in \mathcal{H}_N$ is associated with the positional degree of freedom, and $|i \rightarrow j\rangle \in \mathcal{H}_k$ is associated with the internal degree of freedom. $\psi_{ij}(t)$ is the probability amplitude of $|i \rightarrow j\rangle$. The walk length evolution of the quantum state $|\psi(t)\rangle$ is determined by the coin operator $\hat{C}$ and the shift operator $\hat{S}$:

$$|\psi(t)\rangle = [\hat{S}\hat{C}]^t |\psi(0)\rangle. \tag{2}$$

The coin operator must be a node-dependent since each node has a different number of links, i.e.,

$$\hat{C} = \sum_{i}^{N} |i\rangle \langle i| \otimes \hat{C}_i. \tag{3}$$

The coin operator $\hat{C}_i$ at node i is employed as $\left(|i\rangle \langle i| \otimes \hat{C}_i \right) \sum_{j=1}^{k_i} \psi_{ij}(t) |i\rangle \otimes$ $|i \rightarrow j\rangle = |i\rangle \otimes \hat{C}_i \left(\psi_{ij_1}(t), \psi_{ij_2}(t), .., \psi_{ij_{k_i}}(t) \right)^T$. Here, the symbol T indicates the transposition operation. The matrix size of the coin operator $\hat{C}_i$ is $k_i \times k_i$. Note that the numbering of the neighboring nodes $\{j_1, j_2, \cdots, j_{k_i}\}$ is arbitrary. The shift operator changes the position of the quantum walker based on the movement information of the nearest nodes. We define the shift operator as

$$\hat{S} |i\rangle |i \rightarrow j\rangle = |j\rangle |j \rightarrow i\rangle. \tag{4}$$

The probability of node i is given by

$$P_i(t) = \sum_{j=1}^{k_i} |(\langle i| \otimes \langle i \to j|) |\psi(t)\rangle|^2 = \sum_{j=1}^{k_i} |\psi_{ij}(t)|^2. \tag{5}$$

3 Related Works

Unsupervised feature learning approaches using random walks have been extensively studied. In DeepWalk, which combines Skip-gram with random walks, node embeddings are established by representing the input graph as documents [3,11]. DeepWalk samples nodes from the graph and transforms the graph into ordered node sequences. Node2vec [4], an extension of DeepWalk, introduces search parameters for graphs, proposing a superior sampling strategy.

A comprehensive overview of the field of quantum machine learning is presented by J. Biamonte et al. [12]. For quantum walks, some studies have propose new quantum graph algorithms. F. Mauro et al. [13] show that the initial state of a quantum walk affects the accuracy of community detection. K. Mukae et al. [14] show that implementing a quantum walk on a graph reveals a more explicit community structure than that of a random walk. Y. Wang et al. [15] show that quantum walks exhibit better in graph centrality distinguishing capabilities than classic algorithms.

3.1 Problems

Various sampling strategies have been studied in graph feature representation learning, resulting in different feature representation methods. Indeed, open problems remain regarding the development of a superior sampling strategy that functions effectively across all graphs [4]. Quantum walks demonstrate faster propagation than random walks on graphs [7]. As quantum computing advances, it is natural to consider the node embedding process with quantum walks to realize classification tasks based on new sampling strategies. However, the application of quantum walks for node embedding algorithms remains unexplored. Therefore, we propose QWalkVec, which is a quantum walk-based sampling method.

4 Proposed Method: QWalkVec

For node representation purposes, homophily and structural equivalence are key concepts [16]. To capture these characteristics, using both the BFS and DFS strategies is essential [4]. The homophily hypothesis [17] suggests that nodes with similar attributes should belong to the same community and be embedded closely together. BFS involves sampling nodes restricted to the neighbors of a given node, resulting in embeddings that correspond closely to homophily [4]. On the other hand, the structural equivalence hypothesis [18] suggests that nodes

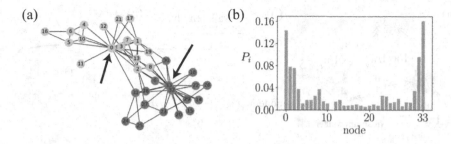

Fig. 1. (a) The visualization of Karate dataset. The node indicated by an arrow is a hub node. (b) The quantum probability distribution of each node visited by a quantum walk with the superposition state at $t = 100$.

with similar structural roles in a graph should be closely embedded. Structural equivalence focuses on the structural roles of nodes in graphs. DFS can sample nodes far from the source node, making it suitable for inferring structural equivalence based on graph roles such as bridges and hubs [4].

To observe how a quantum walk samples nodes, we employ a quantum walk on a Karate [19] as shown in Fig. 1. The initial state of a quantum walk is typically used as a superposition state where the quantum walk starts from all nodes by taking advantage of quantum benefits. When performing a quantum walk in the superposition state, we find that node IDs 0 and 33, which are the hub nodes of the Karate, are almost equally emphasized. Hence, the quantum walk preferentially samples nodes with structural equivalence. On the other hand, the precise information of the neighboring nodes does not appear due to the superposition state. Thus, the BFS effect does not work, and sampling based on a quantum walk tends to have a DFS effect, which extracts structural equivalence.

To overcome this, QWalkVec is proposed. A quantum walk exhibits representation differences depending on the search strategy, such as by changing coin operators [14,20]. Therefore, we propose to leverage coin operators for controlling both BFS and DFS effects. Figure 2(a) shows an overview of the QWalkVec. As shown in this figure, the values on edges defined by 1, $1/w_p$ and $1/w_q$ are weights of coin operators, which can balance BFS and DFS. The return parameter w_p and in-out parameter w_q are created based on each source node, enabling the acquisition of node information around the source node. When $w_p = w_q = 1$, QWalkVec corresponds to sampling using the conventional quantum walk. In QWalkVec, we first initiate the initial state from a superposition state, which leverages the DFS effect. The initial state enables all nodes to be searched, resolving the sample size constraint. We then adjust the priority with which the quantum walker moves to neighboring nodes to characterize the dependencies of nodes. We repeat sampling for all source nodes and create a feature representation of each node Φ^{v_0} shown as Fig. 2(b). From the summation of each node representation Φ^{v_0}, we finally create a feature representation Φ that captures all node information. We describe the detailed procedures in the next section.

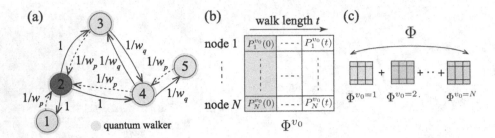

Fig. 2. Overview of QWalkVec. (a) QWalkVec with source node $v_0 = 2$. (b) Φ^{v_0} is the feature representation of node v_0. (c) The feature representations Φ.

4.1 Algorithm

Algorithm 1. QWalkVec (G, t, w_p, w_q)

Input: graph $G(V, E, M)$,
    walk length t
    return probability w_p
    in-out probability w_q
Output: matrix of node representations $\Phi \in \mathcal{R}^{N \times t}$
 1: Initialization : set an $N \times t$-size matrix Φ
 2: **for** $v_0 = 1$ to N **do**
 3:    set parameter $w_{ij}^{v_0}$ for each node and make $\hat{C}_{v_0}, |\psi(0)\rangle_{v_0}$ (i)
 4:    **for** $t_i = 1$ to t **do**
 5:        $|\psi(t)\rangle_{v_0} = [\hat{S}\hat{C}_{v_0}]^{t_i} |\psi(0)\rangle_{v_0}$ (ii)
 6:        **for** $i = 1$ to N **do**
 7:            $\Phi[i, t_i] \leftarrow P_i^{v_0} = \sum_{j=1}^{k_i} |(\langle i| \otimes \langle i \rightarrow j|) |\psi(t)\rangle|^2$ (ii)
 8:        **end for**
 9:    **end for**
10: **end for**
11: **return** Φ

Algorithm 1 is a specific implementation. The algorithm is executed as follows: (i) We set the source node v_0 and determine coin operators and the superposition state with parameters. (ii) We employ quantum walk, and we obtain the quantum probability $P_i(t)$ and store it in a $N \times t$ matrix. We repeat (i) and (ii) for all v_0. For process (i), we define the parameters of the coin operators as

$$w_{ij}^{v_0} = \begin{cases} 1 & \text{for } |i\rangle \otimes |i \rightarrow j\rangle, \ \text{ if } l(v_0, i) = 0 \\ 1/w_p & \text{for } |i\rangle \otimes |i \rightarrow j\rangle, \ \text{ if } l(v_0, i) > l(v_0, j) \\ 1/w_q & \text{for } |i\rangle \otimes |i \rightarrow j\rangle, \ \text{ if } l(v_0, i) = l(v_0, j) \\ 1/w_q & \text{for } |i\rangle \otimes |i \rightarrow j\rangle, \ \text{ otherwise} \end{cases} \tag{6}$$

where $l(a, b)$ is the shortest distance between nodes a and b. We set w_p as the return probability and w_q as the in-out probability based on node2vec [4]. We use the superposition state as the initial state for each source node v_0:

$$|\psi(0)\rangle_{v_0} = \frac{1}{\sqrt{N}} \sum_{i=1}^{N} \frac{1}{\sqrt{\sum_{j=1}^{k_i} w_{ij}^{v_0}}} \sum_{j=1}^{k_i} \sqrt{w_{ij}^{v_0}} |i\rangle \otimes |i \to j\rangle. \qquad (7)$$

$w_{ij}^{v_0}$ is a movement weight between nodes i and j for v_0. In process (ii), the quantum state at t is defined by

$$|\psi(t)\rangle_{v_0} = [\hat{S}\hat{C}_{v_0}]^t |\psi(0)\rangle_{v_0}. \qquad (8)$$

The coin operator is given by $\hat{C}_{v_0} = \sum_i^N |i\rangle \langle i| \otimes \hat{C}_i^{v_0}$, where $\hat{C}_i^{v_0} = 2 |s_i^{v_0}\rangle \langle s_i^{v_0}| - \hat{I}$. $\hat{I}$ is identity operator, and $|s_i^{v_0}\rangle$ is given by

$$|s_i^{v_0}\rangle = \frac{1}{\sqrt{\sum_{j=1}^{k_i} w_{ij}^{v_0}}} \sum_{j=1}^{k_i} \sqrt{w_{ij}^{v_0}} |i \to j\rangle. \qquad (9)$$

The shift operator $\hat{S}$ is given by $\hat{S} |i\rangle |i \to j\rangle = |j\rangle |j \to i\rangle$. We implement the quantum walk based on the coin operator, which is reconstructed for each v_0.

Finally, we define Φ, a feature representation. Each feature representation Φ^{v_0} obtains node information based on each v_0. We assume that the summation of these representations generates an effective composite representation. That is represented as

$$\Phi_{i,t} = \sum_{v_0=1}^{N} \Phi_{i,t}^{v_0}. \qquad (10)$$

The element $\Phi_{i,t}^{v_0}$ is given by Eq. (5) by replacing $|\psi(0)\rangle$ with $|\psi(0)\rangle_{v_0}$. Figures 2(b) and (c) show the representations of Φ, Φ^{v_0} and Eq. (10). In this study, we investigate the effect of feature addition by combining N normalized feature matrices, $(\sum_{i=1}^{N} \Phi_{i,t}^{v_0} = 1)$, to form a new feature matrix Φ $(\sum_{i=1}^{N} \Phi_{i,t} = N)$.

5   Evaluations

We evaluate the performance of QWalkVec under the node classification task by numerical simulation, and analyze the parameter sensitivity for the walk length. Embeddings obtained by each method are fed into a classifier, and it predicts the node labels. As baselines, DeepWalk [3] and node2vec [4] are used.

5.1   Experimental Settings and Dataset

We follow the experimental procedure of DeepWalk [3]. Specifically, we randomly sample a training set of size T_R from the labeled nodes and use the rest of the nodes as testing data. We repeat this process 20 times and evaluate the average micro F_1 and macro F_1 scores as performance scores. We use a one-vs.-rest logistic regression model implemented by LibLinear [21] for the node

Table 1. Node classification results in Karate (34 nodes, 78 edges and 2 labels), Webkb (265 nodes and 479 edges and 5 labels), IIPs (219 nodes, 630 edges and 3 labels) and DD199 (841 nodes and 1,902 edges and 20 labels). We set $(p, q) = (1.0, 2.0)$, $(1.0, 2.0)$, $(0.25, 4.0)$ and $(1.0, 0.5)$, and set $(w_p, w_q) = (0.25, 1.0)$, $(0.5, 4.0)$, $(0.25, 4.0)$ and $(1.0, 1.0)$ for Karate, Webkb, IIPs and DD199, respectively. Numbers in **bold** represent the highest performance.

dataset		Algorithm \ T_R	20%	30%	40%	50%	60%	70%	80%
Karate	micro F_1 (%)	DeepWalk				87±23	91±15	94±30	96±33
		node2vec				98±26	98±28	98±30	99±37
		QWalkVec				**98±16**	**98±18**	**99±21**	**100±22**
	macro F_1 (%)	DeepWalk				93±25	93±23	93±32	96±36
		node2vec				98±31	98±32	98±34	99±38
		QWalkVec				**98±20**	**98±20**	**99±23**	**100±22**
Webkb	micro F_1 (%)	DeepWalk	46±4	48±3	49±2	50±3	51±2	53±3	52±8
		node2vec	48±9	48±9	49±9	51±8	51±8	53±8	52±9
		QWalkVec	**52±8**	**53±8**	**54±6**	**55±6**	**56±7**	**57±8**	58±8
	macro F_1 (%)	DeepWalk	**29±2**	30± 3	31±5	32±4	33±5	35±3	36±9
		node2vec	29±5	29±5	31±5	31±6	32±7	33±8	34±10
		QWalkVec	**29±5**	**32±5**	**33±5**	**34±5**	**36±6**	**37±7**	**39±10**
IIPs	micro F_1 (%)	DeepWalk	**65±3**	**66±1**	**66±2**	**67±4**	**68±3**	69±3	70±5
		node2vec	**65±6**	65±5	**66±5**	**67±5**	**68±6**	**70±8**	**72±10**
		QWalkVec	63±12	64±9	65±6	**67±5**	**68±6**	70±7	71±9
	macro F_1 (%)	DeepWalk	**52±0**	**52±0**	53±0	55±0	56±1	55±1	56±1
		node2vec	49±7	**52±6**	**54±6**	**56±6**	**56±7**	**58±9**	**58±11**
		QWalkVec	40±7	44±7	46±8	49±8	50±9	52±10	57±11
DD199	micro F_1 (%)	DeepWalk	**11.6±3.2**	11.7±2.7	**12.1±2.0**	**12.1±1.4**	12.0±1.9	12.2±2.3	12.4±2.8
		node2vec	11.4±3.1	**12.0±2.7**	12.0±2.0	**12.1±1.6**	**12.5±1.7**	**12.7±1.9**	**12.8±2.8**
		QWalkVec	11.0±2.6	11.2±1.8	12.0±1.4	**12.1±1.5**	12.1±1.8	**12.7±2.3**	12.2±2.6
	macro F_1 (%)	DeepWalk	5.3±1.2	**5.4±0.9**	**5.6±0.8**	5.6±0.6	5.8±0.9	5.9±0.9	6.1±1.4
		node2vec	**5.6±0.9**	5.4±1.2	5.4±1.1	5.5±1.1	5.4±1.4	5.6±1.6	5.9±1.9
		QWalkVec	5.0±1.0	5.2±1.0	**5.6±1.0**	**5.8±1.2**	**6.1±1.5**	**6.2±1.9**	**6.2±2.0**

classification tasks of all methods. We present the results obtained by DeepWalk and node2vec with $\gamma = 80, w = 10$, and $d = 128$. We perform a grid search over $p, q, w_p, w_q \in \{0.25, 0.50, 1, 2, 4\}$ [4]. We use a total of four small-sized real graphs. Karate [19], Webkb [22], Internet Industry Partnerships (IIPs) [22], and DD199 [22]. The statistics of each dataset are shown in the caption of Table 1.

5.2   Overall Results

Node Classification. Table 1 shows the experimental results in Karate, Webkb, IIPs and DD199. For the result of Karate, we set the training ratio T_R from 50% to 80%, which is a different situation from other datasets. We set the training ratio T_R from 20% to 80% for the rest of the datasets. This is because the number of training nodes becomes too small due to the graph size. We run for $t = 400$ to determine the best approach. QWalkVec consistently performs better than DeepWalk and node2vec.

For the result of WebKb, we run for $t = 400$ to determine the best performance among different walk lengths. For the micro F_1 and macro F_1 scores, QWalkVec consistently performs better than DeepWalk and node2vec.

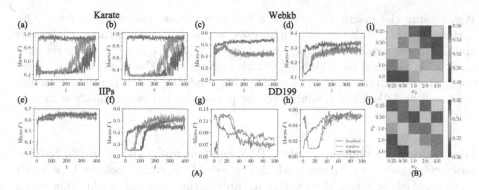

Fig. 3. Parameter sensitivity. (A) Walk length dependence of the F_1 score with $T_R = 50\%$. (B) F_1 scores of the w_p and w_q parameters for Webkb with $T_R = 50\%$.

For the result of IIPs, we run for $t = 400$ to determine the best performance. For the micro F_1 score, DeepWalk, node2vec and QWalkVec have almost the same performance, but node2vec consistently performs slightly better than DeepWalk and QWalkVec. For the macro F_1 score, node2vec consistently performs better than DeepWalk and QWalkVec.

For the result of DD199. We run for $t = 100$ to determine the best performance. For the micro F_1 score, node2vec consistently performs better than Deep-Walk and QWalkVec for $T_R \geq 50\%$. However, for $T_R = 50\%, 70\%$, QWalkVec has the same performance as that of node2vec. For the macro F_1 score, DeepWalk and node2vec perform better than QWalkVec at $T_R = 20\%, 30\%$, but QWalkVec consistently performs better than DeepWalk and node2vec at $T_R \geq 40\%$.

Parameter Sensitivity. QWalkVec involves parameters t, w_p and w_q. We first examine how the value of t affects the resulting performance. Figure 3 shows the parameter sensitivity of QWalkVec. We observe that QWalkVec can reach the best performance faster than the DeepWalk and node2vec algorithms since quantum walks demonstrate faster propagation than random walks on graphs. However, the relation between accuracy and the walk length is not correlated for some graphs. For the macro F_1 score of IIPs, QWalkVec reaches the best performance faster than DeepWalk and node2vec, but its accuracy is lower than those of these algorithms. The dynamics of the F_1 accuracy under different walk lengths exhibit the same pattern for IIPs and DD199 among the three algorithms. For IIPs, we observe that increasing the walk length increases the F_1 performance. On the other hand, for the DD199, we observe that increasing the walk length decreases the micro F_1 score, while the macro F_1 score increases over the walk length t. The dynamics of DeepWalk and node2vec also yield the same results for the quantum walk at $t > 40$. The walk length dependence of F_1 concerns the graph structure. We also examine how different parameter choices affect the performance of the quantum walk on the Webkb dataset using a $T_R = 50$ split between the labeled and unlabeled data, as shown in Fig. 3(B).

We measure the macro F_1 score as a function of parameters w_p and w_q. The performance of QWalkVec tends to improve as the return parameter w_p decreases and the in-out parameter w_q increases, i.e., $w_p < w_q$. While a low w_q encourages DFS, it is balanced by a high w_p, which ensures that the quantum walk does not go too far from the starting node. This tendency of the parameter results produced by QWalkVec is similar to the parameter values of node2vec for the used dataset.

Discussions. In uniform graphs (IIPs), random walks are more effective for measuring the macro F_1 score, which assesses the accuracy achieved for small classes and edges, due to their ability to capture the details of neighboring nodes and connections. On the other hand, in graphs with hubs (Karate and Webkb), the quantum walk, which is more influenced by hubs, attains better results through detailed sampling than those of classic methods. The reason for this is that unlike DeepWalk, which is constrained by the embedding dimensions and window size during the word2vec transformation process, quantum walks directly and probabilistically represent nodes without constraints. In graphs with many missing edges (DD199), there is no significant difference between the proposed method and the existing methods, as the abundance of isolated nodes affects the resulting learning accuracy. The micro F_1 score of QWalkVec tends to decrease with the walk length due to the retention of initial state information. However, for the macro F_1 score, there is no difference between the proposed approach and the existing methods due to the equivalent sampling process performed on isolated nodes. Consequently, QWalkVec performs superior to DeepWalk and node2vec in graphs with hub nodes.

The time complexity of DeepWalk and node2vec is approximately $\mathcal{O}(N \cdot t_c) + m$, where $m = \mathcal{O}(\log N)$ represents the time complexity of Skip-gram [3]. On the other hand, QWalkVec has a time complexity of $\mathcal{O}(N \cdot t_Q \cdot shots)$, where *shots* denotes the number of measurements of quantum states. The actual value of *shots* is not clear at this time. QWalkVec tends to exhibit a shorter optimal walk length than DeepWalk and node2vec, as illustrated in Fig. 3, i.e., $t_Q < t_c$. However, we note that the question of whether QWalkVec will outperform DeepWalk or node2vec with actual quantum computers for sampling time still remains an open problem.

6   Conclusion

In this study, we propose QWalkVec, a quantum walk-based node embedding algorithm. QWalkVec integrates depth-first search and breadth-first search processes by prioritizing node sampling based on the novel coin operators produced for the quantum walker. We use four labeled graph datasets and compare the node embedding accuracies among QWalkVec, DeepWalk, and node2vec. QWalkVec achieves superior performance to that of DeepWalk and node2vec in graphs with hub nodes. For our future work, we will investigate the performance of QWalkVec in other graph-related tasks such as community detection.

References

1. Broder, A., et al.: Graph structure in the web. Comput. Netw. **33**(1–6), 309–320 (2000)
2. Sharma, K., et al.: DeepWalk based influence maximization (DWIM): influence maximization using deep learning. Intell. Autom. Soft Comput. **35**(1) (2023)
3. Perozzi, B., Al-Rfou, R., Skiena, S.: Deepwalk: online learning of social representations. In: Proceedings of the 20th ACM SIGKDD International Conference on Knowledge Discovery and Data Mining, pp. 701–710 (2014)
4. Grover, A., Leskovec, J.: node2vec: scalable feature learning for networks. In: Proceedings of the 22nd ACM SIGKDD International Conference on Knowledge Discovery and Data Mining, pp. 855–864 (2016)
5. Mikolov, T., Chen, K., Corrado, G., Dean, J.: Efficient estimation of word representations in vector space. arXiv preprint arXiv:1301.3781 (2013)
6. Mikolov, T., Sutskever, I., Chen, K., Corrado, G.S., Dean, J.: Distributed representations of words and phrases and their compositionality. In: Advances in Neural Information Processing Systems, vol. 26 (2013)
7. Ambainis, A., Kempe, J., Rivosh, A.: Coins make quantum walks faster. In: Proceedings of the Sixteenth Annual ACM-SIAM Symposium on Discrete Algorithms, SODA 2005, pp. 1099-1108, USA (2005). Society for Industrial and Applied Mathematics
8. Childs, A.M., Farhi, E., Gutmann, S.: An example of the difference between quantum and classical random walks. Quant. Inf. Process. **1**, 35–43 (2002)
9. Koch, D., Hillery, M.: Finding paths in tree graphs with a quantum walk. Phys. Rev. A **97**, 012308 (2018)
10. Ambainis, A., Bach, E., Nayak, A., Vishwanath, A., Watrous, J.: One-dimensional quantum walks. In: Proceedings of the Thirty-third Annual ACM Symposium on Theory of Computing, pp. 37–49 (2001)
11. Ahmed, N.K., et al.: Learning role-based graph embeddings. arXiv preprint arXiv:1802.02896 (2018)
12. Biamonte, J., Wittek, P., Pancotti, N., Rebentrost, P., Wiebe, N., Lloyd, S.: Quantum machine learning. Nature **549**(7671), 195–202 (2017)
13. Faccin, M., Migdał, P., Johnson, T.H., Bergholm, V., Biamonte, J.D.: Community detection in quantum complex networks. Phys. Rev. X **4**(4), 041012 (2014)
14. Mukai, K., Hatano, N.: Discrete-time quantum walk on complex networks for community detection. Phys. Rev. Res. **2**(2), 023378 (2020)
15. Wang, Y., Xue, S., Junjie, W., Ping, X.: Continuous-time quantum walk based centrality testing on weighted graphs. Sci. Rep. **12**(1), 6001 (2022)
16. Hoff, P.D., Raftery, A.E., Handcock, M.S.: Latent space approaches to social network analysis. J. Am. Stat. Assoc. **97**(460), 1090–1098 (2002)
17. Fortunato, S.: Community detection in graphs. Phys. Rep. **486**(3–5), 75–174 (2010)
18. Henderson, K., et al.: RolX: structural role extraction & mining in large graphs. In: Proceedings of the 18th ACM SIGKDD International Conference on Knowledge Discovery and Data Mining, pp. 1231–1239 (2012)
19. Zachary, W.W.: An information flow model for conflict and fission in small groups. J. Anthropol. Res. **33**(4), 452–473 (1977)

20. Wong, T.G.: Coined quantum walks on weighted graphs. J. Phys. Math. Theor. **50**(47), 475301 (2017)
21. Fan, R.E., Chang, K.W., Hsieh, C.J., Wang, X.R., Lin, C.J.: Liblinear: a library for large linear classification. J. Mach. Learn. Res. **9**, 1871–1874 (2008)
22. Rossi, R.A., Ahmed, N.K.: The network data repository with interactive graph analytics and visualization. In: Proceedings of the Twenty-Ninth AAAI Conference on Artificial Intelligence (2015)

Human-Driven Active Verification for Efficient and Trustworthy Graph Classification

Tien-Cuong Bui[1]([✉]) [iD] and Wen-Syan Li[2] [iD]

[1] Department of ECE, Seoul National University, Seoul, South Korea
cuongbt91@snu.ac.kr
[2] Graduate School of Data Science, Seoul National University, Seoul, South Korea
wensyanli@snu.ac.kr

Abstract. Graph representation learning methods have significantly transformed applications in various domains. However, their success often comes at the cost of interpretability, hindering them from being adopted in critical decision-making scenarios. In conventional graph classification, the integration of domain expertise to enhance model training has been underutilized, leading to discrepancies in decision outcomes between humans and models. To address this, we introduce a novel framework involving active human verification in graph classification processes. Our approach features a human-aligned representation learning component, achieved by seamlessly integrating Graph Neural Network architectures and leveraging human domain knowledge and feedback. This framework enhances model transparency and interpretability and fosters collaborative decision-making between humans and AI systems. Extensive evaluations and user studies prove the efficiency of our framework.

Keywords: Graph Classification · Trustworthiness · Explainable AI · Human-in-the-loop Machine Learning

1 Introduction

Graph representation learning methods [23] have revolutionized real-world graph-based applications. While incredibly successful, these techniques are often perceived as mysterious "black boxes," limiting their utility in high-stakes decision-making scenarios. Conventional graph classification models primarily learn the mapping $\mathcal{G} \rightarrow \mathcal{Y}$, where $\mathcal{G}$ represents an input graph, and $\mathcal{Y}$ is the ground truth. The valuable insights that domain experts can provide to enhance model training are frequently overlooked, leading to discrepancies in decision-making between humans and models. Therefore, incorporating domain knowledge into models is beneficial and essential, potentially enhancing both model performance and interpretability.

The collaborative approach between humans and AI [13,15] has garnered significant attention within the research community. This approach holds promise in

© The Author(s), under exclusive license to Springer Nature Singapore Pte Ltd. 2024
D.-N. Yang et al. (Eds.): PAKDD 2024, LNAI 14645, pp. 105–116, 2024.
https://doi.org/10.1007/978-981-97-2242-6_9

enhancing model performance and reliability by harnessing the complementary strengths of both entities. However, integrating human knowledge and feedback into ML models poses significant challenges. Lately, Liu et al. [11] have introduced a case-based reasoning framework that effectively combines the advantages of deep representation learning and case-based decision support, incorporating human feedback through contrastive learning. In the pursuit of making Graph Neural Network (GNN) models more interpretable, several methodologies [14,22] have been proposed. However, these approaches have predominantly prioritized algorithmic assessments and model performance, sometimes overlooking the crucial human dimension of collaborative endeavors.

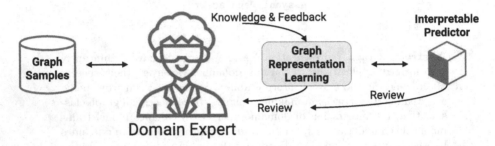

Fig. 1. HVG Framework overview. It centers around a pivotal concept: learning human-aligned graph representations through the empowerment of human active verification.

The motivation for our research originates from the inherent transparency of case-based reasoning processes [17], coupled with the enormous challenges domain experts face when reviewing extensive training samples and offering guidance for the learning process. Practically, they can focus on a select set of prominent examples that effectively represent entire groups of training samples. This intuitive strategy aligns with the idea that training samples should be situated close to at least one of these representative samples in representation learning. Moreover, experts can offer valuable insights, specifying which samples should be close to one another and far from certain others, thus enriching the learning process.

In this paper, we present a novel framework named **HVG**, facilitating efficient and reliable **G**raph classification by active **H**uman **V**erification. Our framework centers around a pivotal concept: developing a human-aligned representation learning component capable of generating graph representations crucial for interpretable predictions. To achieve this, we seamlessly integrate GNN architectures as functions responsible for encoding graph data. Our approach harnesses human knowledge and feedback by imposing dual layers of constraints: class-level knowledge and instance-level feedback. Furthermore, we introduce an iterative human-AI interaction approach for representation learning, a method that significantly elevates predictive performance and model stability. To enhance interpretability, we incorporate two interpretable predictors based on the widely recognized k-nearest-neighbor algorithm. Additionally, we introduce multiple prediction explanation formats based on the resources provided by these interpretable components. Extensive experiments and user studies validate the correctness and efficiency of our framework.

The paper's remainder is as follows. Section 2 presents related work. Section 3 describes our methodology. Experiments are reported in Sect. 4. We discuss possible fairness issues in Sect. 5. The paper is concluded in Sect. 6.

2   Related Work

2.1   Human-in-the-loop Machine Learning

Human-in-the-loop ML [13] is an approach where human expertise collaborates with machine learning algorithms, enabling both iterative feedback loops and improved model performance. Ramos et al., 2020 [15] introduced a comprehensive framework for facilitating human-AI interactions. Lately, Liu et al., 2022 [11] discovered that algorithmic representations might be incompatible with human intuitions requiring human constraints in the training process. Taesiri et al., 2022 [18] proposed a framework for humans and AI to collaborate toward a mutual decision-making process. These approaches leverage the strengths of both humans and machines to build more effective and reliable systems.

2.2   Deep Learning for Case-Based Reasoning

In traditional machine learning, the case-based reasoning paradigm is notable [17], serving as a foundational component in decision-support systems. This approach hinges on utilizing past experiences to address new problems. The inherent ability of deep learning models to recognize patterns and transform data into latent representations significantly enhances the process of retrieving past instances. Recent contributions from Li et al. [9], Chen et al. [2], and Davoudi et al. [4] adopt a prototype-centric method in which prototypes are discovered during the training phase. Our study resembles [4], particularly in differentiating between deep representation learning and the phase of prototype determination.

2.3   Interpretable Graph Neural Networks

Interpretable GNNs [3,6,10,14,22] aim for improved interpretability via mechanisms like node pooling, similarity modules, subgraph aggregation, and prototypes. The method by Dai et al. [3] encountered training difficulties and did not thoroughly address explanation generation methods. Methods by Ragno et al. [14] and Zhang et al. [22] utilize prototype-based prediction techniques but differ in the prototype projection phase. It is pivotal to note that contemporary approaches prioritize prediction accuracy over users' perception of explanations.

3   Methodology

3.1   Problem Formulation and Framework Overview

We formalize the graph classification problem in the context of representation learning and case-based reasoning. The goal is to find a mapping $P : \mathcal{G} \to \mathcal{Y}$. We assume there exists a representation model f, which takes a graph $\mathcal{G} \in \mathcal{D} = \{(\mathcal{G}_1, y_1), ..., (\mathcal{G}_N, y_N)\}$ as an input and outputs a d-dimensional representation $h_\mathcal{G} \in \mathbb{R}^d$. Given a GNN model $g_{\phi,\theta} : \mathcal{G} \to \mathcal{Y}$, the representation model is the last layer before the classifier, referred to as $f = e(g)$, where e is a selection function. For each instance $\mathcal{G}$, a reference policy π selects K-labeled samples from the training set $\mathcal{D}_{\text{train}}$ and presents them to humans. This work's primary focus is the effectiveness of f for case-based graph classification.

As presented in Fig. 1, the essential component is graph representation learning, which turns experts' knowledge and feedback into constraints to achieve human-compatible graph representations. Interpretable predictors utilize these representations to make predictions. Domain experts inspect learned representations and predictions to ensure that AI models and human intuitions are correspondence. Algorithm 1 presents an overview of our framework execution pipeline.

Algorithm 1. HVG framework training and inference

Input: GNN g with ϕ, θ, reference policy π, dataset $\mathcal{D}$, and #epochs T
Output: Representation model f

1: **for** $i = 1$ to T **do**
2:     Execute g on $\mathcal{D}$
3:     Update ϕ, θ via Eq. 4 {Section 3.2}
4:     Periodically suggest new centroids to human {Section 3.2}
5:     Injects knowledge and feedback dynamically {Section 3.2}
6:     Break if early stopping criteria is met
7: **end for**
8: Execute $h_\mathcal{G} = f(\mathcal{G})$ {Obtain representations}
9: Retrieve close references of $\mathcal{G}$ via π {Section 3.3}
10: Execute the interpretable predictor P {Section 3.3}
11: Generate explanations {Section 3.4}

3.2   Human-Compatible Representation Learning

The graph representation learning component incorporates a GNN encoder to transform graph data into a latent space. Our proposed framework is open to various GNN architectures, abstracted as $\mathbf{H}^l = \text{GNN}(\mathcal{G}, \mathbf{A}, \mathbf{H}^{l-1})$, where l is the layer index, $\mathbf{A}$ is the adjacency matrix, and $\mathbf{H}$ is a representation matrix. We perform sum pooling over $\mathbf{H}$ to compute the graph representation vector $h_\mathcal{G}$.

Our goal is to learn representations that are efficient in classification tasks and compatible with human intuitions. First, we adopt the cross entropy loss function to encourage samples to be well-separated in the latent space. This loss function can be replaced in real-world problems with other objective types.

$$\mathcal{L}_{pred} = -\frac{1}{N} \sum_{i=1}^{N} y_i \cdot \log(p_\theta(\hat{y}|h_\mathcal{G})), \tag{1}$$

where p_θ is a variational approximation function estimating class probabilities given a graph representation. Practically, θ represents the weights of a prediction layer in a DL model.

Class-level Knowledge: We hypothesize that domain experts can point out prominent cases with specific features representing diverse groups of samples given a problem. We refer to these representative samples as a prototype set $\mathcal{P} = \{p_1, p_2, ..., p_M\}$. Intuitively, a training sample must be located near at least one of the representative points in the latent space. Moreover, a sample should also be distant from points that do not belong to the same class. Therefore, we introduce the second objective function based on the triplet loss as follows:

$$\mathcal{L}_{ck} = \frac{1}{N} \sum_{i=1}^{N} \min_{j:p_j \in \mathcal{P}_{y_i}} ||f(\mathcal{G}_i) - f(p_j)||_2^2 - \frac{1}{N} \sum_{i=1}^{N} \min_{j:p_j \notin \mathcal{P}_{y_i}} ||f(\mathcal{G}_i) - f(p_j)||_2^2, \tag{2}$$

where $\mathcal{P}_{y_i}$ denotes a subset of $\mathcal{P}$ with the label y_i.

Instance-level Feedback: To enhance discriminative power, domain experts can examine representations in visualization interfaces and provide additional feedback via a contrastive style. Specifically, experts can generate triplets, each consisting of an input graph, a positive graph, and a negative one, denoted as $(\mathcal{G}, \mathcal{G}^+, \mathcal{G}^-)$. Two graphs are considered a positive/negative pair if they are similar/dissimilar to each other on a specific criterion. The instance-level feedback is beneficial for error analysis scenarios or scenarios when experts want to ensure that the reference policy doesn't select human-incompatible examples. The instance-level constraint is presented as follows:

$$\mathcal{L}_{ik} = \sum_{(\mathcal{G}, \mathcal{G}^+, \mathcal{G}^-) \in \mathcal{T}} \max(0, ||f(\mathcal{G}) - f(\mathcal{G}^+)||_2^2 - ||f(\mathcal{G}) - f(\mathcal{G}^-)||_2^2 + \epsilon) \tag{3}$$

Putting everything together, we have the following optimization problem:

$$\min_{\phi, \theta} \mathcal{L}_{pred} + \alpha \mathcal{L}_{ck} + \beta \mathcal{L}_{ik} \tag{4}$$

Iterative Interaction: An iterative representation learning process [15] facilitates human-AI alignment. Specifically, experts can stop the training process anytime and review whether their injected knowledge/feedback is beneficial.

Additionally, AI models can propose new centroid candidates to humans to improve weight adjustments in optimization processes. Humans can either accept or reject the suggestions. For each class, K_c centroids $\mu = \{\mu_i, ..., \mu_{K_c}\}$ are defined using the following equation:

$$\arg\min_{\mu} \sum_{i=1}^{K_c} \sum_{j=1}^{N_i} ||h_{\mathcal{G}_{ij}} - \mu_i||, \tag{5}$$

where N_i is the number of graphs in a cluster i of the class c.

3.3   Interpretable Predictor

Case-based prediction [17] aligns naturally with human cognition, drawing inspiration from our innate ability to tackle new challenges by referencing similar past experiences. Our approach aims to define a predictor $P : h_{\mathcal{G}} \rightarrow \mathcal{Y}$ that outputs a class for an input $\mathcal{G}$, given a representation $h_{\mathcal{G}}$ and a reference policy π. This work focuses on two distinct reference policies rooted in the nearest-neighbor algorithm.

$$\begin{aligned}
\pi_a &= \text{KNN}(\mathcal{G}, f, \mathcal{D}_{\text{train}}) \\
\pi_c &= \{\text{KNN\_CLASS}(\mathcal{G}, f, \mathcal{D}_{\text{train}}^c)\}_{c=1}^C
\end{aligned} \tag{6}$$

As shown in Eq. 6, we employ different subscripts to denote two reference policies. π_a represents a conventional k-nearest-neighbors algorithm, while π_c adopts a strategy that selects an equal number of references for each class based on the corresponding sub-training set D_{train}^c. Strategy selection depends on the characteristics of representation spaces. Practically, π_a is appropriate for scenarios marked by well-separated representations, low noise, and homogeneous neighbors. In contrast, π_c excels in situations with complex decision boundaries, where overlapping representations are prevalent.

$$P(\hat{Y}|\mathcal{G}, \pi) = \sum_{R_i \in \pi} a(\mathcal{G}, R_i) y_i \quad \text{s.t} \quad a(\mathcal{G}, R_i) = \text{softmax}(\text{sim}(\mathcal{G}, R_i)), \tag{7}$$

where y_i is the one-hot ground-truth label, and sim is a similarity function. Practically, we define $\text{sim}(\mathcal{G}, R) = \exp\left(-\frac{||h_{\mathcal{G}} - h_R||^2}{2\sigma^2}\right)$, where $\sigma = 2$.

3.4   Prediction Explanation

Explanations are necessary for improving humans' understanding of model predictions [5]. This module organizes information provided by the interpretable prediction function into user-friendly explanations. Explanations are presented to users in the following types:

- **Comparative Analysis:** Visualizing references enables users to comprehend the model's decisions more easily. Additionally, it facilitates the examination of model errors by comparing incorrect predictions with correct ones in similar cases, aiding model refinement based on instance-level constraints.
- **Reference attribution:** This feature gives users quantitative insights into the decision-making process, revealing the most influential references in shaping the current decision. This feature fosters transparency and interpretability in decision-making.
- **Subgraph visualization:** This function highlights essential components within execution graphs, thus enhancing user understanding. These essential components typically represent common patterns within a group of graphs and are extracted using subgraph extraction techniques like [1,12].

4   Experiments

4.1   Datasets and Baselines

We conducted our experiments using five graph classification datasets: Mutag, IMDB-Binary (IMDB), DD, Proteins [16], and Graph-Twitter (Twitter) [21]. We selected four standard GNN architectures as our baseline models: GCN [8], GraphSage (Sage) [7], GIN [20], and GAT [19]. Each model consists of two GNN layers, followed by a hidden layer and a prediction layer. We applied the objective function in Eq. 4 to train backbones g and used HVG to denote these combinations. To enhance interpretability, we executed the interpretable predictor P on four HVG models, resulting in a group of models denoted as **HVG$^+$**.

4.2   Implementations and Configurations

We adopted an 8:1:1 data-splitting strategy and employed 10-fold cross-validation. Note that in IMDB and DD datasets, node features were represented as one-hot vectors based on node degrees, whereas the features for the Proteins dataset were standardized.

All models underwent training for 100 epochs, with an initial learning rate of 0.01 reduced by a factor of 0.5 after 50 epochs. We also adopted the early stopping technique for model training. Hidden numbers were set to 32, except for the Twitter dataset, which was 16. GAT employed eight attention heads and utilized ReLU activation. Meanwhile, for GraphSage, Mean aggregators were utilized, except for the Twitter dataset, where GCN aggregators were applied. The parameters α and β were selected between 10^{-2} and 10^{-5}.

For inference, we implemented reference selection methods and Eq. 5 based on Faiss v1.7.4. Specifically, we set the number of references K to 10 and 3 in π_a and π_c, respectively.

4.3   Predictive Performance Comparison

Analysis of Table 1 reveals several noteworthy findings. Our proposed HVG method significantly enhances the performance of GNN backbones, achieving up to 8% higher accuracies than baselines. Additionally, the introduction of class-level knowledge constraints results in reduced accuracy variances across all datasets and models. Furthermore, the KNN-based interpretable predictor, which leverages GNN representations, contributes to an overall improvement in predictive performance. Our observations indicate that KNN is particularly effective for the Mutag and Proteins datasets, while KNN_Class exhibits higher performance with these others. This disparity in performance can be attributed to the nature of the node features and graph complexity. Unlike Mutag and Proteins, IMDB and DD solely rely on vertex degrees. Twitter graphs are characterized by their noise and complexity. Consequently, graphs are not well-separated in latent space, thereby hindering the efficiency of KNN.

Table 1. Predictive performance comparison on five datasets

Method	Mutag	Proteins	IMDB	DD	Twitter
BASE-GCN	71.8 ± 9.4	71.4 ± 5.1	71.0 ± 4.9	71.5 ± 4.0	64.2 ± 1.7
BASE-SAGE	73.0 ± 9.6	69.4 ± 4.9	71.5 ± 5.1	74.3 ± 3.8	63.6 ± 2.1
BASE-GIN	86.2 ± 9.6	75.0 ± 5.2	72.6 ± 2.9	69.9 ± 3.5	65.1 ± 1.3
BASE-GAT	75.0 ± 11.2	67.2 ± 12.0	72.6 ± 3.4	69.9 ± 3.5	65.2 ± 1.6
HVG-GCN	74.0 ± 8.5	72.2 ± 3.8	75.5 ± 3.9	72.9 ± 3.7	64.3 ± 1.8
HVG-SAGE	75.1 ± 8.7	73.0 ± 4.6	74.9 ± 3.8	74.0 ± 3.4	65.3 ± 1.4
HVG-GIN	87.3 ± 7.1	77.2 ± 3.6	77.3 ± 4.3	**77.2 ± 3.0**	65.1 ± 1.8
HVG-GAT	78.2 ± 6.8	75.3 ± 4.0	76.2 ± 2.7	70.6 ± 3.2	65.8 ± 1.2
HVG$^+$-GCN	76.6 ± 10.0	74.6 ± 4.1	74.8 ± 3.5	72.8 ± 3.7	64.6 ± 1.8
HVG$^+$-SAGE	76.7 ± 9.0	74.8 ± 3.3	74.4 ± 4.3	72.4 ± 3.4	64.2 ± 1.4
HVG$^+$-GIN	**88.2 ± 6.3**	**78.1 ± 3.3**	**77.7 ± 4.1**	76.1 ± 2.2	65.1 ± 2.1
HVG$^+$-GAT	77.7 ± 7.3	75.3 ± 4.0	76.2 ± 2.6	70.8 ± 3.1	**66.0 ± 1.3**

4.4   Benefits of Human-AI Interactions

This experiment aimed to validate our hypothesis that the proposed interaction strategy could enhance HVG predictive performance and the stability of training processes. In Fig. 2, we present the experiments' results using two different backbones and three datasets. For HVG Interaction, we identified centroid candidates via Eq. 5 and simulated user selections by controlling a rejection threshold. Notably, the HVG Interaction strategy significantly improved the predictive capabilities of GNN backbones compared with Baseline and HVG Random approaches. Since we assumed that datasets possessed IID properties

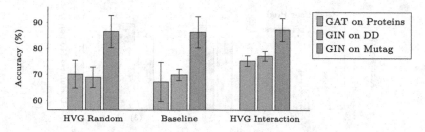

Fig. 2. A comparison of the predictive performance of three training strategies on three datasets and two GNN backbones

and employed random centroid selection, the model accuracies of the HVG Random strategy showed slightly more significant variability than others, and its accuracies were only comparable to those of baseline models. These findings underscore the importance of human-AI interactions as a critical and beneficial factor for achieving superior model performance and alignment between human and AI systems.

4.5 User Perception of Prediction Explanations

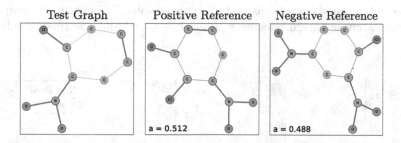

Fig. 3. An example of a test graph and references in Mutag dataset

We organized a small game including 20 participants to study the relationship between user perception and prediction explanations. We gave the winner a gift to boost participants' interests. Participants were asked to predict model outcomes of Twitter graphs based on one of the following explanation formats: (1) Subgraph visualization; (2) Presenting one reference from each class with (1); (3) Showing reference attribution scores with (2), as similar to Fig. 3.

We evaluated user performance based on their prediction accuracy and usefulness score. Results from Fig. 4 revealed that showing only subgraphs extracted from methods like [12] did not help users on this task. The combination of subgraph extraction and reference presentation significantly improved user performance. Additionally, it was notable that attribution scores had a modest impact

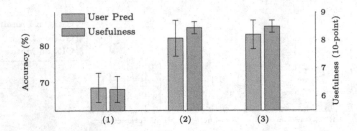

Fig. 4. User predictive performance with different explanation formats

on enhancing users' predictive capability. These findings suggested that incorporating references in explanations is advantageous, as they align with users' inherent ability to learn from similar cases, thereby improving their understanding and performance.

4.6   Is Instance-Level Feedback Helpful in Any Cases?

Based on Sect. 4.5, we investigated the benefits of incorporating instance-level user feedback. We fine-tuned an HVG-GIN model with Eq. 3 and generated triplets from Mutag's validation set. After the task was explained quickly, 19 volunteers guessed model predictions of ten graphs based on supported information like subgraph visualizations and reference graphs. Ultimately, we compared prediction accuracy given references of non-fine-tuned and fine-tuned models.

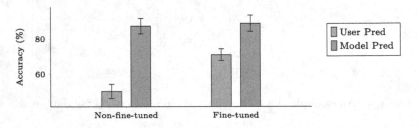

Fig. 5. User predictive performance with references retrieved from non-fine-tuned and fine-tuned models using the instance-level feedback constraint

As depicted in Fig. 5, the fine-tuned model outperformed its non-fine-tuned counterpart, improving both model and user performance. However, participants faced significant challenges due to limited domain knowledge, resulting in relatively low accuracy in both scenarios. Notably, the instance-level feedback adjustments brought target graph representations closer to actual neighbors in the latent space, making references from the fine-tuned model more helpful. This study highlighted the potential of instance-level feedback in specific scenarios, such as human-model alignment enhancement.

5   Discussions of Fairness and Ethical Issues

Our framework holds potential in various applications, but it is crucial to acknowledge the fairness and ethical concerns. Firstly, the prototype selection process, driven by domain experts, can introduce biases, potentially leading to systematic errors. There is also a risk of malicious actors manipulating the system to guide users toward incorrect or harmful decisions by introducing specific prototypes or triplet samples. Second, feedback loops can exacerbate biases, mainly if the system continually receives feedback from a particular viewpoint. Third, reference policies are unintelligible to some extent since graph representations can be different from the original data. Moreover, if certain groups are underrepresented in the training data or among selected prototypes, the system may perform poorly for those groups, resulting in potentially discriminatory outcomes. Lastly, while experts are responsible for addressing ethical concerns, they are also susceptible to their own biases and errors, necessitating ongoing vigilance and mitigation measures.

6   Conclusion and Future Work

We proposed a promising framework bridging the gap between powerful yet opaque graph representation learning models and human decision-making. By incorporating human expertise through an iterative process, we have improved the interpretability and reliability of graph classification models. Our experiments and user studies have shown the effectiveness of this approach, demonstrating the potential for broader adoption in real-world applications where transparency and collaboration between humans and AI systems are crucial.

For future research, we aim to explore more advanced techniques for leveraging human feedback, possibly utilizing reinforcement learning strategies. We also plan to extend our framework to handle more extensive and complex graph datasets, addressing scalability challenges. Lastly, we see opportunities for integrating domain-specific knowledge more effectively and enhancing the adaptability of our framework across various application domains.

Acknowledgments. This work was supported by the National Research Foundation of Korea(NRF) grant funded by the Korean government (MSIT)(No. RS-2023-00222663, RS-2023-00262885).

References

1. Bui, T.C., Le, V.D., Li, W.S.: Generating real-time explanations for GNNs via multiple specialty learners and online knowledge distillation. IEEE Access (2023)
2. Chen, C., Li, O., Tao, D., Barnett, A., Rudin, C., Su, J.K.: This looks like that: deep learning for interpretable image recognition. In: Advances in Neural Information Processing Systems, vol. 32 (2019)

3. Dai, E., Wang, S.: Towards self-explainable graph neural network. In: Proceedings of the 30th ACM International Conference on Information & Knowledge Management, pp. 302–311 (2021)
4. Davoudi, S.O., Komeili, M.: Toward faithful case-based reasoning through learning prototypes in a nearest neighbor-friendly space. In: International Conference on Learning Representations (2021)
5. Doshi-Velez, F., Kim, B.: Towards a rigorous science of interpretable machine learning. arXiv preprint arXiv:1702.08608 (2017)
6. Feng, A., You, C., Wang, S., Tassiulas, L.: KerGNNs: interpretable graph neural networks with graph kernels. In: Proceedings of the AAAI Conference on Artificial Intelligence, vol. 36, pp. 6614–6622 (2022)
7. Hamilton, W., Ying, Z., Leskovec, J.: Inductive representation learning on large graphs. In: Advances in Neural Information Processing Systems, vol. 30 (2017)
8. Kipf, T.N., Welling, M.: Semi-supervised classification with graph convolutional networks. arXiv preprint arXiv:1609.02907 (2016)
9. Li, O., Liu, H., Chen, C., Rudin, C.: Deep learning for case-based reasoning through prototypes: a neural network that explains its predictions. In: Proceedings of the AAAI Conference on Artificial Intelligence, vol. 32 (2018)
10. Li, X., et al.: BrainGNN: interpretable brain graph neural network for FMRI analysis. Med. Image Anal. **74**, 102233 (2021)
11. Liu, H., Tian, Y., Chen, C., Feng, S., Chen, Y., Tan, C.: Learning human-compatible representations for case-based decision support. In: The Eleventh International Conference on Learning Representations (2022)
12. Luo, D., et al.: Parameterized explainer for graph neural network. arXiv preprint arXiv:2011.04573 (2020)
13. Mosqueira-Rey, E., Hernández-Pereira, E., Alonso-Ríos, D., Bobes-Bascarán, J., Fernández-Leal, Á.: Human-in-the-loop machine learning: a state of the art. Artif. Intell. Rev. **56**(4), 3005–3054 (2023)
14. Ragno, A., La Rosa, B., Capobianco, R.: Prototype-based interpretable graph neural networks. IEEE Trans. Artif. Intell. (2022)
15. Ramos, G., Meek, C., Simard, P., Suh, J., Ghorashi, S.: Interactive machine teaching: a human-centered approach to building machine-learned models. Hum. Comput. Interact. **35**(5–6), 413–451 (2020)
16. Rossi, R., Ahmed, N.: The network data repository with interactive graph analytics and visualization. In: Proceedings of the AAAI Conference on Artificial Intelligence, vol. 29 (2015)
17. Slade, S.: Case-based reasoning: a research paradigm. AI Mag. **12**(1), 42–42 (1991)
18. Taesiri, M.R., Nguyen, G., Nguyen, A.: Visual correspondence-based explanations improve AI robustness and human-AI team accuracy. In: Advances in Neural Information Processing Systems, vol. 35, pp. 34287–34301 (2022)
19. Velickovic, P., Cucurull, G., Casanova, A., Romero, A., Lio, P., Bengio, Y.: Graph attention networks. stat **1050**, 20 (2017)
20. Xu, K., Hu, W., Leskovec, J., Jegelka, S.: How powerful are graph neural networks? arXiv preprint arXiv:1810.00826 (2018)
21. Yuan, H., Yu, H., Gui, S., Ji, S.: Explainability in graph neural networks: a taxonomic survey. arXiv preprint arXiv:2012.15445 (2020)
22. Zhang, Z., Liu, Q., Wang, H., Lu, C., Lee, C.: ProtGNN: towards self-explaining graph neural networks. In: Proceedings of the AAAI Conference on Artificial Intelligence, vol. 36, pp. 9127–9135 (2022)
23. Zhang, Z., Cui, P., Zhu, W.: Deep learning on graphs: a survey. IEEE Trans. Knowl. Data Eng. **34**(1), 249–270 (2020)

SASBO: Sparse Attack via Stochastic Binary Optimization

Yihan Meng, Weitao Li, and Lin Shang[✉]

State Key Laboratory for Novel Software Technology, Department of Computer
Science and Technology, Nanjing University, Nanjing, China
{mengyihan,liweitao}@smail.nju.edu.cn, shanglin@nju.edu.cn

Abstract. Deep Neural Networks have shown vulnerability to sparse
adversarial attack, which involves perturbing only a limited number
of pixels. Identifying the coordinates requiring perturbation in sparse
attacks poses a significant computational challenge. Existing solutions
predominantly rely on heuristic methods or relax the ℓ_0-norm to the
ℓ_1-norm. In this paper, we present an efficient algorithm for conducting
sparse attacks. Our algorithm factorizes the perturbation at each pixel to
the product of the perturbation coordinates and the perturbation magnitudes and then optimizes them alternately. We reformulate the ℓ_0-norm
as a stochastic binary optimization problem, assuming that each pixel's
perturbation status is associated with a stochastic binary variable. This
stochastic binary variable follows a Bernoulli distribution, with a parameter value that ranges from 0 to 1, signifying the probability of pixel
disturbance. To tackle this stochastic binary optimization challenge, we
employ an unbiased gradient estimator known as Augment-Reinforce-
Merge (ARM). Once the perturbed coordinates are determined, we optimize the perturbation magnitudes with gradient descent. Furthermore,
we incorporate a binary search algorithm to eliminate redundant pixels
to enhance sparsity. Comprehensive experiments demonstrate the superiority of our proposed method over several state-of-the-art sparse attack
methods.

Keywords: Sparse Attack · Stochastic Binary Optimization ·
Augment-Reinforce-Merge · Sparsity Enhancement

1 Introduction

Deep Neural Networks (DNNs) have made remarkable strides in various computer vision tasks, such as image classification [10,24], facial expression recognition [20], and image retrieval [9]. Despite their notable achievements, DNNs
have been found to exhibit vulnerability to adversarial examples [2,8,19]. Adversarial examples are meticulously crafted images that contain subtle, malicious
alterations, aiming to deceive the targeted model while remaining visually indistinguishable from the original, clean images. The rapid development of adversarial attack has given rise to increased concerns about the security of DNNs,
especially those deployed in real-world applications [12]. Consequently, research

© The Author(s), under exclusive license to Springer Nature Singapore Pte Ltd. 2024
D.-N. Yang et al. (Eds.): PAKDD 2024, LNAI 14645, pp. 117–129, 2024.
https://doi.org/10.1007/978-981-97-2242-6_10

on adversarial attack holds significant importance, given that these adversarial examples are invaluable instruments for uncovering the vulnerabilities of DNNs, illuminating their inner mechanisms, and establishing a foundation for enhancing robustness.

In the realm of adversarial perturbations, the predominant consideration revolves around the constraint of ℓ_p-norm distance, which typically falls into two distinct categories: dense attacks and sparse attacks. Dense attacks adhere to the constraints of either the ℓ_2 or ℓ_∞-norm [8,19]. In contrast, sparse attacks are governed by the ℓ_0-norm [14,15] constraint.

Sparse attacks pose a greater challenge in comparison to dense attacks. Such a circumstance arises from the fact that generating adversarial perturbations under the ℓ_0-norm constraint is an NP-hard problem. To address this problem, a multitude of methods have been proposed, broadly falling into two categories. The first category is to predetermine the perturbed pixels artificially or to find the pixels to be perturbed by some heuristic search methods. Nonetheless, there is no assurance that these heuristic methods can pinpoint the optimal perturbation coordinates, as heuristic search strategies are prone to get trapped in local minima and may inadvertently introduce redundant perturbations. The second category involves the relaxation or transformation of ℓ_0-norm constraint, followed by subsequent problem optimization. However, these methods have the high computational complexity and the complex algorithm implementation.

In this paper, we propose a novel sparse adversarial attack approach **SASBO** (**S**parse **A**ttack via **S**tochastic **B**inary **O**ptimization). Initially, we factorize each perturbation into the product of the perturbation magnitudes and the perturbation coordinates. Subsequently, we alternately optimize the coordinates and the magnitudes in each iteration. Regarding the perturbation coordinates, their values dictate whether the corresponding pixel can be perturbed, where a value of 1 indicates permission and 0 denotes exclusion. Because of the binary nature of the coordinates, we model it as a stochastic binary optimization problem, assuming adherence to a Bernoulli distribution. The parameters's values represent the probability of the gate being activated. Consequently, optimizing the coordinates will be formulated as a stochastic binary optimization problem, which can be tackled by the unbiased gradient estimator Augment-Reinforce-Merge (**ARM**) [21]. Especially, ARM is noteworthy for its impartiality, low variance, and computational efficiency, which enables our method to optimize the perturbation coordinates with precision and efficiency. Regarding perturbation magnitudes, we utilize the gradient descent method for optimization. Furthermore, we convert the perturbation coordinates from continuous values to discrete values by using a dynamic threshold strategy. To enhance sparsity, we employ a binary search algorithm to determine the minimum number of pixels required to generate each adversarial example. Finally, we conduct extensive experiments on two benchmark datasets, CIFAR-10 [11] and ImageNet [5]. These experiments empirically validate the effectiveness and performance of our proposed method.

In summary, our work makes three primary contributions:

- We introduce an innovative sparse attack method, regarding it as a stochastic binary optimization problem and optimizing it with ARM.

- We utilize a binary search algorithm to enhance the sparsity of generated adversarial examples by eliminating redundant pixels.
- We demonstrate our method's outstanding performance in untargeted and targeted attacks through comprehensive experiments.

2 Related Work

The generation of sparse adversarial examples for image classification tasks has been the subject of extensive research in recent years. Existing adversarial attack methods can be primarily categorized into two distinct groups: heuristic strategies and optimized strategies.

Heuristic Strategies. One Pixel Attack [18] leverages the Differential Evolution algorithm to seek an extremely sparse configuration, where only one pixel is perturbed to deceive the target model. JSMA [16] and its extensions select influential pixels based on a saliency map, while CornerSearch [3] employs a heuristic approach that involves traversing all pixels and selecting a subset for perturbation. Additionally, GreedyFool [6] iteratively identifies the most suitable pixels for modification based on gradient information and subsequently applies a greedy strategy to drop as many less critical pixels as possible. However, it is essential to note that these heuristic-based methods share common challenges. Due to the nature of the heuristic search, they are susceptible to local optima, potentially resulting in unnecessary and redundant perturbations.

Optimized Strategies. SparseFool [15] converts the ℓ_0 problem into an ℓ_1 problem. PGD_0 [3] proposes to project the adversarial noise generated by PGD [14] to the ℓ_0-ball to achieve the ℓ_0-version PGD. An ADMM-based method proposed in [22] aims to optimize the sparse attack problem by decoupling the ℓ_0-norm and the adversarial loss. ℓ_1-APGD [4] converts the ℓ_0-norm constrain into the intersection of the ℓ_1-ball and $[0,1]^d$ box constrain. Subsequently, it derives the appropriate steepest descent step for the intersection, which motivates an adaptive sparsity of the chosen descent direction. SAPF [7] reformulates the sparse adversarial attack problem as a mixed integer programming problem, to jointly optimize the binary selection factors and continuous perturbation magnitudes of all pixels, with a cardinality constraint that explicitly controls the degree of sparsity. Homotopy-Attack [23] addresses the sparse attack problem by proposing a homotopy algorithm. In each iteration, Homotopy-Attack optimizes the ℓ_0-regularized adversarial loss by leveraging the nonmonotone Accelerated Proximal Gradient Method for nonconvex programming. Additionally, it incorporates an optional post-attack step designed to escape undesirable local minima.

3 Methods

3.1 Problem Definition

We denote the source image as x and its ground-truth label as y. An input x will be classified as $\arg\max f_c(x) = y$, where f is the target model and $f_c(x)$ is the output logit value for class c. An adversarial example $x^{adv} = x + \delta$ is generated by adding perturbations to the source image.

The untargeted attack aims to ensure $\arg\max f_c(x) \neq y$. Furthermore, targeted attack aims to guarantee $\arg\max f_c(x) = y_{adv}$, where y_{adv} is a designated adversarial label. Moreover, the adversarial perturbation δ should be small enough to guarantee the visual similarity between the adversarial sample and the source image. In this paper, we employ ℓ_0-norm, which means that adversarial perturbations can only be added to a few pixels, known as *sparse attack*. Thus, the goal of the adversarial attack can be formulated as:

$$\min_{\delta} \|\delta\|_0 + \lambda \mathcal{L}\left(f(x + \delta), y_t\right), \quad \text{s.t.} \quad x + \delta \in [0, 1], \tag{1}$$

where $\mathcal{L}$ is the loss function, such as the widely used C&W loss [2], and λ serves as a weighting factor for the regularization.

However, the sparse attack is an NP-hard problem, which can not be solved analytically within polynomial time, primarily due to the non-differentiability of ℓ_0-norm. Previous approaches have employed stochastic search strategies or heuristic methods to determine the perturbed coordinates. In contrast, we propose to transform the discrete optimization problem of perturbed coordinates into a continuous optimization problem, and then alternately optimize the perturbation coordinates and perturbation magnitudes.

3.2 Sparse Adversarial Attack via Stochastic Binary Optimization

We first factorize the perturbation δ into the following two parts:

$$\delta = \mathbf{M} \odot \mathbf{E}. \tag{2}$$

$\mathbf{M} \in \{0, 1\}^N$ denotes the vector of perturbed coordinates, which should be a sparse matrix. $\mathbf{E} \in \mathbb{R}^N$ denotes the vector of perturbation magnitudes. $\odot$ denotes element-wise product. Since $\mathbf{E}$ is continuous, while $\mathbf{M}$ is discrete, it is unfeasible to jointly optimize them using any existing continuous solvers, such as stochastic gradient descent.

Optimize Perturbed Coordinates M with ARM. Following the previous factorization, and given $\mathbf{E}$, we can reformulate the goal of the sparse adversarial attack concerning $\mathbf{M}$ as follows:

$$\min_{\mathbf{M}} \|\mathbf{M} \odot \mathbf{E}\|_0 + \lambda \mathcal{L}\left(f(x + \mathbf{M} \odot \mathbf{E}), y_t\right), \quad \text{s.t.} \quad \mathbf{M} \in \{0, 1\}^N, \tag{3}$$

Note that the Eq. (3) remains computationally intractable since $m_j \in \mathbf{M}$ is a discrete variable. Therefore, we must consider further transformations of the variable $\mathbf{M}$. These transformations should enable us to efficiently optimize Eq. (3) in a continuous manner while maintaining the integrity of the ℓ_0-norm constraints.

The value of $m_j \in \mathbf{M}$ is a discrete value 0 or 1, corresponding to a binary variable that denotes whether the pixel can be perturbed. The ℓ_0-norm represents the number of variables with perturbed permission. To relax this discrete constraint, we assume m_j adheres to Bernoulli distribution with parameter $\pi_j \in [0, 1]$, such as $m_j \sim \text{Bern}(m_j; \pi_j)$. After this transformation, the value of m_j is continuous, and the ℓ_0-norm corresponds to the variable π_j, which determines the probability that gate m_j is equal to one. In addition, according to stochastic variational optimization [1], given any function $\mathcal{F}(z)$ and any distribution $q(z)$, the following inequality holds

$$\min_z \mathcal{F}(z) \leq \mathbb{E}_{z \sim q(z)}[\mathcal{F}(z)], \tag{4}$$

which means the minimum value of a function is upper bounded by the expectation of the function. With Eq. (4) and substituting $\mathbf{M} \sim \text{Bern}(\mathbf{M}; \pi)$ into (3), problem (3) can be rewritten as:

$$\min_\pi \mathcal{R}(\pi) = \mathbb{E}_{M \sim \text{Ber}(M;\pi)} [\lambda \mathcal{L}(f(x + \mathbf{M} \odot \mathbf{E}), y_t)] + \sum_{j=1}^N e_j \cdot \pi_j \tag{5}$$

The second term represents sparse constraint after relaxation, where π_j is the perturbed probability of pixel at m_j, and e_j is the perturbation size of the pixel at m_j. The problem is an expectation minimization problem characterized by introducing parameters π. Note that the second term is differentiable and related to the new parameters. However, the first term remains problematic since the expectation over numerous binary random variables M is intractable and does not allow for efficient gradient-based optimization.

In order to minimize the new problem, we can use the Augment-Reinforce-Merge (ARM) [21] gradient estimator, which is proposed for the optimization of binary latent variables problem. Denote $\sigma(\phi) = 1/(1 + exp(-\phi))$ as a sigmoid function, which shares the same value range as π. $\mathbf{1}_{[\cdot]}$ is an indicator function that equals to one if the argument is true and zero otherwise

Theorem 1 *(ARM). For a vector of V binary random variables $z = (z_1, \ldots, z_V)^T$, the gradient of*

$$\mathcal{E}(\phi) = \mathbb{E}_{z \sim \prod_{v=1}^V \text{Bernoulli}(z_v; \sigma(\phi_v))}[f(z)] \tag{6}$$

with respect to $\phi = (\phi_1, \ldots, \phi_V)^T$, the logits of the Bernoulli probability parameters can be expressed as

$$\nabla_\phi \mathcal{E}(\phi) = \mathbb{E}_{u \sim \prod_{v=1}^V \text{Uni}(u_v; 0, 1)} \left[\left(f\left(\mathbf{1}_{[u > \sigma(-\phi)]}\right) - f\left(\mathbf{1}_{[u < \sigma(\phi)]}\right) \right) \left(u - \frac{1}{2} \right) \right], \tag{7}$$

where $\mathbf{1}_{[u > \sigma(-\phi)]} := \left(\mathbf{1}_{[u_1 > \sigma(-\phi_1)]}, \ldots, \mathbf{1}_{[u_V > \sigma(-\phi_V)]} \right)^T$.

With parameter $\pi_j \in [0, 1]$ as $\sigma(\phi_j)$, problem (5) can be reformulated as

$$\min_\phi \mathcal{R}(\phi) = \mathbb{E}_{M \sim \text{Ber}(M; \sigma(\phi))} [\lambda \mathcal{L}(f(x + \mathbf{M} \odot \mathbf{E}), y_t)] + \sum_{j=1}^N e_j \cdot \sigma(\phi_j) \tag{8}$$

According to Theorem 1, we can compute the gradient of Eq. (8) with respect to ϕ by

$$\nabla_\phi^{ARM} \mathcal{R}(\phi) = \mathbb{E}_{u \sim \prod_{v=1}^V \text{Uni}(u; 0, 1)} \left[\left(f\left(\mathbf{1}_{[u > \sigma(-\phi)]}\right) - f\left(\mathbf{1}_{[u < \sigma(\phi)]}\right) \right) \left(u - \frac{1}{2} \right) \right]$$
$$+ \lambda \sum_{j=1}^N e_j \cdot \nabla_{\phi_j} \sigma(\phi_j) \tag{9}$$

which is an unbiased and low variance estimator as demonstrated in [21].

After completing the optimization steps mentioned above, we can obtain the parameter ϕ. Each element of ϕ represents the perturbation probability of the corresponding pixel. As a result, $\mathbf{M}$ can be obtained by taking the expectation of $m_j \sim \text{Bern}(m_j; \sigma(\phi_j))$

$$m_j = \mathbb{E}[m_j] = \pi_j = \sigma(\phi_j) \qquad j = 1, 2, \cdots, N \tag{10}$$

A conventional approach to convert continuous values to discrete values approximately is specifying a threshold τ for partition. It is worth noting that when utilizing a large threshold, the resulting perturbation coordinates matrix $\mathbf{M}$ exhibits excessive sparsity, rendering it ineffective in generating successful adversarial examples for certain images. Conversely, a smaller threshold leads to perturbations that are overly redundant. Sensitivity to the threshold parameter highlights the necessity for a more nuanced and adaptive discretization strategy within our algorithmic framework.

Hence, we opt for a dynamic thresholding approach by determining the threshold τ based on the k-th largest element in the output vector $\sigma(\phi)$. The parameter k is introduced as a hyperparameter, serving the purpose of regulating sparsity. This strategic selection of the threshold contributes to a nuanced control over the extent of perturbation in the context of adversarial attacks.

$$\tau = \text{argmax}_k(\sigma(\phi)) \tag{11}$$

Although the upper bound is sufficient for successful sparse attacks, it may not be the minimum number of required pixels for the adversary. For sparser perturbations, it is crucial to minimize the number of perturbed pixels as small as possible. Therefore, we adopt a binary search strategy to achieve better sparsity, and the details are introduced in Algorithm 1.

Optimize Perturbation Magnitudes E by Gradient Descent. Given the perturbation coordinates $\mathbf{M}$, the Eq. (3) associated with $\mathbf{E}$ can be reformulated as follows:

$$\min_E \|\mathbf{M} \odot \mathbf{E}\|_2 + \lambda \mathcal{L}(f(x + \mathbf{M} \odot \mathbf{E}), y_t), \quad -\epsilon \leq \mathbf{E} \leq \epsilon, \tag{12}$$

where ϵ is the maximal allowable perturbation magnitudes, and we use ℓ_2-norm as the metric to quantify the magnitudes of the perturbations.

When the perturbation coordinates have been determined, we need to calculate the perturbation magnitudes of the corresponding coordinates. Since the sparse matrix $\mathbf{M}$ satisfies the requirement of the sparsity constraint, ℓ_0-norm is no longer required as a constraint for adversarial perturbations in Eq. (12). To ensure the invisibility of the perturbation, we opt to employ the ℓ_2-norm to constrain the dissimilarity between the adversarial examples and the original images.

After the above reformulation, Eq. (12) is akin to the dense adversarial attack formulation and is differentiable. Hence, we can optimize it by utilizing the gradient descent algorithm as follows:

$$\nabla_{\mathbf{E}} = 2 \cdot (\mathbf{M} \odot \mathbf{M} \odot \mathbf{E}) + \lambda \frac{\partial \mathcal{L}(f(x + \mathbf{M} \odot \mathbf{E}), y_t)}{\partial \mathbf{E}} \tag{13}$$

Algorithm 1. SASBO

Input: source image $\mathbf{x}$, targeted model f.
Parameter: iterative steps N_1, N_2, Bernoulli probability parameter $\sigma(\phi)$, upper bound of pixels k, learning rate η_1, η_2, perturbation threshold ϵ.
Output: adversarial example $\mathbf{x}^{adv}$.

1: **Stage 1: Optimization**
2:    $\phi = \mathcal{N}(0, 0.01)$
3:    $\mathbf{E} = 0.01$
4: **while** not converged **do**
5:      Update $\mathbf{M}$ with ARM.
6:      **for** $0...N_1$ **do**
7:        $\phi = \phi - \eta_1 \cdot \nabla_\phi^{ARM} \mathcal{R}(\phi)$
8:      **end for**
9:      $\tau = \mathrm{argmax}_k(\sigma(\phi))$
10:     $\mathbf{M} = \mathbf{1}_{[\mathbf{m}_j > \tau]}$
11:     Update $\mathbf{E}$ with gradient descent.
12:     **for** $0...N_2$ **do**
13:       $\mathbf{E} = \mathbf{E} - \eta_2 \cdot (2 \cdot (\mathbf{M} \odot \mathbf{E} \odot \mathbf{E}) + \lambda \frac{\partial \mathcal{L}(f(\mathbf{x} + \mathbf{M} \odot \mathbf{E}), y_t)}{\partial \mathbf{E}})$
14:       $\mathbf{E} = clip(\mathbf{E}, -\epsilon, \epsilon)$
15:     **end for**
16: **end while**

17: **Stage 2: Binary Search**
18: $LB = 1$
19: $UB = k$
20: **while** $UB - LB > 1$ **do**
21:     $MI = (LB + UB)/2$
22:     $\tau = \mathrm{argmax}_{MI}(\sigma(\phi))$
23:     $\mathbf{M} = \mathbf{1}_{[\mathbf{m}_j > \tau]}$
24:     $\mathbf{x}^{adv} = \mathbf{x} + \mathbf{M} \odot \mathbf{E}$
25:     **if** $\mathbf{x}^{adv}$ is not adversarial **then**
26:       $LB = MI + 1$
27:     **else**
28:       $UB = MI$
29:     **end if**
30: **end while**
31: **return** $\mathbf{x}^{adv}$
32:

Finally, we clip the perturbation magnitudes $\mathbf{E}$ to the range $[-\epsilon, \epsilon]$ after each gradient step in order to satisfy perturbation threshold ϵ constraints.

4   Experiments and Results

In this section, we conduct experiments on CIFAR-10 [11] and ImageNet [5] datasets. We compare our method with several state-of-the-art sparse adversarial attack algorithms, including PGD-$\ell_0 + \ell_\infty$ [3], SparseFool [15], and GreedyFool [6].

Dataset and Classification Model Setting. For CIFAR-10 dataset, following [6] and [3], we employ a pre-trained network in network [13] model with input size of $32 \times 32 \times 3$. For the classification model on the ImageNet dataset, we opt to utilize the pre-trained VGG19 [17] model. To make an accurate comparison, we generate adversarial examples with 2000 images randomly selected from the ImageNet validation set and 5000 images from the CIFAR-10 test set. For the targeted attack, the targeted label is randomly selected and kept consistent for all methods to make a fair comparison.

Parameter Settings. In SASBO, the trade-off hyperparameter λ is set to 1 to balance the adversarial loss and sparsity. As shown in Algorithm 1, we initialize ϕ by random samples from a normal distribution $\mathcal{N}(0, 0.01)$ and set the initial value of $\mathbf{E}$ to 0.01. During each iteration, $\mathbf{M}$ and $\mathbf{E}$ are updated $N_1 = 5$ and $N_2 = 10$ steps with learning rate $\eta_1 = 0.01$ and $\eta_2 = 0.1$, respectively, so that when one of $\mathbf{M}$ or $\mathbf{E}$ is specified the other can be optimized to the current optimal solution. Besides, the number of upper bound perturbed coordinates k is a key hyper-parameter for our method. We initially run the baseline GreedyFool with 100% ASR and use the average ℓ_0-norm of GreedyFool as the reference value for k. Then, we tune the value of k to ensure successful attacks on all images.

For both targeted attack and non-targeted attacks, we employ the C&W loss function [2] as the adversarial loss function $\mathcal{L}$. As for compared methods, we employ the official implementation and follow default settings in the experiments.

Evaluation Metrics. We report the average ℓ_p norms of generated perturbations, where $p = 0, 1, 2, \infty$, and the attack success rate (ASR) of the experimented methods with both the targeted and non-targeted settings.

4.1 Non-targeted Attack

We evaluate the sparsity and ASR under three different perturbation magnitude thresholds. On both the CIFAR-10 and ImageNet datasets, we compare our methods with recent works: PGD-$\ell_0 + \ell_\infty$, SparseFool, GreedyFool. The non-targeted attack results on both CIFAR-10 and ImageNet datasets are presented in Table 1 and Table 2.

Table 1. Statistics of ASR and average ℓ_p-norms ($p = 0, 1, 2, \infty$) of non-targeted attack on CIFAR-10 dataset.

Threshold	Method	ASR	ℓ_0	ℓ_1	ℓ_2	ℓ_∞
$\epsilon = 255$	PGD-$\ell_0 + \ell_\infty$	100	48.01	38.59	6.05	1.000
	SparseFool	100	22.03	10.61	2.29	**0.808**
	GreedyFool	100	12.29	8.14	1.48	0.986
	SASBO (Ours)	**100**	**3.95**	**3.82**	**1.24**	0.985
$\epsilon = 100$	PGD-$\ell_0 + \ell_\infty$	100	487.82	171.82	8.18	0.389
	SparseFool	100	30.39	9.81	1.71	0.386
	GreedyFool	100	15.56	5.83	1.21	**0.376**
	SASBO (Ours)	**100**	**7.90**	**3.08**	**0.51**	0.382
$\epsilon = 10$	PGD-$\ell_0 + \ell_\infty$	82.72	941.78	45.72	1.88	0.039
	SparseFool	100	315.13	12.25	0.57	0.039
	GreedyFool	100	178.53	6.56	0.49	**0.038**
	SASBO (Ours)	**100**	**144.21**	**5.61**	**0.43**	0.039

For CIFAR-10, as depicted in Table 1, when the perturbation threshold ϵ is 255, it is evident that all methods can achieve a 100% attack success rate. Furthermore, our algorithm exhibits a significant advantage over the other methods

Table 2. Statistics of ASR and average ℓ_p-norms ($p = 0, 1, 2, \infty$) of non-targeted attack on ImageNet dataset.

Threshold	Method	ASR	ℓ_0	ℓ_1	ℓ_2	ℓ_∞
$\epsilon = 255$	PGD-$\ell_0 + \ell_\infty$	100	1315.73	1315.73	114.68	1.000
	SparseFool	100	102.45	49.33	4.56	0.860
	GreedyFool	100	71.65	30.34	3.06	0.867
	SASBO (Ours)	**100**	**50.87**	**24.52**	**1.65**	**0.860**
$\epsilon = 100$	PGD-$\ell_0 + \ell_\infty$	100	1782.94	686.33	16.22	0.385
	SparseFool	100	172.78	45.02	2.96	0.386
	GreedyFool	100	160.46	36.91	2.3	**0.345**
	SASBO (Ours)	**100**	**145.77**	**38.29**	**1.44**	0.379
$\epsilon = 10$	PGD-$\ell_0 + \ell_\infty$	64.92	14750.05	535.25	4.56	0.039
	SparseFool	100	1257.56	39.10	1.04	0.039
	GreedyFool	100	1104.27	38.95	0.84	**0.038**
	SASBO (Ours)	**100**	**1051.24**	**40.86**	**0.57**	0.039

in terms of sparsity. Specifically, the average perturbation number required by our algorithm is only 3.95 pixels to execute a successful attack, while Greedy-Fool needs to perturb 12.29 pixels, which is approximately 3.11 times higher than ours.

As the perturbation threshold ϵ decreases, executing the attack successfully becomes increasingly challenging. All the methods require more perturbation pixels to perform a successful attack. Nonetheless, our method consistently outperforms the compared methods. For instance, when the perturbation threshold ϵ is 10, PGD-$\ell_0 + \ell_\infty$ fail to perform non-targeted attack with 100% attack success rate. In contrast, our algorithm maintains a 100% attack success rate. Besides, our algorithm can still achieve the best sparsity, requiring only 144.21 pixels to be perturbed, which is a notable reduction compared to all other algorithms.

For the ImageNet dataset, it can be observed that our method consistently outperforms other methods in most settings. Under three different perturbation threshold settings, our algorithm can achieve a 100% attack success rate with the lowest average perturbation number. When perturbation threshold ϵ is 255, the average ℓ_0 norm of our algorithm is 50.87 pixels, while the average ℓ_0 norm of GreedyFool is 71.65 pixels, which is 1.41x than us.

As evident from the data presented in Table 1 and Table 2, our algorithm can achieve the lowest ℓ_0, ℓ_1, ℓ_2 and ℓ_∞ norm at the same time under most settings, when compared to the other methods. This indicates that our method can select pixels that contribute more adversarial than other pixels, and all the perturbed pixels have been fully used within the allowable perturbation threshold, resulting in sparser results. Moreover, this also illustrates our method can obtain a better approximate optimal for the adversarial attack problem under ℓ_0-norm.

126   Y. Meng et al.

4.2 Targeted Attack

Since SparseFool can only achieve non-targeted attacks, here we compare targeted attack results with PGD-$\ell_0 + \ell_\infty$, GreedyFool on both CIFAR-10 and ImageNet dataset. Table 3 and Table 4 present the experimental results of the targeted attack on the CIFAR-10 and ImageNet datasets.

Table 3. Statistics of attack success rate (ASR) and average ℓ_p-norms ($p = 0, 1, 2, \infty$) of targeted attack on CIFAR-10 dataset.

Threshold	Method	ASR	ℓ_0	ℓ_1	ℓ_2	ℓ_∞
$\epsilon = 255$	PGD-$\ell_0 + \ell_\infty$	100	54.03	32.59	5.54	1.000
	GreedyFool	100	48.99	16.52	2.61	0.992
	SASBO (Ours)	**100**	**14.88**	**13.95**	**1.87**	**0.981**
$\epsilon = 100$	PGD-$\ell_0 + \ell_\infty$	100	461.91	168.24	8.09	**0.389**
	GreedyFool	100	98.46	15.83	3.41	0.391
	SASBO (Ours)	**100**	**29.94**	**11.34**	**1.55**	0.390
$\epsilon = 10$	PGD-$\ell_0 + \ell_\infty$	54.79	955.35	32.17	1.19	0.039
	GreedyFool	100	412.08	14.41	0.79	0.039
	SASBO (Ours)	**100**	**324.14**	**12.62**	**0.68**	**0.039**

As presented in Table 3, when ϵ is 255, we can observe that all compared methods can achieve a 100% attack success rate. Notably, SASBO demonstrates superior sparsity in comparison to the other methods. In detail, the GreedyFool achieves a sparsity of 1.59% (48.99 pixels) on $32 \times 32 \times 3$ CIFAR-10 images with $\epsilon = 255$, while SASBO can achieve a sparsity of 0.48% (14.88 pixels) on $32 \times 32 \times 3$ CIFAR-10 images, which is 69.6% fewer when compared to GreedyFoll. Although achieving a targeted attack becomes more difficult as the perturbation threshold ϵ decreases to 10, SASBO can still maintain a 100% attack success rate. In contrast, the attack success rate of PGD-$\ell_0 + \ell_\infty$ decreases rapidly to 54.79%. Meanwhile, the ℓ_0-norm of PGD-$\ell_0 + \ell_\infty$ increases rapidly. The average perturbation pixel number of PGD-$\ell_0 + \ell_\infty$ increases from 54.03 to 955.35. However, SASBO requires only 324.14 pixels to achieve the targeted attack successfully.

Regarding the ImageNet dataset, our algorithm still performs best in most settings. This also demonstrates the effectiveness of our algorithm. As shown in Table 4, our algorithm achieves a 100% attack success rate under all three settings. The PGD-$\ell_0 + \ell_\infty$ and GreedyFool algorithms achieve 100% attack success rate under perturbation threshold ϵ is 255, and the average perturbation number is 10015.04 pixels and 503.27 pixels respectively. However, our algorithm can achieve the same 100% attack success rate with the lowest average perturbation number, only 272.95 pixels.

The results presented in Tables 3 and 4 illustrate that SASBO is effective in both targeted and non-targeted attacks that both achieve better sparsity.

Table 4. Statistics of attack success rate (ASR) and average ℓ_p-norms ($p = 0, 1, 2, \infty$) of targeted attack on ImageNet dataset.

Threshold	Method	ASR	ℓ_0	ℓ_1	ℓ_2	ℓ_∞
$\epsilon = 255$	PGD-$\ell_0 + \ell_\infty$	100	10015.04	9873.23	113.13	1.000
	GreedyFool	100	503.27	350.41	27.38	**0.935**
	SASBO (Ours)	**100**	**272.95**	**205.87**	**15.33**	0.995
$\epsilon = 100$	PGD-$\ell_0 + \ell_\infty$	100	32842.92	10867.91	60.23	**0.389**
	GreedyFool	100	1196.42	173.76	10.25	0.391
	SASBO (Ours)	**100**	**600.80**	**142.83**	**6.56**	0.390
$\epsilon = 10$	PGD-$\ell_0 + \ell_\infty$	51.02	49516.20	1826.12	8.72	0.039
	GreedyFool	100	6476.42	225.64	3.12	0.039
	SASBO (Ours)	**100**	**4782.61**	**186.49**	**2.61**	**0.039**

(a) Clean   (b) SASBO   (c) SASBO   (d) GreedyF   (e) GreedyF

Fig. 1. Visualization of nontargeted attack under perturbation threshold $\epsilon = 10$. The images in each row, from left to right, are clean images, our adversarial example, our perturbation coordinates, GreedyFool's adversarial example, and GreedyFool's perturbation coordinates. In the figures of perturbation coordinates, black pixels denote no perturbation; white pixels denote perturbing all three RGB channels; pure red, green, and blue pixels represent perturbing a single channel. (Color figure online)

4.3 Visualization

For adversarial examples, invisibility serves as a pivotal metric. Figure 1 presents visualizations illustrating instances of non-targeted attacks executed by our algorithm and GreedyFool. Scrutinizing columns 1–2 of Fig. 1, it is almost impossible to figure out the difference between clean images and our adversarial examples, proving that our algorithm can guarantee the invisibility of adversarial examples. Furthermore, as depicted in columns 3 and 5 of Fig. 1, our method perturbs only a single channel, while GreedyFool perturbs all three RGB channels. Consequently, the alterations in perturbed pixels are slight, highlighting that our algorithm ensures both sparsity and imperceptibility concurrently.

5 Conclusion

In this paper, we propose a novel sparse adversarial attack algorithm. Our approach commences by converting the discrete variable representing the perturbed coordinates into a continuous random variable. This transformation allows us to frame the sparse adversarial attack as a stochastic binary optimization problem, which can be efficiently and precisely solved using the ARM method. Experiments conducted on two benchmark datasets show that our method achieves optimal performance in terms of attack success rate and sparsity, among other metrics, when compared to existing state-of-the-art sparse adversarial attack methods. In addition, the results of the visual presentation of the adversarial samples also show that the adversarial samples generated by our method have good adversarial perturbation invisibility. In the future, we will further consider how to incorporate the proposed ideas into group sparsity to help achieve better invisibility.

References

1. Bird, T., Kunze, J., Barber, D.: Stochastic variational optimization. arXiv preprint arXiv:1809.04855 (2018)
2. Carlini, N., Wagner, D.: Towards evaluating the robustness of neural networks. In: SP, pp. 39–57. IEEE (2017)
3. Croce, F., Hein, M.: Sparse and imperceivable adversarial attacks. In: ICCV, pp. 4723–4731. IEEE (2019)
4. Croce, F., Hein, M.: Mind the box: l_1-APGD for sparse adversarial attacks on image classifiers. In: ICML, pp. 2201–2211. PMLR (2021)
5. Deng, J., Dong, W., Socher, R., Li, L., Li, K., Fei-Fei, L.: ImageNet: a large-scale hierarchical image database. In: CVPR, pp. 248–255. IEEE (2009)
6. Dong, X., et al.: GreedyFool: distortion-aware sparse adversarial attack. In: NeurIPS, pp. 11226–11236 (2020)
7. Fan, Y., et al.: Sparse adversarial attack via perturbation factorization. In: Vedaldi, A., Bischof, H., Brox, T., Frahm, J.-M. (eds.) ECCV 2020. LNCS, vol. 12367, pp. 35–50. Springer, Cham (2020). https://doi.org/10.1007/978-3-030-58542-6_3
8. Goodfellow, I.J., Shlens, J., Szegedy, C.: Explaining and harnessing adversarial examples. In: ICLR (2015)
9. Gordo, A., Almazán, J., Revaud, J., Larlus, D.: Deep image retrieval: learning global representations for image search. In: Leibe, B., Matas, J., Sebe, N., Welling, M. (eds.) ECCV 2016. LNCS, vol. 9910, pp. 241–257. Springer, Cham (2016). https://doi.org/10.1007/978-3-319-46466-4_15
10. He, K., Zhang, X., Ren, S., Sun, J.: Deep residual learning for image recognition. In: CVPR, pp. 770–778. IEEE (2016)
11. Krizhevsky, A., Hinton, G., et al.: Learning multiple layers of features from tiny images. In: Handbook of Systemic Autoimmune Diseases (2009)
12. Kurakin, A., Goodfellow, I.J., Bengio, S.: Adversarial machine learning at scale. In: ICLR (2017)
13. Lin, M., Chen, Q., Yan, S.: Network in network. In: ICLR (2014)
14. Madry, A., Makelov, A., Schmidt, L., Tsipras, D., Vladu, A.: Towards deep learning models resistant to adversarial attacks. In: ICLR (2018)

15. Modas, A., Moosavi-Dezfooli, S.M., Frossard, P.: SparseFool: a few pixels make a big difference. In: CVPR, pp. 9087–9096. IEEE (2019)
16. Papernot, N., McDaniel, P., Jha, S., Fredrikson, M., Celik, Z.B., Swami, A.: The limitations of deep learning in adversarial settings. In: EuroSP, pp. 372–387 (2016)
17. Simonyan, K., Zisserman, A.: Very deep convolutional networks for large-scale image recognition. In: ICLR (2015)
18. Su, J., Vargas, D.V., Sakurai, K.: One pixel attack for fooling deep neural networks. TEVC (2019)
19. Szegedy, C., et al.: Intriguing properties of neural networks. In: ICLR (2014)
20. Yang, H., Ciftci, U.A., Yin, L.: Facial expression recognition by de-expression residue learning. In: CVPR, pp. 2168–2177. IEEE (2018)
21. Yin, M., Zhou, M.: ARM: augment-reinforce-merge gradient for stochastic binary networks. In: ICLR (2019)
22. Zhao, P., Liu, S., Wang, Y., Lin, X.: An ADMM-based universal framework for adversarial attacks on deep neural networks. In: ACM MM. ACM (2018)
23. Zhu, M., Chen, T., Wang, Z.: Sparse and imperceptible adversarial attack via a homotopy algorithm. In: ICML, pp. 12868–12877. PMLR (2021)
24. Zoph, B., Vasudevan, V., Shlens, J., Le, Q.V.: Learning transferable architectures for scalable image recognition. In: CVPR, pp. 8697–8710. IEEE (2018)

LEMT: A Label Enhanced Multi-task Learning Framework for Malevolent Dialogue Response Detection

Kaiyue Wang, Fan Yang, Yucheng Yao, and Xiabing Zhou[✉]

School of Computer Science and Technology, Soochow University, Suzhou, China
{kywang2022,fyangoct,ycyao}@stu.suda.edu.cn, zhouxiabing@suda.edu.cn

Abstract. Malevolent Dialogue Response Detection has gained much attention from the NLP community recently. Existing methods have difficulties in effectively utilizing the conversational context and the malevolent information. In this work, we propose a novel framework, the **L**abel **E**nhanced **M**ulti-**t**ask learning (LEMT), which incorporates a structured representation of malevolence description information and exploits malevolence shift detection as an auxiliary task. Specifically, we introduce a hierarchical structure encoder based on prior probability knowledge to capture the semantic information of different malevolent types and integrate it with utterance information. In addition, the malevolence shift detection is modeled to improve the ability of the model to distinguish between different malevolent information. Experimental results show that our LEMT outperforms state-of-the-art methods and verifies the effectiveness of the modules.

Keywords: Malevolent Dialogue Response Detection · Dialogue Safety · Multi-Task Learning · Graph Convolutional Network

1 Introduction

Current dialogue systems can easily produce toxic, biased, or offensive content [7,12], which has raised concerns about the security of these systems and become a new research hot [10,14,16,18]. In this work, we focus on the *Malevolent dialogue response detection and classification* (MDRDC) task, which aims to identify malevolent utterances in multi-turn dialogues [15,16]. Unlike most traditional tasks that concentrate on the presence of malevolent speech or mainly detect instances related to personality abuse, the MDRDC task considers more comprehensive malevolence-related aspects. These include arrogance, disgust, jealousy, and blame, etc., as shown in Fig. 1.

Previous works [5,13] apply CNNs and LSTMs for aggressive and offensive detection. Recently, pre-trained language models such as BERT and RoBERTa have been employed in ad hominem attacks, malevolence, and safety concerns detection [9,10,16]. However, these methods have limitations in exploiting label information and struggle to extract deep and robust semantic representations, resulting in poor performance.

© The Author(s), under exclusive license to Springer Nature Singapore Pte Ltd. 2024
D.-N. Yang et al. (Eds.): PAKDD 2024, LNAI 14645, pp. 130–142, 2024.
https://doi.org/10.1007/978-981-97-2242-6_11

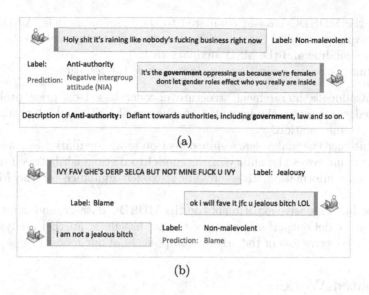

Fig. 1. Examples of the MDRDC. The prediction results are generated by BERT [4].

For example, as demonstrated in Fig. 1(a), the corresponding gold label to the second utterance is *Anti-authority*. However, the BERT model misclassifies the utterance as *Negative intergroup attitude (NIA)* due to its failure to fully understand the meaning of malevolent type. When the description of the malevolence label is fused, it becomes evident that the semantic representation of the utterance containing the term "government" aligns more closely with the label *Anti-authority*. Therefore, fusing such detailed description information associated with these labels is crucial for the MDRDC task. Additionally, as illustrated in Fig. 1(b), since previous work [16] concatenate dialogue history and current response as the input of the classification model, which may cause the model to fail to distinguish the target utterance to be classified, and malevolent information in the history may interfere with the judgment of current utterance.

To tackle the above issues, we propose a novel Label Enhanced Multi-task Learning Framework (LEMT) for Malevolent Dialogue Response Detection, which fully incorporates the descriptions of different malevolence labels to help better understand dialogue content and detects malevolence shift in dialogue to eliminate interference from irrelevant historical discourse. Specifically, we first adopt a structure encoder based on Graph Neural Networks to model the hierarchical dependencies between different malevolence labels, which combines with prior hierarchical probability knowledge. Then, the hierarchy-aware malevolence label representations and the dialogue response are fused to obtain the label-aware utterance representation for MDRDC. Thirdly, we construct an auxiliary Malevolence Shift Detection (MSD) task to enhance the ability to capture and distinguish malevolence semantics between the target dialogue response and the dialogue context. Our proposed method obtains significant improvements at all levels on the MDRDC dataset compared to the baseline models. Additionally, ablation experiments and case studies prove the effectiveness of the hierarchy-aware label module and the MSD auxiliary task.

To summarize, our main contributions include the following:

- We introduce a hierarchical structure encoder based on prior probability knowledge to capture the semantic information of labels and integrate it with utterance information.
- We construct the malevolence shift detection as an auxiliary task, which significantly improves the ability of the model to distinguish between different malevolent information, thus providing explicit guidance for the MDRDC task.
- We conduct extensive experiments on the MDRDC dataset, and demonstrate that our model outperforms state-of-the-art baselines. Further analyses validate the effectiveness of the critical components of our model.

2   Related Work

Detecting malevolent dialogue response is a challenging natural language understanding (NLU) task. The early rule-based work [8] distinguished malevolent online content by keywords. Additionally, traditional machine learning methods with feature engineering are widely used in the field of malevolent dialogue response detection, such as Logistic Regression [11] and SVM with effective n-gram features [1]. Recently, with the development of deep learning and large pre-training language models, many advanced models have been proposed and achieved outstanding performance in malevolent dialogue response detection. CNNs and LSTMs are applied to aggressiveness detection [5] and offensiveness detection [13]. Additionally, Sheng et al. [9], Zhang et al. [9], and Sun et al. [10] leveraged pre-trained models such as BERT and RoBERTa for ad hominem, malevolence, and safety-related tasks, respectively. However, the effectiveness of these models is limited, as malevolence detection demands a more deep understanding of dialogue semantics.

3   Method

3.1   Problem Definition

Given a dialogue response u_n and its previous dialogue context $\{u_1, u_2, \cdots, u_{n-1}\}$, the purpose of the MDRDC task is not only to identify whether the dialogue response u_n is malevolent but also to identify the malevolence label $y_n \in \mathcal{Y}$, where $\mathcal{Y}$ represents the set of possible malevolence labels, such as obscenity, jealousy, etc. Following prior work [16], the former goal is formulated as a binary classification task over the first-level categories, and the latter goal is formulated as multi-classification tasks over the second-level and third-level categories.

3.2 Overall Framework

The overall architecture of our LEMT is shown in Fig. 2. The shared Bert-based utterance encoder provides utterance features for both MSD and MDRDC. For the main task, various malevolent interpretations and the associated information between malevolence labels are integrated into the comprehension of dialogue content. This integration enhances the accurate understanding of implicit or explicit semantic information related to malevolence detection. For the auxiliary task, we employ a malevolence shift classifier to capture discourse correlations between dialogues. This approach enables a more in-depth exploration of semantic information, ultimately enhancing the optimization of the utterance content encoder. The details of each part are as follows.

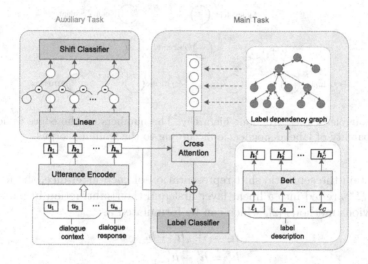

Fig. 2. The overall framework of our model.

3.3 Utterance Encoder

As shown in Fig. 2, we adopt a shared utterance encoder to extract utterance representations. Specifically, the utterance u_i is tokenized into a token sequence with two special tokens [CLS] and [SEP] added as the first and the last token. Subsequently, all utterances in dialogue context and dialogue response are concatenated together and fed into a pre-trained model (i.e., BERT [4]). The final hidden state corresponding to [CLS] token is regarded as the feature representation $h_i \in \mathbb{R}^{d_u}$ of u_i. Here, d_u represents the dimension of the feature representation. The feature representations for all utterances are represented as $H^u \in \mathbb{R}^{n \times d_u}$.

3.4 Malevolence Shift Detection

Malevolence shift detection is designed to detect whether each utterance in a conversation has a different malevolence label than its previous utterance. There-fore, the auxiliary task needs to assign a shift label $\tilde{y}$ to each utterance, where $\tilde{y} \in \{\texttt{shift}, \texttt{unshift}, \texttt{start}\}$. The malevolence shift label of the current utterance is $\texttt{unshift}$ if the malevolence label of the current utterance is consistent with that of its previous utterance, otherwise, it is $\texttt{shift}$. In particular, the malevolence shift label for the first utterance in a conversation is $\texttt{start}$.

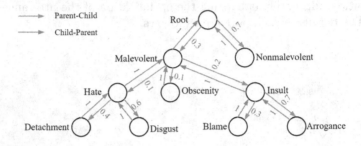

Fig. 3. Example of the taxonomic hierarchy. The numbers on the edges indicate the prior probability of label dependencies according to the training corpus.

To obtain the semantic shift representation of each utterance, the utterance features $\boldsymbol{H}^u$ are fed into a linear layer first and then subtraction in pairs in line with previous methods [2], which can be formulated as:

$$h'_i = \boldsymbol{W}_h \boldsymbol{h}_i + \boldsymbol{b}_h, \qquad (1)$$
$$h^s_i = h'_i - h'_{i-1}, \qquad (2)$$

where $\boldsymbol{W}_h \in \mathbb{R}^{d_s \times d_u}$, $\boldsymbol{b}_h \in \mathbb{R}^{d_s}$ are learnable parameters, $\boldsymbol{h}_i \in \mathbb{R}^{d_u}$ denote the ith representation in $\boldsymbol{H}^u$ and h'_{i-1} is randomly initialized as $h'_{start} \in \mathbb{R}^{d_s}$ when $i = 0$. The final representation is classified via a fully connected softmax layer:

$$\boldsymbol{p}_i = \text{softmax}(\boldsymbol{W}_s \boldsymbol{h}^s_i + \boldsymbol{b}_s), \qquad (3)$$

where $\boldsymbol{W}_s \in \mathbb{R}^{n_s \times d_s}$, $\boldsymbol{b}_s \in \mathbb{R}^{n_s}$ are learnable parameters, and n_s denotes the number of malevolence shift labels.

3.5 Hierarchy-Aware Label Encoder

Given a set of label description information $\{\ell_1, \ell_2, \cdots, \ell_C\}$, where C denotes the total number of labels at all hierarchies. We employ BERT and a linear layer to obtain label representation $\boldsymbol{H}^\ell \in \mathbb{R}^{C \times d_l}$ as follows:

$$h^i_{[\texttt{CLS}]} = \text{BERT}_{[\texttt{CLS}]}(\{[\texttt{CLS}], x_1, x_2, \ldots, x_{n_i}\}), \qquad (4)$$
$$h^\ell_i = \boldsymbol{W}_\ell h^i_{[\texttt{CLS}]} + \boldsymbol{b}_\ell, \text{ for } i = 1, 2, \ldots, C, \qquad (5)$$

where $h_i^\ell \in \mathbb{R}^{d_\ell}$, d_ℓ is the dimension of the label representation, $\{x_1, x_2, \ldots, x_{n_i}\}$ represents the token sequence of the ith label description information ℓ_C, n_i denotes the length of the token sequence, BERT$_{[CLS]}$ is the output of BERT corresponding to [CLS], W_ℓ and b_ℓ are learnable parameters.

Following the previous work [17], we further introduce a hierarchy-GCN module to encode fine-grained label hierarchy information. As illustrated in Fig. 3, the hierarchical label structure can be regarded as a directed graph $G = (V, \overrightarrow{E}, \overleftarrow{E})$, where $V = \{v_1, v_2, \ldots, v_C\}$ indicates the set of hierarchy structure nodes and v_i can be represented by h_i^ℓ, $\overrightarrow{E} = \{(v_i, v_j) \mid v_i \in V, j \in child(i)\}$ are built from the top-down hierarchy paths representing the prior statistical probability from parent nodes to children nodes, $\overleftarrow{E} = \{(v_i, v_j) \mid v_i \in V, j \in parent(i)\}$ are built from the bottom-up hierarchy paths representing the connection relationship from children nodes to parent nodes. Update the label representation of node v_i based its associated neighbourhood $N_i = \{i, child(i), parent(i)\}$ as:

$$a_{i,j} = \begin{cases} 1, & \text{if } i = j \text{ or } j \in child(i) \\ \frac{p_i}{p_j}, & \text{if } j \in parent(i) \end{cases}, \tag{6}$$

$$u_{i,j} = a_{i,j} h_j^\ell + b_\ell^i, \tag{7}$$

$$g_{i,j} = \sigma(W_g^{r_{j,i}} h_i^\ell + b_g^i), \tag{8}$$

$$h_i^v = \text{ReLU}(\sum_{j \in N_i} g_{i,j} \odot u_{i,j}), \tag{9}$$

where p_i denotes the prior probability that the ith label appears in the training corpus. $r_{j,i}$ is the relation type of the edge from label node v_j to v_i, including *Parent-Child*, *Child-Parent* and *Self-Loop*. $W_g^{r_{j,i}} \in \mathbb{R}^{d_\ell}$, $b_\ell \in \mathbb{R}^{C \times d_\ell}$, and $b_g \in \mathbb{R}^C$ are learnable parameters.

To sum up, the semantic information of the label description information is explicitly combined with the prior hierarchical probability through the graph convolutional network to obtain the final label representation $\{h_i^v \in \mathbb{R}^{d_\ell}\}_{i=1}^C$.

3.6 Malevolence Detection in Dialogues

We obtain label representation $\{h_j^v \mid j \in \text{level}_k\}$ of the corresponding level, where k denotes the level of classification and level_k denotes the set of node indices of the level k. Then we use cross-attention to fuse utterance and label information as:

$$\mu_{nj} = \frac{e^{(W_Q h_n)(W_K h_j^v)}}{\sum_{j \in \text{level}_k} e^{(W_Q h_n)(W_K h_j^v)}}, \tag{10}$$

$$h_{attn} = \sum_{j \in \text{level}_k} \mu_{nj} W_V h_j^v, \tag{11}$$

where W_Q, W_K, W_V are trainable parameters.

We concatenate the original utterance representation h_n and the utterance-wise label representation h_{attn} together for the final classification:

$$H = [h_n; h_{attn}],   \quad (12)$$
$$p = \text{softmax}(W_{MD}H + b_{MD}),   \quad (13)$$

where W_{MD} and b_{MD} are trainable parameters.

3.7   Multi-task Learning

Since malevolent dialogue response detection suffers from the problem of data imbalance, we use focal loss to assign lower weights to well-classified samples, the loss function is as follows:

$$\mathcal{L}^{MD} = -\sum_{i=1}^{M}\sum_{j=1}^{C}(\alpha_j \cdot y_{ij} \cdot (1 - p_{ij})^{\gamma} \cdot \log(p_{ij}))   \quad (14)$$

where M is the number of dialogue responses, γ is a non-negative tunable focusing parameter to differentiate between easy and difficult samples, $\alpha_j \in [0, 1]$ is a weighting factor, p_{ij} represents the probability of sample i belonging to class j, y_{ij} indicates whether sample i belongs to class j (0 or 1).

For the auxiliary task, we use cross-entropy loss to supervise the training process of the MSD task, the loss function is as follows:

$$\mathcal{L}^{SD} = -\sum_{i=1}^{N}\sum_{j=1}^{T_i} [\mathbf{y}_{i,j}^{SD}] \log(\mathbf{p}_{i,j}^{SD}),   \quad (15)$$

where N is the number of the dialogues, T_i is the number of utterances in dialogue i, $\mathbf{p}_{i,j}^{SD}$ is the probability distribution of shift labels for utterance j of dialogue i, $[\mathbf{y}_{i,j}^{SD}]$ represents the one-hot vector of the ground truth shift label.

The overall loss function for our proposed method is calculated as the total sum of losses from the auxiliary task and the main task:

$$\mathcal{L}_{total} = \mathcal{L}^{MD} + \beta\mathcal{L}^{SD},   \quad (16)$$

where β is the weight for loss of auxiliary malevolence shift detection.

4   Experiments

4.1   Datasets and Evaluation Metrics

To verify the effectiveness of our model, we conduct experiments on MDRDC [16], which is a benchmark dataset for the task. There are 6000 dialogues containing 10299 malevolent utterances and 21081 non-malevolent utterances, respectively. For the malevolent utterances, it devises the second and the third levels of malevolent categories, which include 10 categories for the second level based on negative psychological behavior, unethical issues, and negative emotion. The third level label is subdivided under the second level label, forming 17 different malevolent types. Each label contains a description of the relevant definition. Following prior work [16], we create training, validation and test splits with a ratio of 7: 1: 2. We use precision, recall, and F1 as the evaluation metrics, and report the macro scores as since the data is imbalanced in terms of labels.

4.2 Compared Baselines

We compare the performance of our network with the following baselines:

- **BERT** [4], a pre-trained model that can be fine-tuned to solve the MDRDC task.
- **HiAGM-LA** [17], an end-to-end hierarchical text classification model with a multi-label attention module, which learns label-aware embeddings through the structure encoder and conducts inductive fusion of label features.
- **DialogGCN-BERT** [3], a graph-based network to leverage inter and intra-speaker dependency relationship. We use BERT as the context encoder.
- **CoG-BART** [6], a BART-based emotion recognition model in conversation, which conducts contrastive learning and response generation as the auxiliary tasks to enhance the dialogue comprehension ability of BART.

4.3 Implementation Details

We adopt the uncased base version of BERT[1] as our utterance encoder and label encoder. The hidden dimension of the context feature representation, semantic shift representation, and label representation are set to 768, 300, and 300, respectively. We set the number of the hierarchy-GCN layer as 2. During training, we use AdamW to optimize the model parameters. The dropout rate of the shift classifier, the label classifier, and the hierarchy-GCN layer are set to 0.4, 0.4, and 0.05, respectively. We set $\beta = 0.3$ for the multi-task learning. To avoid overfitting, we conduct validation every 100 steps, and the patience for early stopping is set to 13. All models were trained on NVIDIA A100 GPU, and the reported results are based on the average performance of 5 random runs on the test set.

4.4 Main Results

The overall performance of all the compared baselines and our proposed model on the MDRDC dataset is reported in Table 1. Our model consistently obtains the best macro-F1 score over others on three levels. Compared to HiAGM-LA, the performance of LEMT is significantly improved. Since HiAGM-LA may introduce noises from the randomly initialized label embedding, while we obtain the label embedding by utilizing the label description information that is partly derived from the dataset and partly constructed by ourselves. For the Dialog-GCN model, which mainly relies on the dependency of the speakers, lacks the

---

[1] https://huggingface.co/bert-base-uncased.

capture of malevolent information expression. Compared with the modeling of the speakers, the capture of malevolent information in the dialogue context is more important. It is worth noting that LEMT obtains 2.26%, 2.98%, and 1.38% absolute improvements of the F1 scores on all levels compared to CoG-BART, respectively. This is probably because the contrastive learning module employed in CoG-BART can not gain a clear understanding of different malevolent labels, and compared with the response generation as an auxiliary task, exploring malevolence shift explicitly considers malevolent information in the context.

4.5   Ablation Study

To further investigate the role of the modules added to our model, we conduct ablation studies. The results are reported in Table 2. Obviously, removing any of the two modules makes the overall performance worse. We can also find that the MSD auxiliary task has a greater impact on fine-grained levels, and there is little difference in the results at the second and third levels. However, due to the relatively simple MDRDC at the second level, the improvement at the second level is more obvious. Specifically, the macro f1 at the second and the third level is increased by 3.05% and 2.44% after adding the shift module, respectively. These observations indicate that the MSD auxiliary task effectively captures and distinguishes malevolence semantics in the dialogue context and the target dialogue response, thus significantly improving the performance of the main task. Meanwhile, removing the label module degrades the macro f1 on the first level by 0.74 points, which demonstrates the effectiveness of the label module.

Table 1. Main results on the MDRDC dataset. The previous best results are underlined, and results better than state-of-the-art are in **bold**.

model	level = 3			level = 2			level = 1		
	p	r	f1	p	r	f1	p	r	f1
BERT [4]	53.44	55.65	53.63	59.32	60.15	58.87	80.69	83.07	81.69
HiAGM-LA [17]	43.60	44.53	43.29	50.74	48.65	49.08	77.00	77.42	77.18
DialogGCN-BERT [3]	55.71	54.03	53.80	60.84	59.84	59.43	80.97	83.03	81.89
CoG-BART [6]	56.56	52.66	53.85	61.66	56.97	58.92	81.11	81.31	81.18
LEMT	56.23	**57.46**	**56.11**	**61.75**	**63.17**	**61.90**	**81.71**	**83.61**	**82.56**

Table 2. Experimental results of ablation study, where w/o shift denotes remove the MSD module and w/o label denotes remove the label module.

model	level = 3			level = 2			level = 1		
	p	r	f1	p	r	f1	p	r	f1
ours	56.23	**57.46**	**56.11**	**61.75**	**63.17**	**61.90**	**81.71**	**83.61**	**82.56**
w/o shift	52.76	56.48	53.67	58.91	60.62	58.84	81.20	82.99	81.99
w/o label	**56.57**	57.14	56.07	61.27	62.50	61.46	80.95	82.91	81.82

4.6 Analysis of Malevolence Shift Detection

As shown from Table 3, the performance of the MSD task in the shift category at level 2 and level 3 is higher than that at level 1. This is because the label space of the first level is relatively smaller, which makes it more possible to misclassify *shift* as *unshift* due to the incorrect interpretation of malevolence labels in utterances. In contrast, the label spaces at the second and third levels are relatively broader. Even if a malevolence label is misinterpreted as another incorrect one, it does not result in the misclassification of *shift* as *unshift*.

4.7 Case Study

Table 4 shows a few cases predicted by our LEMT model and the BERT model. In the first example, LEMT predicts the correct label "Phobia" for the dialogue response, while BERT mistakenly predicts it as "Non-malevolent". It can be found that the label of the previous utterance is "Phobia", and the shift type predicted by the MSD auxiliary task is "unshift". This observation highlights the valuable contribution of MSD to LEMT, enhancing its ability to accurately predict dialogue response labels by considering the malevolent type shift between a dialogue response and its preceding utterances. Correspondingly, the second example can also prove the effectiveness of the MSD auxiliary task and the label module. At the same time, as shown in Fig. 4, the dialogue response and corresponding label "Phobia" have the highest attention score (0.201), suggesting that incorporating label description information can assist the malevolent prediction well.

Table 3. Performance of MSD under different label levels

label	level = 3			level = 2			level = 1		
	p	r	f1	p	r	f1	p	r	f1
start	99.96	99.99	99.98	99.99	100.00	100.00	99.99	100.00	100.00
shift	68.94	73.07	70.89	65.69	76.53	70.70	60.28	64.50	62.31
unshift	87.38	84.91	86.11	88.70	82.16	85.30	87.22	85.07	86.13
Macro	85.43	85.99	85.66	84.80	86.23	85.33	82.50	83.19	82.81

Table 4. Case studies of our LEMT model compared with the BERT model.

Dialogue History	Dialogue Response	BERT	LEMT	Golden Label	Predicted Shift Type
young men aged 18 35 are now invading europe. . millions of them! they are not refugees, they are invaders! Refugee crisis	do you get the point of pumped up zombie movies of hollywood now? zombies are the migrants and they won't stop	Non-malevolent	Phobia	Phobia	unshift
Romo is legit the worst quarterback in the league; neither interception was his fault?	this guy has 9 fantasy points and that is his fault	Arrogance	Blame	Blame	shift

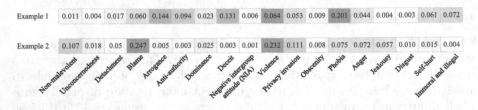

Fig. 4. Attention score visualization for examples in Table 4.

4.8   Analysis of LLMs

With the advent of large language models (LLMs), there is a great potential for their employment on the MDRDC task. We evaluate the most mainstream LLMs (i.e., ChatGPT[2]) in zero-shot mode with appropriate prompts. Unexpectedly, Table 5 shows the results obtained from ChatGPT are not as good as expected. The performance of ChatGPT is lower than our model on all levels. Although it has 73% of macro f1 on the first level, it has poor results on the second level and the third level. We believe the following reasons might lead to the result: 1) The powerful text comprehension capabilities enable LLMs a jack of all trades. However, when dealing with complex linguistic phenomena, LLMs struggle to understand nuances in text, such as the semantic difference between arrogance, deceit, and dominance. 2) In such a zero-shot learning setting, ChatGPT lacks prior knowledge of the feature distribution within the target space of the MDRDC task. Sufficient training data and fine-tuning are necessary for achieving state-of-art performance. In general, though ChatGPT exhibits impressive text generation and conversation ability, it still can not serve as a well-behaved malevolent dialogue response detector.

Table 5. Experimental results of 100 stratified random samples on ChatGPT and our LEMT model.

model	level = 3			level = 2			level = 1		
	p	r	f1	p	r	f1	p	r	f1
ChatGPT	42.18	37.04	33.22	58.89	42.42	42.95	80.35	75.00	73.85
LEMT	72.37	59.26	60.53	69.70	62.63	63.27	86.78	85.00	84.82

5   Conclusion

In this paper, we propose a novel Label Enhanced Multi-task Learning Framework (LEMT) for Malevolent Dialogue Response Detection, which makes full use of label description information in combination with hierarchical prior probability and exploits malevolenc shift detection as an auxiliary task. Experimental

---

[2] https://chat.openai.com/.

results show that our LEMT outperforms previous state-of-the-art baselines, and further analysis validates the effectiveness of critical modules in LEMT. In the future, we will explore the introduction of external knowledge to enhance the detection performance of implicit malevolent dialogue responses.

Acknowledgement. The work is supported by National Nature Science Foundation of China (No. 62176174).

References

1. Davidson, T., Warmsley, D., Macy, M., Weber, I.: Automated hate speech detection and the problem of offensive language. In: Proceedings of the International AAAI Conference on Web and Social Media, vol. 11, pp. 512–515 (2017)
2. Gao, Q., et al.: Emotion recognition in conversations with emotion shift detection based on multi-task learning. Knowl.-Based Syst. **248**, 108861 (2022)
3. Ghosal, D., Majumder, N., Poria, S., Chhaya, N., Gelbukh, A.: DialogueGCN: a graph convolutional neural network for emotion recognition in conversation. In: Proceedings of the 2019 Conference on Empirical Methods in Natural Language Processing and the 9th International Joint Conference on Natural Language Processing (EMNLP-IJCNLP), pp. 154–164 (2019)
4. Kenton, J.D.M.W.C., Toutanova, L.K.: BERT: pre-training of deep bidirectional transformers for language understanding **1**, 2 (2019)
5. Kumar, R., Ojha, A.K., Malmasi, S., Zampieri, M.: Benchmarking aggression identification in social media. In: Proceedings of the First Workshop on Trolling, Aggression and Cyberbullying (TRAC-2018), pp. 1–11 (2018)
6. Li, S., Yan, H., Qiu, X.: Contrast and generation make BART a good dialogue emotion recognizer. In: Proceedings of the AAAI Conference on Artificial Intelligence, vol. 36, pp. 11002–11010 (2022)
7. Perez, E., et al.: Red teaming language models with language models. In: Proceedings of the 2022 Conference on Empirical Methods in Natural Language Processing, pp. 3419–3448 (2022)
8. Roussinov, D., Robles-Flores, J.A.: Applying question answering technology to locating malevolent online content. Decis. Support Syst. **43**(4), 1404–1418 (2007)
9. Sheng, E., Chang, K.W., Natarajan, P., Peng, N.: "Nice try, kiddo": investigating ad Hominems in dialogue responses. In: Proceedings of the 2021 Conference of the North American Chapter of the Association for Computational Linguistics: Human Language Technologies, pp. 750–767 (2021)
10. Sun, H., et al.: On the safety of conversational models: taxonomy, dataset, and benchmark. In: Findings of the Association for Computational Linguistics: ACL 2022, pp. 3906–3923 (2022)
11. Waseem, Z., Hovy, D.: Hateful symbols or hateful people? Predictive features for hate speech detection on twitter. In: Proceedings of the NAACL Student Research Workshop, pp. 88–93 (2016)
12. Wolf, M.J., Miller, K., Grodzinsky, F.S.: Why we should have seen that coming: comments on microsoft's tay "experiment," and wider implications. ACM SIGCAS Comput. Soc. **47**(3), 54–64 (2017)
13. Zampieri, M., Malmasi, S., Nakov, P., Rosenthal, S., Farra, N., Kumar, R.: Predicting the type and target of offensive posts in social media. In: Proceedings of

the 2019 Conference of the North American Chapter of the Association for Computational Linguistics: Human Language Technologies, Volume 1 (Long and Short Papers), pp. 1415–1420 (2019)

14. Zhang, M., Jin, L., Song, L., Mi, H., Chen, W., Yu, D.: SafeConv: explaining and correcting conversational unsafe behavior. In: Proceedings of the 61st Annual Meeting of the Association for Computational Linguistics (Volume 1: Long Papers), pp. 22–35 (2023)

15. Zhang, Y., Ren, P., Deng, W., Chen, Z., Rijke, M.: Improving multi-label malevolence detection in dialogues through multi-faceted label correlation enhancement. In: Proceedings of the 60th Annual Meeting of the Association for Computational Linguistics (Volume 1: Long Papers), pp. 3543–3555 (2022)

16. Zhang, Y., Ren, P., de Rijke, M.: A taxonomy, data set, and benchmark for detecting and classifying malevolent dialogue responses. J. Am. Soc. Inf. Sci. **72**(12), 1477–1497 (2021)

17. Zhou, J., et al: Hierarchy-aware global model for hierarchical text classification. In: Proceedings of the 58th Annual Meeting of the Association for Computational Linguistics, pp. 1106–1117 (2020)

18. Zhou, J., et al.: Towards identifying social bias in dialog systems: framework, dataset, and benchmark. In: Findings of the Association for Computational Linguistics: EMNLP 2022, pp. 3576–3591 (2022)

Two-Stage Knowledge Graph Completion Based on Semantic Features and High-Order Structural Features

Xiang Ying[1,2,3], Shimei Luo[1], Mei Yu[1,2,3], Mankun Zhao[1,2,3], Jian Yu[1,2,3], Jiujiang Guo[1], and Xuewei Li[1,2,3(✉)]

[1] College of Intelligence and Computing, Tianjin University, Tianjin, China
[2] Tianjin Key Laboratory of Cognitive Computing and Application, Tianjin, China
[3] Tianjin Key Laboratory of Advanced Networking, Tianjin, China
lixuewei@tju.edu.cn

Abstract. Recently, multi-head Graph Attention Networks (GATs) have incorporated attention mechanisms to generate more enriched feature embeddings, demonstrating significant potential in Knowledge Graph Completion (KGC) tasks. However, existing GATs based KGC approaches struggle to update entities with few neighbors, making it challenging to obtain structured semantic information and overlooking complex and implicit information in distant triples. To this effect, we propose a novel model named the Two-Stage KGC model with integrated High-Order Structural Features (HOSAT), designed to enhance the learning process of GATs. Initially, we leverage the conventional GATs module to acquire embeddings encapsulating local semantic intricacies. Subsequently, we introduce a global biased random walk algorithm, strategically amalgamating graph topology, entity attributes, and relationship attributes. This algorithm aims to extract high-order structured semantic neighbor sequences from multiple perspectives and construct nuanced reasoning paths. By propagating the embedding along this path, it is ensured that with an increasing number of iterations, the aggregated information of each node becomes an almost perfect combination of local and global features. Evaluation on two public benchmark datasets using entity prediction methods demonstrates that HOSAT achieves substantial performance improvements over state-of-the-art methods.

1 Introduction

As a type of semantic knowledge base, Knowledge Graphs (KGs) aim to store various entities, semantic types, concepts and relations between entities in the real-world. Usually, KGs comprise collections of triples (e_i, r_k, e_j), where e_i and e_j represent the head entity and the tail entity and r_k the corresponding relation between them. Over the past decade, existing KGs (e.g., YAGO [17], DBpedia [1], FreeBase [3], NELL [5], etc.) have provided powerful support for artificial intelligence applications and have been well used in the field of question answering [6,10], recommendation systems [9,24], sentiment analysis [12], and cultural protection [8]. However, the practical application of existing KG is hindered by

© The Author(s), under exclusive license to Springer Nature Singapore Pte Ltd. 2024
D.-N. Yang et al. (Eds.): PAKDD 2024, LNAI 14645, pp. 143–155, 2024.
https://doi.org/10.1007/978-981-97-2242-6_12

their incompleteness and noise, caused by the constantly emerging new knowledge. Therefore, Knowledge Graph Completion (KGC), which attempts to predict missing facts based on existing triples.

Most KGC models are based on KG embeddings. The classical KGC models (e.g., TransE [4], ComplEx [20], ConvE [7], etc.) use a low-dimensional embedding space to map entities and relations and evaluate the plausibility of triples based on scores computed from the embeddings. These models have strong generalization capabilities but ignore meaningful graph contexts and are limited in updating the embeddings of individual triplets. In recent years, Graph Convolutional Networks (GCNs) (e.g., R-GCN [16], CompGCN [22], and PathCon [23], etc.) update the embedding of the target node by recursively aggregating the potentially relevant information of the nodes. GCNs treat neighboring nodes equally and do not fully utilize the information contained in the network structure. In contrast, GATs assign different weights to different neighbors of entities in KGs, which can better capture neighborhood features. As a result, some works have introduced GATs into their embeddings to further enhance their effectiveness.

However, we consider that these GAT-based models cannot well solve the following problems: (1) **Deeper high-order structural features were not successfully acquired.** GAT-based models mainly capture local features up to three-hop contexts around an entity. Yet, their potential to graph long-distance contexts beyond three hops remains largely untapped. For example, the path in Fig. 1: $Plato \xrightarrow{student} Aristotle \xrightarrow{advocate} Realism \xrightarrow{origins\_in} 1850s$, which has more complex and richer semantics. (2) **The method of global neighborhood aggregation exhibits inadequacies.** The global information aggregation module of the model [14] integrates all multi-hop neighbor information into the learning process of GAT. For instance, all three-hop neighbors in Fig. 1 participate in the aggregation. However, it is evident that this approach is flawed, as some neighbors introduce noise and result in an adverse effect. Moreover, one model [27] tries to embed the global importance as the weight update of neighboring nodes, we think this aggregation method is still not enough because it ignores the importance of adjacent nodes and relation embedding itself.

For the Deeper High-Order Structural Features PProblem, in order to better capture long-distance contexts, we enhance the weighting mechanism for a globally biased random walk, considering entity and relation importance, as well as structural similarity. This allows us to reach up to eight-hop connections centered on key entities. We assign weights based on semantic and relational ties, adjusting the walk's direction based on the graph's structure. For highly homogeneous nodes, we prioritize a depth-first search (DFS), whereas for structurally similar nodes, we lean towards a breadth-first search (BFS). **For the global neighborhood aggregation problem**, we selectively choose the most relevant neighbor to the target node from the k-hop neighbors in each layer, and employ a two-stage framework to implicitly integrate local and long-range contexts to better distinguish their importance in the KGs. During the training process, we utilize the entity and relation embeddings generated in the first stage as the input of the second stage, thus ensuring that the aggregated information of each node is almost a perfect combination of local features and global features as the number of iterations increases.

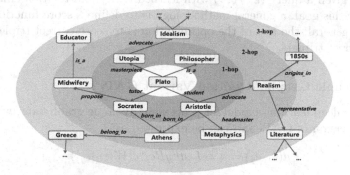

Fig. 1. A subgraph of philosopher *'Plato'*. The concentric circles denotes the ripple sets with different hops.

In summary, our contributions are as follows:

1) We introduce a two-stage graph neighborhood aggregation strategy, enhancing GAT's learning and information aggregation by implicitly integrating local and long-distance contextual information, ensuring superior expressive capabilities.
2) To enhance global information aggregation, we optimized the global biased random walk method. Comprehensive consideration of graph structure, entity attributes, and relationship attributes ensures the selection of k-hop neighbors most relevant to the central entity in each hop.
3) Experimental results on benchmark datasets show that HOSAT significantly outperforms other baseline methods and shows significant improvements in exploiting the structural and semantic features of KGs.

2 Preliminary

2.1 Knowledge Graph

Let $\mathcal{E}$ and $\mathcal{R}$ denote a set of entities and relations, respectively. A KG $\mathcal{G}$ can be considered as a directed graph with a collection of factual triples $(e_i, r_k, e_j) \in \mathcal{E} \times \mathcal{R} \times \mathcal{E}$, where $(e_i, e_j) \in \mathcal{E}$ denotes the head and tail entities respectively and $r_k \in \mathcal{R}$ represents the relation linking from e_i to e_j.

2.2 Knowledge Graph Completion

The aim of KGC is to learn the embeddings and relations of entities with valid triples, and then predict missing head entities e_i given a query $(?, r_k, e_j)$ or tail entities e_j given a query $(e_i, r_k, ?)$ with learned entity and relation embeddings. To achieve this goal, a general methodology is to define a score function for the triplet. In general, the aim of the optimization is to score correct triplets higher than incorrect triplets.

2.3 Dynamic Graph Attention Variant GATv2

GATv2 is a deep learning model for processing dynamic graph data, which is an extended version of GATs and can be expressed as:

$$a(e_i, e_j) = a^T \cdot LeakyReLu(W[e_i||e_j]). \tag{1}$$

They propose a method for self-learning edge weights, Where $a(e_i, e_j)$ is the absolute attention weight between node e_i and e_j, W is a parametric linear transformation matrix that maps input features to a high-dimensional output feature space, and a is any attention function we choose.

3   Methodology

We now present the proposed HOSAT model. Figure 2 shows the workflow of the overall architecture.

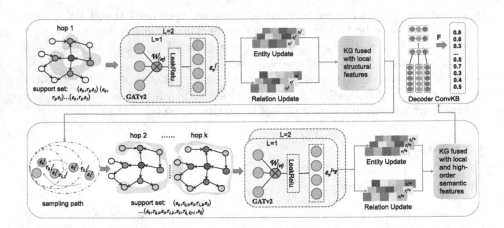

Fig. 2. Overview of the proposed HOSAT, which consists of three parts: (1) Local structural features aggregation; (2) Global high-order structural features aggregation; (3) Decoder.

3.1   Structural Local Contexts Aggregation

For an entity in KGs, it is often linked with many other entities that can enhance its information. However, different entities are of different importance to the prediction task and need to be given different attention. For example, the relationship between nodes "$Plato$" and "$Socrates$" is teacher-student, and the neighbor contexts of $Socrates$ are $(born\_in, Athens)$ and $(propose, Midwifery)$. Neighborhood context $(born\_in, Athens)$ is more important when we need to predict $(Plato, born\_in, ?)$. To obtain more descriptive entity embeddings, GATv2 aggregates the feature information of each neighbor of the central entity by assigning different attention weights to different neighbors.

Here, we denote by $N_{e_i}^l = e_j | (e_i, r_k, e_j) \in \mathcal{G}$ the local neighbors of an entity e_i. To incorporate relations into the attention mechanism, we learn these embeddings by linearly transforming the concatenation of entity and relation feature vectors for a particular triplet $t_{ij}^k = (e_i, r_k, e_j)$ integrating neighborhood entities by $e_j \in N_{e_i}^l$ the following linear transformation Embedding and embedding of correspondences:

$$v_{ijk} = W_1 \cdot [e_i || r_k || e_j], \qquad (2)$$

among them, v_{ijk} is a vector representation of triplet. The vectors e_i, r_k and e_j denote the embeddings of entities and relation, respectively. W_1 is a linear transformation matrix. Then, we utilize self-attention to learn attention values of neighboring nodes, which represent the importance of one node to another. For the triplet t_{ij}^k, the attention weight of the neighborhood node e_j to the node e_i under the relation r_k can be expressed as:

$$b_{ijk} = a^T \cdot LeakyReLU(W_2 \cdot v_{ijk}). \qquad (3)$$

To get the relative attention weight of a single neighbor entity to the central entity, apply softmax normalization to obtain the calculation result:

$$\alpha_{ijk} = softmax(b_{ijk}) = \frac{exp(b_{ijk})}{\sum_{n \in N_{e_i}^l} \sum_{r \in R_{in}} exp(b_{inr})}, \qquad (4)$$

where R_{ij} represents the relation set connecting entities e_i and e_j. To aggregate information from local neighbors $N_{e_i}^l$, the embedding of entity e_i is the sum of attention-weighted sums of each triplet, whose features are updated as:

$$e_i^l = \sigma(\sum_{e_j \in N_{e_i}^l} \alpha_{ijk} \cdot v_{ijk}). \qquad (5)$$

3.2   High-Order Connected Contexts Aggregation

During the local context aggregation stage, we model the local structure and semantic information of KGs. Next, we model the global structural and semantic information of KGs by finding paths with high-order correlations with central

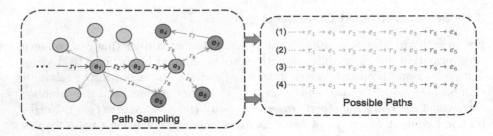

Fig. 3. An example for the path generation by biased random walk.

entities, thereby enhancing the expressiveness and completeness of embedded nodes. Inspired by the node2vec model, we use a more complex global biased random walk to obtain high-order paths that can implicitly represent the global context. Figure 3 depicts the probability of jumping process. The better the random walk, the better the context the algorithm finds, and the more efficient it is. Specifically, assuming that the current sampling path is p_i, we sample the next neighbor (r_j, e_j) with the following probability for path growth:

$$P(r_j, e_j)|p_i) = \frac{\Phi(p_i, r_j, e_j)}{\sum_{(r_k, e_k) \in \mathcal{N}_{e_p}} \Phi(p_i, r_k, e_k)}, \tag{6}$$

where e_p denotes the last entity of path p_i, $\mathcal{N}_{e_p}$ the relation-entity pair neighbor of the target entity e_p. As Fig. 3 shows, the current path has the last entity e_3 and its neighbors include (r_5, e_4), (r_8, e_5), (r_9, e_6) and (r_6, e_7). Unlike node2vec which chooses the next entity in a uniform probability distribution. We use a parameter representing the importance of the entity and a parameter representing the importance of the relationship, which improves the accuracy and efficiency of the random walk. Φ is the function to calculate the unnormalized jump probability of neighbors, which is defined as:

$$\Phi(p_i, r_j, e_j) = \alpha_{bd}(e_{p_{i-1}}, e_j) \times \phi(p_i, e_j) \times \phi(r_{p_i}, r_j), \tag{7}$$

where $\phi(p_i, e_j)$ and $\phi(r_{p_i}, r_j)$ represents the semantic correlation weight and relational correlation weight between the current path and the sample neighborhood, respectively. Formally, $\phi(p_i, e_j) = \cos(p_i, e_j)$ and $\phi(r_{p_i}, r_j) = \cos(r_{p_i}, r_j)$, where both r_{p_i} and r_j represent the embedding of the current path and neighborhood relations, respectively. When new neighbors are added for a path, the path and relation path embeddings are updated to:

$$p_i = \theta_1 \cdot p_{i-1} + (1 - \theta_1) \cdot (e_p \circ r_j), \tag{8}$$

$$r_{p_i} = \theta_2 \cdot r_{p_{i-1}} + (1 - \theta_2) \cdot r_j, \tag{9}$$

among them, $\theta_1, \theta_2 \in [0,1]$ represents the proportion of retained information of the previous sampling path. In addition,

$$\alpha_{bd}(e_{p_{i-1}}, e_j) = \begin{cases} \frac{1}{p} & d_{e_{p_{i-1}}e_j} = 0 \\ 1 & d_{e_{p_{i-1}}e_j} = 1, \\ \frac{1}{q} & d_{e_{p_{i-1}}e_j} \geq 2 \end{cases} \tag{10}$$

where p and q are used to adjust DFS or BFS walking respectively. $d_{e_{p_{i-1}}e_j}$ represents the minimum distance between e_i and e_j. As Fig. 3 shows, the current neighbors e_5 and e_6 are connected to the sampling nodes, which can also be regarded as BFS neighbors. The larger the p value, the more attention will be paid to e_5 and e_6 during the sampling process, indicating that the model usually pays more attention to the local structure of the current node. Instead, the parameter q determines the probability of DFS occurring. Through the above steps, the global biased random walk can correctly sample the path $l = e_0 \xrightarrow{r_{0,1}} e_1 \xrightarrow{r_{1,2}} \ldots e_{n-1} \xrightarrow{r_{n-1,n}} e_n$, which incorporates randomness to ensure that all features of a graph are sampled. To aggregate the distant neighbors, we introduce an auxiliary edge for each distant neighbor using the relation composition. For n-hop neighbors $(e_0, r_{0,1}, e_1) \rightarrow (e_1, r_{1,2}, e_2) \rightarrow \cdots \rightarrow (e_{n-1}, r_{n-1,n}, e_n)$, we can represent them as $\mathcal{G}^g = (e_1, r_1 \oplus r_{1,2} \oplus \cdots \oplus r_{n-1,n}, e_n)$, where $\oplus$ is an addition composition operator. The neighborhood after sampling is denoted by $N_{e_i}^g = \{(e_i, r_k, e_j)\} \in \mathcal{G}^g$. The sampled neighbors are also input into the GATv2 network in the form of triplets, and other global features are further aggregated on the basis of the locally updated embedding:

$$e_u = \sum_{j \in N_{e_i}^g} \alpha_{ijk} v_{ijk}]. \tag{11}$$

With the above equation, we can get the global embedding of entity e_i, which captures the global neighbor information in the background graph. To make the learning process more stable and the captured features more complete, we add the multi-head self-attention mechanism and concatenate the features of different perspectives as follows:

$$e_i^{(l+g)} = \sigma \left(\sum_{m=1}^{\mathcal{M}} \sum_{n \in N_{e_i}^g} \sum_{r \in R_{in}} \alpha_{ijk}^m v_{ijk}^m \right). \tag{12}$$

However, while learning new embeddings, entities lose their initial embedding information. To resolve this issue, we obtain the final entity embedding e^* by combining the transformed initial embeddings e_i and the output entity embedding $e_i^{(l+g)}$, as follows:

$$e_i^* = W^* \cdot e_i + e_i^{(l+g)}, \tag{13}$$

where W^* is a projecting matrix. Furthermore, within every layer of HOSAT, we incorporate a linear transformation to revise the relational embedding, expressed in the form:

$$r_k^* = W^r \cdot r_k, \tag{14}$$

where W^r is a learnable weight matrix that projects relations into the same embedding space as entities.

3.3 Decoder

Within the extant KGC model, the ConvKB [15] network model is employed as the decoder, whereby 3×3 convolutional filters and 1×3 convolutional filters are utilized to extract feature representations from triplets (e_i, r_k, e_j). The decoder model generates a scoring function to calculate the effective probability of a given triple, as follows:

$$f(t_{ijk}) = (\sum_{m=1}^{\Omega} ReLU([\boldsymbol{e_i}, \boldsymbol{r_k}, \boldsymbol{e_j}]) * w^m) \cdot W, \tag{15}$$

where Ω represents a hyper-parameter, denoting the number of filters used. Ω and W are shared parameters and independent of e_i, r_k and e_j. $*$ denotes a convolution operator. The decoder is trained using soft-margin loss as:

$$L_2 = \sum_{t_{ij}^k \in \{\mathcal{G} \cup \mathcal{G}'\}} log(1 + exp(l_{t_{ij}^k} \cdot f(t_{ij}^k))) + \lambda \|W\|_2^2, \tag{16}$$

in which, $l_{t_{ij}^k} = \begin{cases} 1 & t_{ij}^k \in \mathcal{G} \\ -1 & t_{ij}^{k'} \in \mathcal{G}' \end{cases}$, $\mathcal{G}$ and $\mathcal{G}'$ are the sets of positive triples and negative triples, respectively.

$$\mathcal{G}' = \{t_{i'j}^k | e_i' \in \varepsilon \; e_i\} \bigcup \{t_{ij'}^k | e_j' \in \varepsilon \; e_j\}. \tag{17}$$

4 Experiment

4.1 Datasets and Metrics

To evaluate the efficacy of HOSAT, we have selected two benchmark datasets, WN18RR [7]and FB15k-237 [19], which have been previously divided into training, validation, and test sets.

Analogous to typical baselines, we adhere to the customary convention of examining all models in a filtered setting, whereby any corrupted triples existing within the training, validation, or test sets are eliminated during ranking. Our approach is evaluated using three distinct metrics, namely mean reciprocal rank (MRR), mean rank (MR), and the proportion of accurately ranked entities within the top N (HITS@N, with N = 1, 3, 10). Superior performance is denoted by lower MR, higher MRR, and higher HITS@N values.

Table 1. Experimental results on FB15K-237 and WN18RR.

	FB15K-237					WN18RR				
	MR	MRR	hit@1	hit@3	hit@10	MR	MRR	hit@1	hit@3	hit@10
TransE [4]	323	0.279	19.8	37.6	44.1	3384	0.226	4.27	42.1	53.2
DistMult [25]	512	0.281	19.9	30.1	44.6	5110	0.43	39.2	44.1	49.3
ComplEx [20]	546	0.278	19.4	29.7	45	5261	0.44	41.6	46.1	51.3
RotatE [18]	177	0.338	24.1	37.5	53.3	3340	0.476	42.8	49.2	57.1
TuckER [2]	–	0.358	26.6	39.4	54.4	–	0.47	44.3	48.2	54.6
ConvE [7]	245	0.312	22.5	34.1	49.7	4187	0.43	40.7	44.3	52.8
ConvKB [15]	216	0.289	19.8	32.4	47.1	3324	0.265	5.82	44.5	55.8
R-GCN [16]	600	0.164	10	18.1	30	6700	0.123	8	13.7	20.7
HRAN [11]	156	0.355	26.3	38	54.1	2322	0.471	44.1	48.4	54.2
InteractE [21]	172	0.354	26.3	–	53.5	5202	0.46	43	–	52.8
RAGAT [13]	199	0.365	27.3	40.1	54.7	2390	0.489	45.2	49.3	56.2
RGHAT [26]	196	0.516	46	53.1	62.5	1846	0.47	42.2	49.1	57.5
KBGAT [14]	210	0.518	46.1	54	62.6	1940	0.44	36.1	48.3	58.1
HOSAT	**152**	**0.528**	**46.3**	**56.4**	**65.9**	**1891**	0.492	45.8	49.6	**58.4**

4.2 Results and Analysis

We conduct extensive experiments on the KGC to demonstrate the effectiveness of HOSAT. We compare our results with various SOTA methods. Table 1 shows the comparison results for all benchmark datasets. We can observe that our proposed method on HOSAT has comparable performance to the SOTA baseline, validating the effectiveness of exploiting the structural and semantic features of KG.

Examining Attention Values Across Training Epochs. We investigate the distribution of attention values for a particular node as the number of training epochs increases. Figure 4 illustrates this distribution on FB15k-237. During the first stage of training, the attention values are randomly distributed in the initial stage, and there are only attention values at one-hop. By the second stage of training, HOSAT will have collected more information from more distant neighbors as training progresses. Once the model converges, it learns to gather multi-hop and clustering relationship information from the multi-hop neighborhood of nodes. The results demonstrate that it is necessary to pay attention to some important long-distance contexts in KGs.

Effect of Walk Length on Performance. Finally, we assessed the impact of different walk lengths on model performance. We find that performance improves considerably as the walk length increases initially, but stabilizes around $l = 8$. This aligns with our intuition that entities farther away have less influence. It is important to note that training time also increases with walk length. Therefore, it is crucial to strike a balance between performance gains and time overhead.

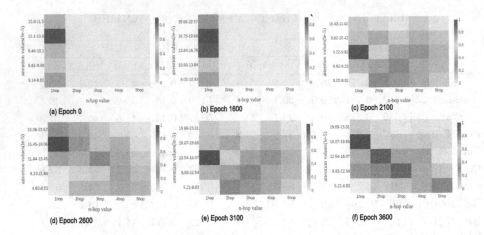

Fig. 4. The two stage learning process of HOSAT on FB15K-237 dataset.

4.3   Ablation Study

Global importance Analysis. To accurately assess the importance of long-distance contexts in HOSAT, we remove the global biased random walk module. The comparative analysis presented in Table 2 demonstrates that HOSAT achieves better improvements than HOSAT-Remove-Global on all metrics. In addition, the results suggest that GATv2 is more suitable for KGC than GATs. Results show that by considering global high-order connected contexts, the model can better understand the relationships between different nodes in KGs and make more informed predictions. Moreover, replacing GATv2 with GAT can improve the model's ability to capture the nuances of the data, resulting in more nuanced and accurate predictions.

Table 2. Result of ablation study.

	FB15K-237					WN18RR				
	MR	MRR	hit@1	hit@3	hit@10	MR	MRR	hit@1	hit@3	hit@10
HOSAT	**152**	**0.528**	**46.3**	**56.4**	**65.9**	**1891**	**0.472**	**44.8**	**49.6**	**58.4**
Eq. (3) w/o *GATv2*	157	0.518	43.7	55.2	64.4	1922	0.463	43.2	48.4	57.6
Eq. (12) w/o *global*	210	0.513	42.2	53.4	63.6	3277	0.432	36.2	43.3	40.2
w/o *DFS sampling*	186	0.505	45	55.3	64.7	2210	0.442	40.9	45.1	53.9
w/o *BFS sampling*	216	0.488	40.8	51.8	64.1	3260	0.425	39.9	42.4	41.8
w/o *relation contexts*	166	0.521	45.8	56.2	65.3	2155	0.448	42.8	46.2	55.1
w/o *entity features*	580	0.459	36.4	48.6	54.2	4205	0.411	34.2	40	38.7

Effects of the Random Walk Strategy. Our global biased random walk model incorporates entity importance, relation importance, and structural similarity walk strategies in order to evaluate the contribution of different walk strategies. As shown in Table 2, we found that if we remove the HOSAT relevance-weighted walking module, all performance metrics are significantly degraded. This proves that entities have a critical role to play in message propagation. After conducting our analysis, we observed that taking away the weighted walk module that represents the graph's structure also has a noticeable impact on the model. Therefore, it is crucial to make use of the graph structure information to identify high-order contexts.

5   Conclusions

In this work, we focus on integrating structured local and long-distance contexts to enhance the learning process of GATs through a two-stage approach. Due to the challenges in directly fusing this complex contexts, we pursue an indirect approach. Specifically, we introduce a comprehensive random walk algorithm to generate paths to capture the importance of global structural information from multiple perspectives. This enables us to obtain high-order relevant contexts through path transformation. HOSAT is simple yet effective and can be combined with many existing KGC methods. Our experiments demonstrate that the structural expressivity of random walks can improve the performance of KGC and provide additional insights into how to set the walk length. We believe our work can inspire more KGC methods and provide a deeper understanding of the role of random walks in KGC research. Future work includes designing logic rules to infer better high-order contexts for integration into HOSAT.

References

1. Auer, S., Bizer, C., Kobilarov, G., Lehmann, J., Cyganiak, R., Ives, Z.: DBpedia: a nucleus for a web of open data. In: Aberer, K., et al. (eds.) ASWC/ISWC -2007. LNCS, vol. 4825, pp. 722–735. Springer, Heidelberg (2007). https://doi.org/10. 1007/978-3-540-76298-0_52
2. Balazevic, I., Allen, C., Hospedales, T.M.: TuckER: tensor factorization for knowledge graph completion, pp. 5184–5193. Association for Computational Linguistics (2019). https://doi.org/10.18653/v1/D19-1522
3. Bollacker, K.D., Evans, C., Paritosh, P.K., Sturge, T., Taylor, J.: FreeBase: a collaboratively created graph database for structuring human knowledge, pp. 1247–1250. ACM (2008). https://doi.org/10.1145/1376616.1376746
4. Bordes, A., Usunier, N., García-Durán, A., Weston, J., Yakhnenko, O.: Translating embeddings for modeling multi-relational data, pp. 2787–2795 (2013)
5. Carlson, A., Betteridge, J., Kisiel, B., Settles, B.Jr., Hruschka, E., Mitchell, T.M.: Toward an architecture for never-ending language learning. AAAI Press (2010)
6. Chen, X., Hu, Z., Sun, Y.: Fuzzy logic based logical query answering on knowledge graphs, pp. 3939–3948. AAAI Press (2022)

7. Dettmers, T., Minervini, P., Stenetorp, P., Riedel, S.: Convolutional 2d knowledge graph embeddings, pp. 1811–1818. AAAI Press (2018)
8. Fan, T., Wang, H.: Research of Chinese intangible cultural heritage knowledge graph construction and attribute value extraction with graph attention network. Inf. Process. Manag. **59**(1), 102753 (2022). https://doi.org/10.1016/j.ipm.2021.102753
9. Geng, S., Fu, Z., Tan, J., Ge, Y., de Melo, G., Zhang, Y.: Path language modeling over knowledge graphs for explainable recommendation, pp. 946–955. ACM (2022). https://doi.org/10.1145/3485447.3511937
10. Huang, X., Zhang, J., Li, D., Li, P.: Knowledge graph embedding based question answering, pp. 105–113. ACM (2019). https://doi.org/10.1145/3289600.3290956
11. Li, Z., Liu, H., Zhang, Z., Liu, T., Xiong, N.N.: Learning knowledge graph embedding with heterogeneous relation attention networks. IEEE Trans. Neural Networks Learn. Syst. **33**(8), 3961–3973 (2022). https://doi.org/10.1109/TNNLS.2021.3055147
12. Li, Z.X., Li, Y.J., Liu, Y.W., Liu, C., Zhou, N.X.: K-CTIAA: automatic analysis of cyber threat intelligence based on a knowledge graph. Symmetry **15**(2), 337 (2023)
13. Liu, X., Tan, H., Chen, Q., Lin, G.: RAGAT: relation aware graph attention network for knowledge graph completion. IEEE Access **9**, 20840–20849 (2021). https://doi.org/10.1109/ACCESS.2021.3055529
14. Nathani, D., Chauhan, J., Sharma, C., Kaul, M.: Learning attention-based embeddings for relation prediction in knowledge graphs, pp. 4710–4723. Association for Computational Linguistics (2019). https://doi.org/10.18653/v1/p19-1466
15. Nguyen, D.Q., Nguyen, T.D., Nguyen, D.Q., Phung, D.Q.: A novel embedding model for knowledge base completion based on convolutional neural network, pp. 327–333. Association for Computational Linguistics (2018). https://doi.org/10.18653/v1/n18-2053
16. Schlichtkrull, M., Kipf, T.N., Bloem, P., van den Berg, R., Titov, I., Welling, M.: Modeling relational data with graph convolutional networks. In: Gangemi, A., et al. (eds.) ESWC 2018. LNCS, vol. 10843, pp. 593–607. Springer, Cham (2018). https://doi.org/10.1007/978-3-319-93417-4_38
17. Suchanek, F.M., Kasneci, G., Weikum, G.: YAGO: a core of semantic knowledge, pp. 697–706. ACM (2007). https://doi.org/10.1145/1242572.1242667
18. Sun, Z., Deng, Z., Nie, J., Tang, J.: RotatE: knowledge graph embedding by relational rotation in complex space. OpenReview.net (2019)
19. Toutanova, K., Chen, D.: Observed versus latent features for knowledge base and text inference, pp. 57–66. Association for Computational Linguistics (2015). https://doi.org/10.18653/v1/W15-4007
20. Trouillon, T., Welbl, J., Riedel, S., Gaussier, É., Bouchard, G.: Complex embeddings for simple link prediction. In: JMLR Workshop and Conference Proceedings, vol. 48, pp. 2071–2080. JMLR.org (2016)
21. Vashishth, S., Sanyal, S., Nitin, V., Agrawal, N., Talukdar, P.P.: InteractE: improving convolution-based knowledge graph embeddings by increasing feature interactions, pp. 3009–3016. AAAI Press (2020)
22. Vashishth, S., Sanyal, S., Nitin, V., Talukdar, P.P.: Composition-based multi-relational graph convolutional networks. OpenReview.net (2020)
23. Wang, H., Ren, H., Leskovec, J.: Relational message passing for knowledge graph completion, pp. 1697–1707. ACM (2021). https://doi.org/10.1145/3447548.3467247

24. Wang, X., Liu, K., Wang, D., Wu, L., Fu, Y., Xie, X.: Multi-level recommendation reasoning over knowledge graphs with reinforcement learning, pp. 2098–2108. ACM (2022). https://doi.org/10.1145/3485447.3512083

25. Yang, B., Yih, W., He, X., Gao, J., Deng, L.: Embedding entities and relations for learning and inference in knowledge bases (2015)

26. Zhang, Z., Zhuang, F., Zhu, H., Shi, Z., Xiong, H., He, Q.: Relational graph neural network with hierarchical attention for knowledge graph completion, pp. 9612–9619. AAAI Press (2020)

27. Zhao, Y., et al.: EIGAT: incorporating global information in local attention for knowledge representation learning. Knowl. Based Syst. **237**, 107909 (2022). https://doi.org/10.1016/j.knosys.2021.107909

Instance-Ambiguity Weighting for Multi-label Recognition with Limited Annotations

Daniel Shrewsbury[1], Suneung Kim[1], Young-Eun Kim[1], Heejo Kong[2], and Seong-Whan Lee[1(✉)]

[1] Department of Artificial Intelligence, Korea University, Seoul, Republic of Korea
{daniel_shrewsbury,se_kim,ye_kim,sw.lee}@korea.ac.kr
[2] Department of Brain and Cognitive Engineering, Korea University, Seoul, Republic of Korea
hj_kong@korea.ac.kr

Abstract. Multi-label recognition with limited annotations has been gaining attention recently due to the costs of thorough dataset annotation. Despite significant progress, current methods for simulating partial labels utilize a strategy that uniformly omits labels, which inadequately prepares models for real-world inconsistencies and undermines their generalization performance. In this paper, we consider a more realistic partial label setting that correlates label absence with an instance's ambiguity, and propose the novel Ambiguity-Aware Instance Weighting (AAIW) to specifically address the performance decline caused by such ambiguous instances. This strategy dynamically modulates instance weights to prioritize learning from less ambiguous instances initially, then gradually increasing the weight of complex examples without the need for predetermined sequencing of data. This adaptive weighting not only facilitates a more natural learning progression but also enhances the model's ability to generalize from increasingly complex patterns. Experiments on standard multi-label recognition benchmarks demonstrate the advantages of our approach over state-of-the-art methods.

Keywords: Partial Label Learning · Multi-Label Recognition

1 Introduction

Multi-label image recognition(MLR) is the task of identifying multiple semantic labels in an image, which benefits several applications such as content-based image retrieval and recommendation systems [1, 22, 30]. MLR methods are dependent on extensive, cleanly labeled datasets which are both difficult and expensive to create [23], leading to the study of MLR with partial labels as it diminishes the need for comprehensively labeled datasets.

The primary challenge for partial label methods lies in managing the noise introduced when a proportion of class labels are unavailable [5]. The prevalent methods for multi-label recognition with partial labels [6, 10, 16, 17, 25, 31, 37] predominantly operate under the unrealistic assumption that the absence of labels

© The Author(s), under exclusive license to Springer Nature Singapore Pte Ltd. 2024
D.-N. Yang et al. (Eds.): PAKDD 2024, LNAI 14645, pp. 156–167, 2024.
https://doi.org/10.1007/978-981-97-2242-6_13

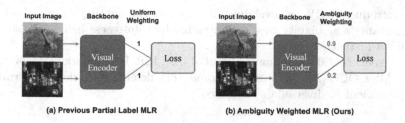

(a) Previous Partial Label MLR   (b) Ambiguity Weighted MLR (Ours)

Fig. 1. A conceptual comparison between other multi-label recognition (MLR) with partial label methods and ours. Previous works weight all instances equally, regardless of the instance's ambiguity and difficulty, which leads to a higher susceptibility to noise from ambiguous data. In contrast, we initially assign lower weights to high ambiguity instances, reducing their influence on early model updates and promoting a more stable learning trajectory.

occurs uniformly, neglecting the real-world complexities of label absence. To simulate partial labels, prevailing methods assign an equal chance of absence to each label across the dataset. This approach fails to account for the varying clarity of instances, making models particularly susceptible to noise from ambiguous images.

We consider a more realistic partial label setting, where the likelihood of a label being missed is correlated with image ambiguity. In the real world, labels are frequently missed due to the complexity and ambiguity of instances rather than a uniform randomness. Intuitively, an annotator is unlikely to miss annotations on a simple, easy instance, but are more likely to miss labels on difficult instances. Recognizing this, our approach assigns a higher probability of label omission to ambiguous instances, which more accurately mirror real-world data scenarios.

Motivated by the above consideration, we propose an Ambiguity-Aware Instance Weighting (AAIW) method that employs entropy as a measure of an instance's ambiguity to dynamically reweight each instance's loss according to its clarity and reliability as seen in Fig. 1. We estimate this instance ambiguity-based weighting by calculating the entropies of known labels through the model's confidence scores and use derived neighbor-inferred scores from positively labeled nearest neighbor predictions for the unavailable labels. We employ a dynamic instance weighting approach, initially emphasizing more reliable instances and incrementally integrating ambiguous instances as training progresses. Our strategy ensures all data, regardless of complexity, contribute to the model's learning, fostering a nuanced understanding of the data distribution and enhancing noise robustness.

To summarize our primary contributions:

1. We investigate a new partial label learning scenario, which more accurately reflects real-world situations, and highlights the need of accounting for instance specific factors in the partial label scenario.

2. We introduce AAIW, a novel dynamic instance reweighting strategy based on an instance's ambiguity that enhances model robustness against noise from difficult instances.
3. Our experimental validations, executed on MLR Benchmarks (Pascal VOC [11], MS-COCO [20] and Visual Genome [18]), demonstrate clear performance improvement in diverse label settings.

2   Related Works

2.1   Multi-label Recognition with Full Annotations

Multi-label recognition is evolving to more accurately reflect the complexity inherent in real-world visual perception, recognizing the critical role of label interrelationships. Early methods focused on independent binary classifiers for each class, an approach that overlooks label correlations [9,14,29]. In contrast, contemporary strategies incorporate these relationships by employing graph neural networks [7,8,33]. However, the reliance of these methods on costly, labor-intensive fully annotated datasets has spurred research into multi-label recognition with limited annotations.

2.2   Multi-label Recognition with Limited Annotations

In multi-label recognition with limited annotations, only a few labels need to be annotated for each image. Earlier approaches simply regard the missing labels as negative [2,19], while alternative approaches predict the missing labels via probabilistic modeling [15,32,34]. Further research focus on utilizing features from known labels to enhance pseudo label generation via feature blending [27] or label co-occurrences [6]. Moreover, new loss functions have been developed to account for missing labels by utilizing the maximum likelihood criterion under an approximate distribution of missing labels [3,37] and leveraging statistical information from existing datasets [36,38].

Recently, vision-language models have seen significant success in computer vision. Through extensive pretraining, vision language models e.g. CLIP [28], provides rich prior representations for downstream tasks. DualCoOp [31] utilizes CLIP's rich semantic representations prompt tuning to address the partial label problem. SCPNet [10] makes use of CLIP's rich label-to-label correspondences through a calculated semantic prior to reason unknown labels from annotated labels.

While these approaches demonstrate promising results, when simulating partial labels they assume uniform random label omission, which overlooks the nuances of instance-specific difficulty and ambiguity. This oversight can compromise the efficacy of models, as it fails to consider the uneven distribution of noise introduced by varying image complexities-an issue our method addresses.

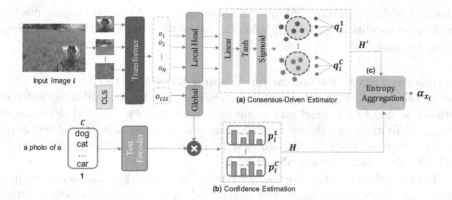

Fig. 2. Overview of our proposed instance weighting strategy. We propose to dynamically weight an instance using dual-method entropy estimation. (a) A consensus-driven estimator is employed for the missing classes, estimating the entropy scores with a nearest neighborhood approach with the class features of positively labeled instances. (b) For existing classes, we use a confidence estimation, utilizing probabilities from a vision-language backbone. (c) The instance-specific weighting factor is then obtained through the Entropy Aggregation module.

3   Methodology

3.1   Problem Definition

In multi-label recognition with partial labels, a dataset consists of N images $\{x_i\}_{i=1}^N \in X$. Each image x_i is associated with a corresponding label vector $\{y_i\}_{i=1}^N \in Y$, containing multiple classes $y_i = \{y^c\}_{c=1}^C$ for C total classes. y^c can take a value of 1, 0, or -1 indicating a positive, negative, or missing label, respectively. The objective is to predict the semantic labels from every image. A key difficulty stems from labels missed due to instance ambiguity, adding noise to the learning process. Our method, outlined in (Fig. 2), dynamically weights an instance according to its entropy, which we use as an estimate of its ambiguity [26]. The dynamic weighting guides model learning, by emphasizing easy instances initially, while gradually increasing the importance of difficult instances as training progresses.

3.2   Ambiguity-Aware Instance Weighting

Confidence Estimation for Existing Labels. Existing class labels are estimated using the model's confidence score, enabling a direct validation of predicted probabilities and an accurate ambiguity reflection. Treating each label as a separate binary classification, the entropy H for a given label is calculated as follows:

$$H(x_i^c) = -p_i^c \cdot \log(p_i^c) - (1 - p_i^c) \cdot \log(1 - p_i^c), \qquad (1)$$

where x_i^c is the ith image containing class c, and $p = P(y = 1|x)$ represents the model's probability score with p_i^c being the probability score for the labeled class of the ith image.

Consensus-Driven Estimator for Missing Labels. In the presence of missing labels, the reliability of model predictions diminishes, complicating entropy estimation due to potential over or underestimations, which could skew model accuracy. To mitigate this, we employ a consensus-driven estimator, drawing on the consistency of class visual features within class objects [6] to enhance entropy estimation reliability amidst absent labels.

To derive distinct class features we implement a class-specific memory bank, storing M unique class features and corresponding model confidences for positive labels, extracted through transformer patch-level embeddings. Following [12], we update the bank with a slow changing momentum model. A dedicated linear layer weights these patches, refined via tanh and sigmoid activations, to derive distinct class features.

Our strategy hinges on cosine similarity between the features of absent labels and similar class features, initiating a soft voting process [24] to compute the neighbor-inferred score q^c for each label:

$$q^c = \frac{1}{\mathcal{K}} \sum_{n \in \mathcal{N}} p'^c_n, \tag{2}$$

where $\mathcal{N}$ represents neighboring indices, $\mathcal{K}$ the selected samples with the most similar features, and p'^c_n the confidence score of the nth neighbor for class c. Unlike [6,21], we do not use these estimates for pseudo-labeling as it would sidestep direct model confidence, instead we focus on applying the estimate towards the entropy estimation only.

Using the derived neighbor inferred score q^c, we calculate the entropy H' of each missing label using the same formula as existing labels:

$$H'(x_i^c) = -q_i^c \cdot \log(q_i^c) - (1 - q_i^c) \cdot \log(1 - q_i^c), \tag{3}$$

where q_i^c represents the neighbor-inferred score for class c of the ith image when the corresponding label is unavailable. We use this feature similarity derived score as the probability estimate for the entropy.

Entropy Aggregation and Weighting. We derive instance entropy $\hat{H}(x_i)$ by aggregating entropies across all classes c as follows:

$$\hat{H}(x_i) = \sum_{c=1}^{C} I(x_i^c) \cdot H(x_i^c) + (1 - I(x_i^c)) \cdot H'(x_i^c), \tag{4}$$

where I is an indicator function that equals 1 if class c is labeled and 0 for unlabeled. $\hat{H}(x_i)$ reflects the total uncertainty per instance x, serving as a comprehensive ambiguity metric. It enables refined interaction with each training instance, mitigating noise from instance-specific variances.

We then normalize $\hat{H}(x_i)$ to $\bar{H}(x_i)$ within a $[0,1]$ range, enhancing comparative uncertainty analysis across instances:

$$\bar{H}(x^i) = \frac{\hat{H}(x_i) - \hat{H}_{\min}}{\hat{H}_{\max} - \hat{H}_{\min}}. \tag{5}$$

Subsequently, we determine each sample x_i's weight α via:

$$\alpha_{x_i} = \exp(-\bar{H}(x_i)). \tag{6}$$

The formula applies an exponential function to ensure a positive, gradual scaling of instance influence, with lower entropy garnering more weight. This approach ensures that the learning emphasis placed on each instance is proportionate to its informative value and reliability.

Adaptive Focus Adjustment. Recognizing that ambiguous instances still contain valuable information, our model employs an adaptive focus adjustment strategy inspired by self-paced curriculum principles [4]. It begins by assigning greater significance to low-entropy instances to ensure the initial learning is stable and less prone to noise. As the model matures, it is important to expose it to more complex instances; thus, as training advances, the entropy weighting is adaptively modified to increase the importance of ambiguous instances as follows:

$$\alpha'_{x_i} = \alpha_{x_i} + (1 - \alpha_{x_i}) \times \left(1 - \exp\left(-s \times \frac{e}{E}\right)\right), \tag{7}$$

where s acts as an annealing rate hyperparameter gradually increasing the importance of high-ambiguity instances into the model's training focus across epochs e out of total E.

3.3   Total Training Loss

For training, we use the focal margin loss from [37] as the classification loss denoted as $\mathcal{L}_{\text{cls}}$. The loss function is designed specifically for the partial label scenario and avoids over-reliance on easy positives to improve model generalization. We denote this loss as $\mathcal{L}_{\text{cls}}$.

Additionally, we adopt a consistency regularization loss with dynamic thresholding from [35] to stabilize pseudo-label generation amidst sample re-weighting, thus ensuring more reliable entropy calculations, denoted as $\mathcal{L}_{\text{cst}}$.

The total loss $\mathcal{L}_{\text{total}}$ is as follows:

$$\mathcal{L}_{\text{total}} = \alpha'_{x_i}(\mathcal{L}_{\text{cls}} + \mathcal{L}_{\text{cst}}), \tag{8}$$

where our weighting factor α'_{x_i} modulates the impact of an instance's contribution to model learning, ensuring a stable learning trajectory.

4 Experiments

4.1 Experiment Settings

Implementation Details. Our experiments span standard MLR benchmarks with limited annotations-MS-COCO (COCO) [20], PASCAL-VOC (VOC) [11], and Visual Genome (VG-200) [18]-in an instance-ambiguity context. For VG-200, we follow [6] and choose the 200 most frequent classes. We use the ViT-B/16 based visual encoder from MKT [13] as the backbone for all methods. We use CLIP's transformer as our text encoder, with both encoders initialized with CLIP's pretrained weights. We freeze the text encoder and all visual encoder layers except the last.

Instance Ambiguity Setting. To emulate environments with incomplete annotations, we apply a selective masking process to the dataset's full label set. The fraction of labels masked varies from 10% to 90%, according to the label omission rate r (e.g., at $r = 0.10$, we mask 10% of labels). Unlike other methods that naively uniformly mask labels [6,10,27,31], our partial label setting provides a more realistic scenario by using each instance's ambiguity to calculate the probability of dropping a label. This ambiguity is assessed by training a neural network on the fully labeled dataset, from which we derive the entropy H (Eq. (1)) and normalize it to $\bar{H}$ (Eq. (5)). To differentiate label certainty we calculate ambiguity thresholds b_{low} and b_{high} defined as:

$$b_{\text{low}} = (1 - r)\left(\frac{H_{\text{max}} - H_{\text{min}}}{H_{\text{max}}}\right), \tag{9}$$

$$b_{\text{high}} = (1 - r)\left(1 + \frac{H_{\text{max}}}{\log(C)}\right), \tag{10}$$

where r is the average label omission rate for the overall dataset, C the total number of classes, and H is the instance's entropy. Notably, $\log(C)$ serves as the theoretical maximum entropy. Using these bounds, we derive the initial probability d' of a label being masked out as:

$$d' = (b_{\text{low}} + \bar{H}(b_{\text{high}} - b_{\text{low}})). \tag{11}$$

This aligns label omission with instance ambiguity, offering a more realistic partial label simulation. To ensure that on average $1 - r$ labels will be retained we align the probabilities with the label retention rate to obtain the final label omission probability d for an instance as follows:

$$d = d' \times \left(\frac{1 - r}{\mu_{d'}}\right), \tag{12}$$

where $\mu_{d'}$ is the mean of all label omission probabilities.

Table 1. Multi-label recognition results on PASCAL VOC2012, MS-COCO, and Visual Genome with partial labels.

Methods	10%	20%	30%	40%	50%	60%	70%	80%	90%	Avg.
PASCAL VOC 2012 [11]										
MKT [31]	82.3	86.6	92.8	93.0	93.2	93.6	93.8	93.9	94.0	84.9
SST [6]	76.1	79.6	81.3	82.0	84.1	84.5	84.9	85.4	85.8	82.6
SARB [27]	75.2	75.8	76.5	76.9	77.1	77.2	78.3	78.9	79.4	77.2
DualCoOp [31]	**87.0**	**88.4**	90.3	91.4	91.5	91.5	91.7	92.0	92.2	90.7
SCPNet [10]	86.5	88.1	88.9	90.4	91.2	93.1	93.4	93.6	94.1	91.0
AAIW(ours)	85.7	85.7	**93.1**	**93.4**	**94.1**	**94.2**	**94.2**	**94.4**	**94.5**	**92.1**
MS-COCO [20]										
MKT [31]	65.7	70.2	70.5	72.1	76.4	79.9	80.1	81.1	81.3	75.2
SST [6]	54.6	58.4	59.3	61.4	63.5	64.2	67.9	68.1	69.4	63.0
SARB [27]	55.2	59.7	60.3	62.4	64.0	65.2	69.1	69.2	70.4	63.9
DualCoOp [31]	**71.6**	**71.8**	71.9	75.1	76.3	76.5	77.1	77.9	78.1	75.1
SCPNet [10]	63.3	66.7	70.4	73.1	76.4	79.8	80.8	81.1	81.5	74.8
AAIW(ours)	66.6	70.9	**72.9**	**76.1**	**78.9**	**80.3**	**80.9**	**81.3**	**81.7**	**76.6**
VG-200 [18]										
MKT [31]	33.4	37.2	40.9	43.3	44.1	44.6	44.8	45.2	45.4	42.1
SST [6]	30.8	32.3	33.1	33.9	35.9	37.2	38.1	39.8	40.5	35.7
SARB [27]	31.2	33.4	34.7	35.5	37.9	38.4	39.8	41.0	41.2	37.0
SCPNet [10]	33.7	36.8	41.8	43.5	44.4	45.0	45.3	45.6	45.7	42.0
AAIW(ours)	**34.5**	**38.3**	**42.2**	**44.0**	**44.7**	**45.1**	**45.4**	**45.7**	**45.9**	**42.8**

Comparison Methods. We compare our model with the following baselines: SCPNet [10], DualCoOp [31], SARB [27], SST [6], and MKT [13]. We initialize SST [6] and SARB [27] with CLIP's pretrained weights. MKT is implemented with the focal margin loss [37] for fairness.

Evaluation. With all datasets we report the mean average precision (mAP) for each proportion of available labels (from 10% to 90%) and the overall average value.

4.2 Results

Table 1 highlights AAIW's comparative performance against state-of-the-art methods under our instance ambiguity paradigm across VOC, MS-COCO, and VG-200. Notably, AAIW excels in all settings, barring the 10% and 20% label availability scenarios. DualCoOp is designed for unsupervised scenarios, whereas AAIW, designed for real-world partial label applications, assumes a small amount of labeled data which is typically available. In conditions with more than 20% label availability, AAIW's advantages become evident. It registers a 1.5% mAP increase on MS-COCO over DualCoOp and outperforms SCPNet by 1.1% on VOC. In the VG-200 dataset, AAIW achieves a 1% edge

Table 2. Effect of different modules in the proposed AAIW method for the instance ambiguity setting under different label proportions(%).

	L_{cls}	α'_{cls}	L_{cst}	α'_{cst}	s	30%	50%	70%	90%	Avg.
backbone	✓					92.2	93.0	93.1	93.5	92.9
AAIW	✓	✓				92.8	93.7	93.8	94.1	93.6
	✓		✓			92.4	93.5	93.6	93.8	93.3
	✓		✓	✓		92.6	93.6	93.8	93.9	93.5
	✓	✓	✓	✓		93.0	93.7	94.0	94.5	93.8
	✓	✓	✓	✓	✓	**93.3**	**93.9**	**94.2**	**94.5**	**94.0**

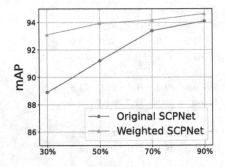

Fig. 3. Comparison between both the original and an ambiguity-weighted SCPNet under different label proportions on VOC

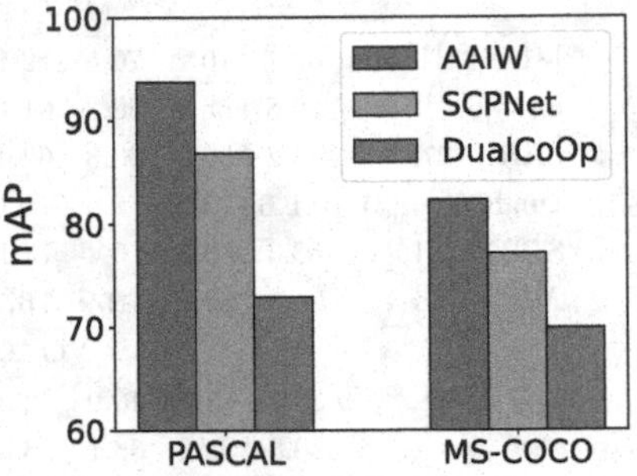

Fig. 4. Performance comparison on the top 1000 most ambiguous instances of each dataset's test set

over SCPNet. These figures affirm the robustness of our approach in multi-label recognition under partial labeling.

4.3 Ablation Studies

We conduct an ablation study with results shown in Table 2. We introduce a ViT-B/16 based MLR model with the L_{cls} loss as our baseline and train each component under the 30%, 50%, 70%, 90% label proportion settings in VOC. Incremental additions to the baseline yields improvements: α'_{cls} alone enhances performance by 0.7%; $\mathcal{L}_{cst}$ contributes 0.4%, further rising to 0.6% when weighted; combined, both weighted losses increase gains to 0.9%, with an overall potential enhancement up to 1.1%, affirming the robustness of our method against noise from ambiguity.

4.4 Model Analysis

In this section we perform further experiments demonstrating the overall applicability of our proposed method.

Adaptability to Other Methods. To validate the adaptability of our instance ambiguity weighting, we integrate our dynamic weighting with SCPNet. We compare this new weighted SCPNet with the original on VOC under our instance ambiguity setting. As seen in Fig. 3, the mAP increases across all settings, highlighting our method's compatibility with different architectures and potential as a plug-and-play enhancement for other multi-label recognition models.

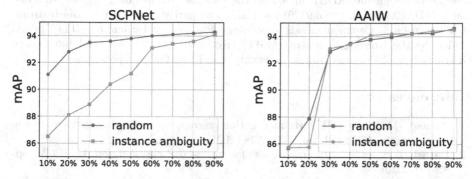

Fig. 5. Comparison of SCPNet and AAIW's performance on both the random and instance ambiguity partial label settings on VOC

Effectiveness on Ambiguous Instances. Using the network trained on full labels in Sect. 4.1, we select the top 1000 highest entropy images from the test sets of VOC/COCO, evaluating them with AAIW, SCPNet, and DualCoOp. The results are showcased in Fig. 4. The dynamic instance weighting mechanism allows our model to learn subtle features from the ambiguous instances without overfitting on the noise introduced by these ambiguous instances.

Partial Label Setting Comparison. We train SCPNet [10] and our AAIW model under both the original random partial label setting as in [6,10,27,31], where labels are uniformly dropped, and our proposed instance ambiguity setting with the results shown in Fig. 5. SCPNet shows significant reduced performance under the instance ambiguity setting, highlighting its susceptibility to instance noise. In contrast, AAIW shows little change between settings, demonstrating our model's robustness and ability to generalize under different settings.

5   Conclusion

This work presents a significant step forward in multi-label recognition with limited annotations, addressing the challenges of real-world instance ambiguity. We introduce AAIW, which enhances robustness through entropy-based adaptive weighting within a novel partial label setting. Our approach effectively balances the learning process, reducing the influence of initial weights over time to harness the full spectrum of data, including ambiguous instances. Extensive experiments on benchmark datasets underscore its effectiveness, marking it as a robust solution for improving multi-label recognition in the presence of ambiguous instances.

Acknowledgement. This research was supported by the Challengeable Future Defense Technology Research and Development Program through the Agency For Defense Development (ADD) funded by the Defense Acquisition Program Administration (DAPA) in 2024 (No.912911601) and was partly supported by the Institute of Information & Communications Technology Planning & Evaluation (IITP) grant, funded by the Korea government (MSIT) (No. 2019-0-00079, Artificial Intelligence Graduate School Program (Korea University)) and NCSOFT corporation.

References

1. Ahmad, M., Lee, S.W.: Human action recognition using multi-view image sequences. In: ICAFGR, pp. 523 – 528 (2006)
2. Baruch, E.B., et al.: Asymmetric loss for multi-label classification. In: ICCV, pp. 82–91 (2020)
3. Ben-Baruch, E., et al.: Multi-label classification with partial annotations using class-aware selective loss. In: CVPR, pp. 4754–4762 (2021)
4. Bengio, Y., Louradour, J., Collobert, R., Weston, J.: Curriculum learning. In: ICML, pp. 41–48 (2009)
5. Chen, M., Zheng, A.X., Weinberger, K.Q.: Fast image tagging. In: ICML (2013)
6. Chen, T., Pu, T., Wu, H., Xie, Y., Lin, L.: Structured semantic transfer for multi-label recognition with partial labels. In: AAAI, pp. 339–346 (2022)
7. Chen, T., Xu, M., Hui, X., Wu, H., Lin, L.: Learning semantic-specific graph representation for multi-label image recognition. In: ICCV, pp. 522–531 (2019)
8. Chen, Z.M., Wei, X.S., Wang, P., Guo, Y.: Multi-label image recognition with graph convolutional networks. In: CVPR (2019)
9. Cole, E., Mac Aodha, O., Lorieul, T., Perona, P., Morris, D., Jojic, N.: Multi-label learning from single positive labels. In: CVPR (2021)
10. Ding, Z., et al.: Exploring structured semantic prior for multi label recognition with incomplete labels. In: CVPR, pp. 3398–3407 (2023)
11. Everingham, M., Van Gool, L., Williams, C.K.I., Winn, J., Zisserman, A.: The PASCAL Visual Object Classes Challenge 2012 (VOC2012) Results. http://www.pascal-network.org/challenges/VOC/voc2012/workshop/index.html
12. He, K., Fan, H., Wu, Y., Xie, S., Girshick, R.: Momentum contrast for unsupervised visual representation learning. In: CVPR (2020)
13. He, S., Guo, T., Dai, T., Qiao, R., Shu, X., Ren, B., Xia, S.T.: Open-vocabulary multi-label classification via multi-modal knowledge transfer. In: AAAI (2023)
14. Huynh, D., Elhamifar, E.: Interactive multi-label CNN learning with partial labels. In: CVPR (2020)
15. Kapoor, A., Jain, P., Viswanathan, R.: Multilabel classification using bayesian compressed sensing. In: NeurIPS, pp. 2645–2653 (2012)
16. Kim, Y., Kim, J., Akata, Z., Lee, J.: Large loss matters in weakly supervised multi-label classification. In: CVPR, pp. 14136–14145 (2022)
17. Kim, Y., Kim, J.M., Jeong, J., Schmid, C., Akata, Z., Lee, J.: Bridging the gap between model explanations in partially annotated multi-label classification. In: CVPR, pp. 3408–3417 (2023)

18. Krishna, R., et al.: Visual genome: connecting language and vision using crowd-sourced dense image annotations. Int. J. Comput. Vis. **123**(1), 32–73 (2017)
19. Lee, M.S., Yang, Y.M., Lee, S.W.: Automatic video parsing using shot boundary detection and camera operation analysis. Pattern Recogn. **34**, 711–719 (2001)
20. Lin, T.Y., et al.: Microsoft coco: common objects in context. In: ECCV, pp. 740–755 (2014)
21. Litrico, M., Del Bue, A., Morerio, P.: Guiding pseudo-labels with uncertainty estimation for source-free unsupervised domain adaptation. In: CVPR (2023)
22. Liu, F., Xiang, T., Hospedales, T.M., Yang, W., Sun, C.: Semantic regularisation for recurrent image annotation. In: CVPR, pp. 4160–4168 (2016)
23. Liu, W., Wang, H., Shen, X., Tsang, I.W.H.: The emerging trends of multi-label learning. IEEE Trans. Pattern Anal. Mach. Intell. **44**, 7955–7974 (2020)
24. Mitchell, H.B., Schaefer, P.A.: A "soft" k-nearest neighbor voting scheme. Int. J. Intell. Syst. **16**(4), 459–468 (2001)
25. Nam, W.J., Gur, S., Choi, J., Wolf, L., Lee, S.W.: Relative attributing propagation: interpreting the comparative contributions of individual units in deep neural networks. In: AAAI, pp. 2501–2508 (2020)
26. Park, L.A.F., Simoff, S.: Using entropy as a measure of acceptance for multi-label classification. In: Advances in Intelligent Data Analysis XIV, pp. 217–228 (2015)
27. Pu, T., Chen, T., Wu, H., Lin, L.: Semantic-aware representation blending for multi-label image recognition with partial labels. In: AAAI, pp. 2091–2098 (2022)
28. Radford, A., et al.: Learning transferable visual models from natural language supervision. In: ICML (2021)
29. Rajeswar, S., López, P.R., Singhal, S., Vázquez, D., Courville, A.C.: Multi-label iterated learning for image classification with label ambiguity. In: CVPR, pp. 4773–4783 (2021)
30. Simonyan, K., Zisserman, A.: Very deep convolutional networks for large-scale image recognition. In: ICLR (2015)
31. Sun, X., Hu, P., Saenko, K.: Dualcoop: fast adaptation to multi-label recognition with limited annotations. In: NeurIPS (2022)
32. Vasisht, D., Damianou, A., Varma, M., Kapoor, A.: Active learning for sparse bayesian multilabel classification. In: SIGKDD, pp. 472–481 (2014)
33. Wang, Y., et al.: Multi-label classification with label graph superimposing. In: AAAI, vol. 34, pp. 12265–12272 (2020)
34. Wu, B., Liu, Z., Wang, S., Hu, B.G., Ji, Q.: Multi-label learning with missing labels. In: ICPR, pp. 1964–1968 (2014)
35. Zhang, B., et al.: Flexmatch: boosting semi-supervised learning with curriculum pseudo labeling. In: NeurIPS, vol. 34, pp. 18408–18419 (2021)
36. Zhang, X., Song, Y., Zuo, F., Wang, X.: Towards imbalanced large scale multi-label classification with partially annotated labels. In: SERA, pp. 195–200 (2023)
37. Zhang, Y., et al.: Simple and robust loss design for multi-label learning with missing labels. arXiv abs/2112.07368 (2021)
38. Zhou, D., Chen, P., Wang, Q., Chen, G., Heng, P.A.: Acknowledging the unknown for multi-label learning with single positive labels. arXiv abs/2203.16219 (2022)

Chaotic Neural Oscillators with Deep Graph Neural Network for Node Classification

Le Zhang[ID] and Raymond S. T. Lee[✉][ID]

Guangdong Provincial Key Laboratory of Interdisciplinary Research and Application for Data Science, Beijing Normal University-Hong Kong Baptist University United International College, Zhuhai, China
raymondshtlee@uic.edu.cn

Abstract. Node classification is a pivotal task in spam detection, community identification, and social network analysis. Compared with traditional graph learning methods, Graph Neural Networks (GNN) show superior performance in prediction tasks, but essentially rely on the characteristics of adjacent nodes. This paper proposed a novel Chaotic Neural Oscillator Feature Selection Graph Neural Network (CNO_FSGNN) model integrating Lee Oscillator which serves as a chaotic memory association to enhance the processing of transient information and transitions between distinct behavioral patterns and synchronization of relevant networks, and a Feature Selection Graph Neural Network to address the limitations. Consequently, the synthesis can improve mean classification accuracy across six homogeneous and heterogeneous datasets notably in Squirrel dataset, and can mitigate over-smoothing concerns in deep layers reducing model execution time.

Keywords: Node Classification · Graph Neural Network · Lee Oscillator · Oversmoothing

1 Introduction

Graph-structured databases are ubiquitous in social media, recommendation systems, and intelligent question-answering (QA) chatbots. Knowledge graphs use entities and edges to represent a broad knowledge base. They have predefined patterns but result from weak relationships or missing node labels due to incomplete information. Compared with traditional graph learning methods, Graph Neural Networks (GNN) revolutionize graph-level tasks by leveraging graph topology to assume the independence of individual examples. Among these tasks, node classification is of great significance in graph analysis due to its wide range of applications such as community detection and traffic flow prediction.

Node classification algorithms have three major categories: convolutional mechanism, attention mechanism, and autoencoder mechanism. Convolutional mechanism is an information aggregation paradigm widely uses convolutional or pooling operations to derive ample representations for each node. Attention mechanism, by contrast, assigns vary contributions from different neighbors to the target node, enhancing model to focus

© The Author(s), under exclusive license to Springer Nature Singapore Pte Ltd. 2024
D.-N. Yang et al. (Eds.): PAKDD 2024, LNAI 14645, pp. 168–180, 2024.
https://doi.org/10.1007/978-981-97-2242-6_14

Table 1. Common models of three types of node classification algorithms

Categories	Models
Convolutional mechanism	GCN [1], GraphSAGE [2]
Attention mechanism	GAT [3], AGNN [4]
Autoencoder mechanism	VAGE [5], DGI [6]

on relevant aspects. Autoencoder mechanism uses unsupervised technology to learn low-dimensional embeddings. These categories and their models are listed in Table 1.

GNN enables precise and analysis and efficient graph-structured data processing across various tasks. For instance, Wang et al. [7] leveraged GNN's capabilities to introduce the Graph Explicit Neural Network (GENN) framework, enhancing natural language inference (NLI) within features-constrained and latency-constrained environments. This innovation not only improved interpretability but also facilitated implementation. While additional methods have enhanced GNNs for industrial contexts, but excess dependence on neighbor node features poses limitations for applications particularly in heterogeneous datasets where nodes may not share similar features between neighbors.

GNN makes aggregation of information from neighbor nodes is a crucial characteristic distinct from other neural networks as many propagation methods are proposed. It only uses homogenous datasets where nodes are assumed to share similar features with neighbor nodes. Although some aggregation algorithms use graph topology [8] for datasets with strong heterophily but they may lead to meager learning. Since nodes in heterogeneous datasets have dissimilar features from their neighbor nodes, predicting a single node's outcome by aggregating all its neighbor information can produce neighbor explosion and high inference latency. Hence, Chaotic Neural Oscillator Feature Selection Graph Neural Network (CNO_FSGNN) is proposed by integrating Lee oscillator into a Feature Selection Graph Neural Network (FSGNN) to address these limitations. The key contributions of the model are to:

1. introduce CNO_FSGNN model for node classification to achieve the highest mean classification accuracy across nine datasets compared with baseline models.
2. attain the latest performance on Squirrel dataset.
3. mitigate over-smoothing issues at deep layers to reduce the divergence between 8 hops and 16 hops models.

1.1 Node Classification

Industrial data are often expressed in the form of graphs to provide a universal representation of the data, solving numerous computing tasks including social networks, transportation networks, protein-protein interaction networks, knowledge graphs, and brain networks. Node classification involves inferring node properties, detecting anomalies, identifying disease-related genes, and providing personalized medication recommendations. However, extensive graph databases in many nodes and edges such as Freebase,

DBpedia, and Wikipedia often encounter incomplete or missing information due to data scarcity.

This is a significant challenge due to privacy concerns [9], making it difficult to create large shared datasets with complete labels for model training. Node classification becomes crucial to predict node labels on incomplete graphs such as identifying fraudulent users in a social network diagram by analyzing their neighbors' labels. Traditional machine learning techniques are inadequate for graph computing tasks due to the assumption of independent and equally distributed samples. Gori [10] introduced the concept of Graph Neural Networks (GNN) to process graph-structured data by splitting the whole graph into subgraphs, input a graph embedding model, update node information iteratively through different aggregation algorithms based on node features and graph topology, and output a predisposed label.

1.2 Graph Neural Network (GNN)

Graph Neural Networks (GNN) are renowned for processing graph-structured, non-Euclidean data. Unlike standard Neural Networks, GNN integrates information from neighboring nodes to enrich a node's features and facilitate message passing throughout the graph as follows:

$$H^{(l+1)} = \sigma(\tilde{D}^{-\frac{1}{2}} \tilde{A} \tilde{D}^{-\frac{1}{2}} H^{(l)} w^{(l)}) \tag{1}$$

A graph is represented as $G = (V, E)$ where V represents vertices and E represents edges, which can be directed or undirected, with or without weights. An essential characteristic of GNN is its adjacency matrix which represent the relationship between graph vertices. Set V as the vertices, the size of A is $|V| \times |V|$. Normally, the adjacency matrix is written as a sparse tensor, suitable for describing a dense graph with high space utilization and query efficiency. In 2013, Bruna [11] first proposed Spectral domain and Spatial-domain based convolutional neural networks on graphs based on Graph Signal Processing by using eigenvalues and eigenvectors of the Laplacian matrix for the spectral domain-based method and adjacency matrix and node features for the spatial domain-based method. Gilmer et al. introduced Message Passing Neural Network (MPNN) [12] to solve various problems, provide the latest learning capabilities to address issues in urban traffic planning and capture spatial-temporal relationships of road networks. Since then, researchers have proposed numerous variants to overcome weaknesses in model training and enhance prediction capabilities with models to address challenges across heterogeneous datasets.

Feature Selection Graph Neural Network (FSGNN)

Feature Selection Graph Neural Network (FSGNN) is a simple architecture designed to enhance the performance of graph neural networks by separating feature aggregation and propagation learning. FSGNN effectively addresses over-reliance problem on neighbor node features by mapping each feature to a distinct linear layer thereby mitigating this dependence. Additionally, FSGNN introduces a novel weighting scheme known as soft-selection, which assigns higher weights to relevant features, thereby providing a robust and flexible approach compared to binary feature selection.

Meanwhile, FSGNN addresses over-smoothing node features that commonly occurs in deep neural networks. When features propagate through numerous stacked layers, they discern the significance of features and the overall fluency of the neural networks becomes difficult. While researchers often use skip connections or residual connections, these methods contend to identify features that are truly meaningful. FSGNN has introduced the Hop-Normalization strategy, involving the normalization of activations from different hops after linear transformation using row-wise L2-normalization as follows:

$$z_{ij} = \frac{z_{ij}}{\|z_i\|_2} \tag{2}$$

where z_i represents the activation function of ith row and z_{ij} represents single values.

The depth of GNN is often surfaced contrary to traditional neural networks is primarily due to the introduction of noise during information propagation with increasing layer depth. In FSGNN, it proposes 3-hop, 8-hop, 16-hop, and 32-hop GNNs with the same problem. It uses a new framework and feature selection scheme but the model's performance diminishes as the number of layers increases specifically exceeding 8 hops.

Directed Graph Neural Network (Dir-GNN)

Dir-GNN is an innovative general framework extending from MPNN on directed graphs by implementing separate aggregations of incoming and outgoing edges. The emergence of MPNN disrupts the assumption of most GNNs by assuming the input graph is undirected. Thus, Dir-GNN is introduced to extract valuable information from both directions, while other GNNs only propagate information in a single direction. It updates features by incorporating both directions, thereby becoming the latest model on datasets such as Squirrel, Chameleon, and arXiv-year etc.

1.3   Chaotic Neural Oscillator (CNO)

Chaotic Neural Oscillator (CNO) is artificial intelligence-based machine learning technology neural network that simulates the human brain's neural network. The neural network in the human brain is a complex system containing approximately 100 billion neurons. Initially, research focused on small neural networks, such as multiple neurons (2–3) described by differential or difference equations. The solutions to these equations may change over time given initial hyperparameters. However, when the neural network is large, the solution will randomly fall within a certain range instead of presenting a single or consistent solution, leading to chaotic behavior.

In 1990, Aihara et al. first proposed a Chaotic Neural Network (CNN) based on previous derivations and animal experiments [13]. Many biological experiments have confirmed that the neural system exhibits bifurcation, chaotic attractors and multiple wings. Consequently, CNN mimics this characteristic to become one of the most effective models for processing temporal information, where the system encodes and processes stimuli. Most CNNs are designed to emulate biological neural activities based on models like Hodgkin and Huxley [14], or Wilson and Cowan [15]. However, some of these models are either too oversimplified to simulate real-world problems accurately, or are too complex to be integrated into artificial neural networks.

Falcke et al. [16] proposed chaotic oscillators, leading to several subsequent extensions developed for three primary tasks: long-term memory, temporal information processing, and dynamic pattern retrieval. LEE [17] summarized the two primary focuses in neuroscience: *1) spiking neural dynamic behavior and 2) interactive triggering between excitatory and inhibitory neurons.* Thus, Lee oscillator [17] was proposed with four neurons to maintain continuous changes and transient progressive growth of a sigmoid curve with chaotic properties in the transition region, showing how Lee oscillator processes temporal information within neural oscillatory networks. Shi [18] has demonstrated it with a bifurcation transfer unit (BTU) that can be used as an activation function to enhance contemporary neural network efficiency and induce chaotic growth in neural dynamics. This paper will further explore the application of Lee oscillator for information processing and auto-association within deep graph neural networks.

Table 2. Eight sets of parameters verified repeatedly by the Lee oscillator

a_1	a_2	b_1	b_2	θ_E	θ_I	k	s
5.	5.	5.	5.	0.	0.	500	1
5.	5.	5.	5.	1.	1.	500	1
5.	5.	5.	−5.	0.	0.	500	1
5.	5.	5.	−5.	1.	1.	500	1
1.	1.	1.	−1.	0.	0.	500	1
1.	1.	1.	−1	1.	1.	500	1
5.	6.	1.	1.	0.	0.	500	1
5.	6.	1.	1.	1.	1.	500	1

The chaotic dynamics of the dynamic neural model with Lee oscillator's dynamic equations are expressed as follow:

$$E(t+1) = f[a_1 E(t) - a_2 I(t) + S(t) - \theta_E] \tag{3}$$

$$I(t+1) = f[b_1 E(t) - b_2 I(t) - \theta_I] \tag{4}$$

$$W(t+1) = f[S(t)] \tag{5}$$

$$Lee(t) = [E(t) - I(t)]e^{-kS^2(t)} + W(t) \tag{6}$$

In the equations, E, I, W, and Lee represent the excitatory, inhibitory, input, and output neurons, where a_i and b_i denote the weights for the excitatory and inhibitory neurons, and θ_E and θ_I are the respective threshold values. Additionally, S(t) represents the external input stimulus, and f is the sigmoid function given by:

$$f = \frac{1}{1+e^{-st}} \tag{7}$$

The Lee Oscillator displays a sigmoid-like function with chaotic properties within the transition region, making it an ideal candidate as a Chaotic Transfer Unit (CTU) for modeling chaotic systems on Complex Sets. Table 2 outlines eight sets of parameters verified repeatedly by Lee oscillator.

2 Methodology

Hypothetically, a neural network deepness determines a favorable effect due to the increased abstraction level of input features resulting from a greater number of hidden layers in the neural network. Surfaced neural networks represent features with a low degree of abstraction whereas deeper counterparts achieve a higher degree of feature abstraction. This scenario occurs only in specific tasks. Generally, adding multiple layers to a neural network leads to over-smoothing of features, rendering each layer indistinguishable from the others. Thus, FSGNN incorporated a feature selection scheme and tested it on 3-hop, 8-hop, 16-hop, and 32-hop networks with only 3-hop and 8-hop networks are resolved. It was observed that the mean classification accuracy decreased as the number of layers increased from 8 to 16, suggesting that FSGNN is suitable for application to surfaced graph neural networks and GNNs with fewer than 8 layers.

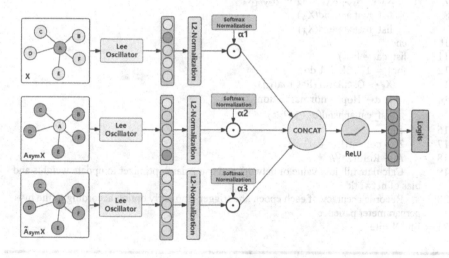

Fig. 1. Model Architecture of CNO_FSGNN

Consequently, the Lee Oscillator is introduced to the FSGNN to leverage the oscillator's remarkable memory-recalling scheme for improved information processing. Further, it also serves as an auto-associative function due to the progressive memory association scheme in the oscillator's structure, to capture pattern associations during chaotic memory association [19]. More importantly, Chaotic Neural Oscillators enable models to rapidly transit between different behavior patterns or synchronize relevant networks. The Lee Oscillator is placed in front of a linear transformation for each feature as depicted in Fig. 1.

Table 3. Parameters of Lee Oscillator

a_1	a_2	b_1	b_2	θ_E	θ_I	k	s
5.0	5.0	1.0	1.0	0	0	500	5

Table 4. CNO_FSGNN Algorithm

Algorithm 1 CNO_FSGNN

Input: Features X; Adjacency matrix without self-loop A_{sym}; Adjacency matrix $\tilde{A}_{sym}$; number of hops H; weights W_l for each layer l
Output: Accuracy

1   Initialize weights and bias of network through hybrid searching, and pretrain parameters of Lee Oscillator
2   Load the training data, build adjacency and normalize features
3   While (training times < Epochs)
4        $\alpha \leftarrow softmax([1.0 \text{ for i in range}(H)])$
5        $X_A \leftarrow X; X_{\tilde{A}} \leftarrow X$
6   For k $= 1 \dots$ K do
7           $X_A \leftarrow A_{sym}X_A; \quad X_{\tilde{A}} \leftarrow \tilde{A}_{sym}X_{\tilde{A}}$
8   list_mat.append(X_A)
9   list_mat.append($X_{\tilde{A}}$)
10   end
11   list_cat = list()
12   for j $= 1 \dots 2K + 1$ do
13        $X_f \leftarrow$ Oscillator(list_mat[j])
14   Out$\leftarrow$ Hop $-$ normalization$(X_f W_j)$
15   list_cat.append($\alpha_j \odot Out$)
16   end
17    $Z \leftarrow$ concat(list_cat)
18    $A \leftarrow$ Relu(Z)W
19   Calculate nll_loss value of network and use Adam optimizer to update weights and bias of network
20   Records accuracy of each epoch and trigger stop early option according to the hyperparameter patience
21   End While

A graph is input in three formats: the node itself, neighboring nodes, and both, significantly alleviate the heterophily of graph datasets. Once each feature at different hops is processed by the Lee Oscillator, it introduces temporal patterns and transient chaotic characteristics. Subsequently, each feature is mapped through a single linear layer and can be distinctly learned by all features. To overcome over-smoothing caused by deep layers, a row-wise L2-normalization is designed. To identify useful features, a soft-selection strategy is incorporated into CNO_FSGNN instead of binary selection. Higher weights are assigned to features contributing to lower loss, and then constraints

α_i are imposed on the weights through softmax normalization, where $\sum_{i=1}^{K} \alpha = 1$. Finally, all features concatenated from different layers are transformed by ReLU, and the likelihood of each class is obtained from the transformation of a linear layer and a nonlinear activation function. The CNO_FSGNN algorithm is outlined in Table 4. An alternate set of parameters is used for experiment listed in Table 3 which are different from Table 2 and the f function has changed from a sigmoid to a tanh function as follows:

$$f = \frac{e^{st} - e^{-st}}{e^{st} + e^{-st}} \tag{8}$$

3 Experiment

CNO_FSGNN model is introduced by incorporating the Lee Oscillator into a feature selection graph neural network. Initially, the model's performance on heterogeneous datasets are evaluated in comparison to other GNNs for node classification tasks. Subsequently, the model's performance at different hops are compared and ablation studies are conducted to demonstrate the influence of Lee Oscillator's components and the feature selection scheme.

3.1 Datasets

The experiments are conducted on standard benchmark datasets: Cora, Citeseer, Pubmed, Wisconsin, Cornell, Texas, Chameleon, Squirrel, and Actor for evaluation. The first three are citation networks categorized as homophily datasets. Wisconsin, Cornell, Texas contain webpage linkage data, while Actor, Chameleon, and Squirrel belong to the Wikipedia dataset. The Actor dataset represents actor cooccurrences in Wikipedia pages, and the last two are related to topics of web pages in Wikipedia. The last six datasets are considered as heterophily datasets. These datasets statistics are listed in Table 5.

Table 5. Statistics of datasets

Datasets	# Nodes	# Edges	# Features	# Classes
Cora	2,708	5,429	1,433	7
Citeseer	3,327	4,732	3,703	6
Pubmed	19,717	44,338	500	3
Chameleon	2,277	36,101	2,325	4
Wisconsin	251	499	1,703	5
Texas	183	309	1,703	5
Cornell	183	295	1,703	5
Squirrel	5,201	198,353	2,089	5
Actor	7,600	26,659	932	5

There are several publicly available data splits taken from [20] are used, where the dataset splitting ratio is training: validation: testing = 3:1:1 for each class, amounting to a total of 10 random split files. Thus, model performances are compared by using the mean classification accuracy of 10 sub-datasets. All models are tested on the same dataset files.

3.2 Settings and Baselines

Further, a grid search on hyperparameters are conducted to determine the best model performance on hyperparameter settings as listed in Table 6. CNO_FSGNN uses 8-hop for accuracy performance comparing other GNNs with popular GCN architectures such as GCN, GAT, FSGNN, and Dir-GNN, which is a latest model on several heterogeneous datasets for experiment.

Table 6. Hyperparameter of CNO_FSGNN

DATASETS	WD_{sca}	LR_{sca}	WD_{fc1}	WD_{fc2}	LR_{fc}	Dropout	Hidden	agg_per
CORA	0.0005	0.02	0.0001	0.0001	0.01	0.7	128	sum
CITESEER	0.0005	0.02	0.0001	0.001	0.01	0.6	256	sum
PUBMED	0.0005	0.02	0.001	0.001	0.01	0.7	256	cat
CHAMELEON	0.1	0.04	0.0	0.0001	0.005	0.5	64	cat
WISCONSIN	0.0005	0.02	0.001	0.0	0.01	0.7	256	sum
TEXAS	0.0005	0.02	0.001	0.001	0.005	0.7	512	sum
CORNELL	0.0005	0.02	0.001	0.001	0.01	0.5	256	cat
SQUIRREL	0.1	0.002	0.0	0.0001	0.01	0.5	64	cat
ACTOR	0.01	0.02	0.0001	0.0001	0.005	0.5	64	cat

3.3 Results

CNO_FSGNN model's performance is evaluated and compared the mean classification accuracy with 6 baselines: GCN, GAT, Dir-GNN, FSGNN (3-hop), FSGNN (8-hop) and FSGNN (16-hop) as listed in Table 7.

The experiment results have showed that CNO_FSGNN demonstrated high accuracy at the majority of datasets and has achieved satisfactory classification performance on Squirrel dataset. CNO_FSGNN model's performance results at different hops are listed in Table 8. Model performance at different hops which indicates that the Lee Oscillator's ability to capture pattern associations and enhance information processing during chaotic memory association. FSGNN is designed to mitigate the over-smoothing that may occur in deep layers. However, it is noted that the model's learning ability diminishes as the hop count increases, its performance initially improves from 3-hop to 8-hop networks and then declines from 8-hop to 16-hop networks. On the contrary, a similar trend is

Table 7. Model performance compared with baselines

MODEL	DATASET								
	Cora	CiteSeer	PubMed	Chameleon	Wisconsin	Texas	Cornell	Squirrel	Actor
GCN	76.38 ±2.98	53.11 ±2.15	81.97 ±1.21	55.39 ±3.67	55.88 ±7.66	61.35 ±4.99	57.84 ±3.67	35.95 ±1.38	24.99 ±1.32
GAT	77.12 ±2.67	45.27 ±4.38	78.11 ±1.61	47.5 ±4.04	48.43 ±5.56	58.92 ±4.32	58.92 ±3.15	26.98 ±1.81	25.03 ±1.40
DIR-GNN	83.53 ±1.23	73.27 ±1.61	84.08 ±0.56	79.42 ±1.30	64.16 ±5.73	60.95 ±4.62	57.32 ±3.03	74.53 ±0.79	25.06 ±0.96
FSGNN (3-HOP)	87.73 ±1.36	77.19 ±1.35	89.73 ±0.39	78.14 ±1.25	88.43 ±3.22	87.30 ±5.55	87.03 ±5.77	73.48 ±2.13	35.67 ±0.69
FSGNN (8-HOP)	87.93 ±1.00	**77.40** ±**1.93**	89.75 ±0.39	78.27 ±1.28	87.84 ±3.37	**87.30** ±**5.28**	**87.84** ±**6.19**	74.10 ±1.89	35.75 ±0.96
FSGNN (16-HOP)	87.79 ±1.36	77.18 ±1.55	89.29 ±0.58	78.31 ±1.1	86.47 ±4.34	85.41 ±5.82	85.68 ±6.17	73.95 ±1.61	35.51 ±1.29
OURS	**88.07** ±**1.24**	77.36 ±1.61	**90.21** ±**0.51**	**79.67** ±**1.06**	**88.63** ±**3.01**	86.76 ±4.59	87.03 ±5.24	**75.90** ±**1.66**	**35.93** ±**0.79**

not observed in the CNO_FSGNN model. It also noted that pattern associations can be readily identified by Chaotic Neural Oscillators by integrating the Lee Oscillator to better process chaotic information, enabling CNO_FSGNN model to swiftly transition between different behavior patterns or synchronize relevant networks. For models with fewer than 8 hops, there is no substantial difference in running time between CNO_FSGNN model and FSGNN. However, as the model becomes deeper, specifically for models with 8 hops, the running time of CNO_FSGNN model is, faster than FSGNN on average. Additionally, unlike FSGNN, the mean classification accuracy of CNO_FSGNN model with 16 hops does not exhibit significant changes compared with the model with 8 hops and surpasses FSGNN with 16-hops on the majority of datasets.

3.4 Ablation Study

The findings in Table 9 present the results of the ablation study focusing on key components: Lee Oscillator and the feature selection scheme. In the absence of the feature selection framework, we replaced the FSGNN framework with popular GNN architectures GCN and GAT. It is evident that the FSGNN framework significantly enhances model performance, particularly on heterogeneous datasets, where the models lacking feature selection framework exhibited performance declines ranging from 15% to 54%.

Regarding the Lee oscillator functionality, models incorporating the Lee oscillator outperformed those without it on most datasets, with only three exceptions. The potential limitation of the Lee oscillator's functionality may be associated with the model's depth. During the ablation study, we employed 8-hop networks—a relatively shallow graph neural network—where the Lee oscillator significantly contributes to processing chaotic transient information. In contrast, for deeper graph neural networks, such as the 16-hop networks outlined in Table 8, the Lee oscillator effectively captures chaotic patterns, facilitates transitions between patterns, mitigates over-smoothing issues, and reduces model execution time.

Table 8. Model performance at different hops

MODEL	DATASET								
	CORA	CITESEER	PUBMED	CHAMELEON	WISCONSIN	TEXAS	CORNELL	SQUIRREL	ACTOR
FSGNN (3-HOP)	87.73 ±1.36	77.19 ±1.35	89.73 ±0.39	78.14 ±1.25	88.43 ±3.22	**87.30** ±5.55	87.03 ±5.77	73.48 ±2.13	35.67 ±0.69
OURS (3-hop)	**88.05** ±1.12	**77.36** ±1.61	**90.21** ±0.51	79.43 ±0.88	88.63 ±3.01	86.76 ±4.90	87.03 ±5.24	**74.82** ±1.90	35.78 ±0.79
FSGNN (8-HOP)	87.93 ±1.00	**77.40** ±1.93	89.75 ±0.39	78.27 ±1.28	87.84 ±3.37	**87.30** ±5.28	**87.84** ±6.19	74.10 ±1.89	35.75 ±0.96
Time(s)	187.44	165.00	219.31	**334.50**	158.54	169.51	192.53	764.81	**72.36**
OURS (8-HOP)	88.07 ±1.24	77.36 ±1.61	90.21 ±0.51	79.67 ±1.06	88.63 ±3.01	86.76 ±4.59	87.03 ±5.24	75.90 ±1.66	35.93 ±0.79
Time(s)	65.08	58.95	164.46	717.07	41.52	48.55	92.66	3025.22	315.45
FSGNN (16-HOP)	87.79 ±1.36	77.18 ±1.55	89.29 ±0.58	78.31 ±1.10	86.47 ±4.34	85.41 ±5.82	85.68 ±6.17	73.95 ±1.61	**35.51** ±1.29
Time(s)	329.79	360.95	432.19	**997.65**	335.34	400.71	250.64	**1163.08**	109.64
OURS (16-HOP)	88.07 ±1.2	77.36 ±1.61	90.21 ±0.51	79.32 ±1.44	88.63 ±3.01	86.76 ±4.90	87.03 ±5.24	75.92 ±1.95	34.92 ±0.95
Time(s)	76.23	53.63	163.35	2365.32	39.79	104.07	98.41	2378.07	55.69

Table 9. Ablation study

MODEL		DATASET								
		Cora	CiteSeer	PubMed	Chameleon	Wisconsin	Texas	Cornell	Squirrel	Actor
OURS		88.07 ±1.24	77.36 ±1.61	90.21 ±0.51	79.67 ±1.06	88.63 ±3.01	86.76 ±4.59	87.03 ±5.24	75.90 ±1.66	35.93 ±0.79
Without Lee Oscillator		87.93 ±1.00	77.40 ±1.93	89.75 ±0.39	78.27 ±1.28	87.84 ±3.37	87.30 ±5.28	87.84 ±6.19	74.10 ±1.89	35.75 ±0.96
Without Feature Selection Framework	GAT	81.45 ±1.91	65.15 ±2.19	68.11 ±1.61	51.62 ±2.04	56.22 ±4.15	61.08 ±5.95	56.49 ±3.07	32.82 ±1.16	24.46 ±0.63
	GCN	80.12 ±1.33	63.30 ±2.68	81.66 ±0.74	60.57 ±2.70	58.24 ±5.26	62.16 ±7.35	51.62 ±5.73	63.30 ±2.68	23.76 ±0.93

4 Conclusion

This paper has introduced a CNO_FSGNN model for the node classification task integrating the Lee Oscillator with a simple feature selection framework. The graph is initially transformed into three formats for each node: a single node, its neighbor nodes, and both. This provides three data modes that can be chosen based on the homophily and heterophily of the dataset. These nodes are then input to the Lee oscillator for transient information processing, capturing chaotic patterns and characteristics. Subsequently, each feature is separately mapped through a single linear layer and normalized to overcome over-smoothing produced by deep layers. Generally, higher weights are assigned to features contributing to lower loss. In the CNO_FSGNN, a constraint α_i is added to the weights through softmax normalization, where $\sum_{i=1}^{K} \alpha = 1$. Finally, two activation functions and one linear layer are concatenated and used to obtain the final likelihood. The experiment results showed that CNO_FSGNN model outperformed other models on six datasets and achieved satisfactory results on the Squirrel dataset. Additionally, CNO_FSGNN model has reduced running time and maintained performance with increasing model depth compared with the original framework FSGNN.

Acknowledgements. This paper was supported in part by the Guangdong Provincial Key Laboratory IRADS (2022B1212010006, R0400001-22), and Guangdong Province F1 project grant UICR0400050-21CTL.

References

1. Kipf, T.N., Welling, M.: Semi-supervised classification with graph convolutional networks. In: Proceedings of the 4th International Conference on Learning Representations (2016)
2. Hamilton, W., Ying, Z., Leskovec, J.: Inductive representation learning on large graphs. In: Proceedings of the 31st Conference on Neural Information Processing Systems, pp. 1024–1034 (2017)
3. Veličković, P., Cucurull, G., Casanova, A., Romero, A., Lio, P., Bengio, Y.: Graph attention networks. In: Proceedings of the 5th International Conference on Learning Representations (2017)
4. Thekumparampil, K.K., Wang, C., Oh, S., Li, L.J.: Attention based graph neural network for semi-supervised learning (2018). arXiv:1803.03735
5. Kipf, T.N., Welling, M.: Variational graph auto-encoders (2016). arXiv:1611.07308
6. Veličković, P., Fedus, W., Hamilton, W.L., Liò, P., Bengio, Y., Hjelm, R.D.: Deep graph infomax. In: Proceedings of the 6th International Conference on Learning Representations (2018)
7. Wang, Y., Hooi, B., Liu, Y., Shah, N.: Graph explicit neural networks: explicitly encoding graphs for efficient and accurate inference. In: Proceedings of the Sixteenth ACM International Conference on Web Search and Data Mining, February 2023. ACM, Singapore, pp. 348–356 (2023). https://doi.org/10.1145/3539597.3570388
8. Duong, C.T., Hoang, T.D., Dang, H.T.H., Nguyen, Q.V.H., Aberer, K.: On node features for graph neural networks. arXiv preprint arXiv:1911.08795 (2019)
9. Fu, X., King, I.: FedHGN: a federated framework for heterogeneous graph neural networks. In: International Joint Conference on Artificial Intelligence (2023)
10. Gori, M., Monfardini, G., Scarselli, F., A new model for learning in graph domains. In: 2005 IEEE International Joint Conference on Neural Networks, 2005. IJCNN'05. Proceedings, vol. 2, pp. 729–734. IEEE (2005)
11. Bruna, J., et al.: Spectral networks and locally connected networks on graphs. CoRR abs/1312.6203 (2013). n. pag
12. Gilmer, J., et al.: Neural message passing for quantum chemistry. In: International Conference on Machine Learning (2017)
13. Aihara, K., et al.: Chaotic neural networks. Phys. Lett. A **144**, 333–340 (1990)
14. Goldwyn, J.H., Shea-Brown, E.T., The what and where of adding channel noise to the Hodgkin-Huxley equations. PLoS Comput. Biol. **7** (2011)
15. Wilson, H.R., Cowan, J.D.: Excitatory and inhibitory interactions in localized populations. Biophys. J. **12**, 1–24 (1972)
16. Falcke, M., Huerta, R., Rabinovich, M., et al.: Modeling observed chaotic oscillations in bursting neurons: the role of calcium dynamics and IP3. Biol. Cybern. **82**, 517–527 (2000). https://doi.org/10.1007/s004220050604

17. Lee, R.S.T.: A transient-chaotic autoassociative network (TCAN) based on Lee oscilla-
    tors. IEEE Trans. Neural Netw. **15**(5), 1228–1243 (2004). https://doi.org/10.1109/TNN.2004.
    832729
18. Lee, R.S.T.: Lee-associator—a chaotic auto-associative network for progressive memory
    recalling. Neural Netw. **19**(5), 644–666 (2006). https://doi.org/10.1016/j.neunet.2005.08.017
19. Shi, N., Chen, Z., Chen, L., Lee, R.S.T.: CNO-LSTM: a chaotic neural oscillatory long short-
    term memory model for text classification. IEEE Access **10**, 129564–129579 (2022). https://
    doi.org/10.1109/ACCESS.2022.3228600
20. Pei, H., et al.: Geom-GCN: geometric graph convolutional networks. In: ICLR (2020)

Adversarial Learning of Group and Individual Fair Representations

Hao Liu[✉] [iD] and Raymond Chi-Wing Wong[iD]

The Hong Kong University of Science and Technology, Kowloon, Hong Kong
{hliubs,raywong}@cse.ust.hk

Abstract. Fairness is increasingly becoming an important issue in machine learning. Representation learning is a popular approach recently that aims at mitigating discrimination by generating representation on the historical data so that further predictive analysis conducted on the representation is fair. Inspired by this approach, we propose a novel structure, called GIFair, for generating a representation that can simultaneously reconcile utility with both group and individual fairness, compared with most relevant studies that only focus on group fairness. Due to the conflict of the two fairness targets, we need to trade group fairness off against individual fairness in addition to considering the utility of classifiers. To achieve an optimized trade-off performance, we include a focal loss function so that all the targets can receive more balanced attention. Experiments conducted on three real datasets show that GIFair can achieve a better utility-fairness trade-off compared with existing models.

Keywords: Fairness · Adversarial Learning · Learning Representation

1 Introduction

Fairness is increasingly becoming an important issue in machine learning. Many studies have shown that using unfair historical datasets that are biased against some groups of people to train accurate machine learning models for decision-making can lead to discrimination of those groups. We refer to groups that are often discriminated against as *protected groups* (e.g., women and African-Americans), and the corresponding attributes that define them as *protected attributes* (e.g., gender and race). For instance, when evaluating loan applications, a bank officer may use applicant information such as age, gender, and credit history to determine creditworthiness, leading to a lower likelihood of approval for applications from women [1]. Motivated by this, we want to propose a fair classification model to help alleviate discrimination in decision-making systems.

To assess the fairness of various classification models, many fairness notions have been proposed and most of them can be divided into *group fairness* [2,3] and *individual fairness* [4,5]. Group fairness requires treating different groups defined by protected attributes equally. Individual fairness requires *similar* individuals should be treated *similarly* by classifiers. Based on these fairness notions, many

© The Author(s), under exclusive license to Springer Nature Singapore Pte Ltd. 2024
D.-N. Yang et al. (Eds.): PAKDD 2024, LNAI 14645, pp. 181–193, 2024.
https://doi.org/10.1007/978-981-97-2242-6_15

approaches [6–9] have been proposed to solve the fair classification problem. Among them, representation learning [8,9] is a common approach, which transforms the original datasets into new representations that obfuscate the information about the protected attributes in the representations. Then, different groups have similar representations and will be treated similarly by any classifier, which satisfies group fairness. However, most existing studies only focus on group fairness, which may harm individual fairness and create discrimination. For example, in hiring decision, some unqualified people in the protected group (e.g., females) are interviewed deliberately [7], which is, in fact, biased against the individuals in the unprotected group. Individual fairness can alleviate such discrimination by ensuring that individuals who are similar in terms of attributes/background (e.g., similar academic experience) are treated similarly.

Only a handful of studies on fair classification [7,10] consider both individual and group fairness in their designs. In LFR [7], a loss function is defined that combines accuracy, group fairness and individual fairness. However, the three terms are trained at the same time, but not well reconciled *at the same time*. Besides, the loss function in LFR enforces fairness *indirectly*, so the fairness performance of learned representation is not guaranteed. DualFair [10] explores an alternative formation of individual fairness called *counterfactual fairness* [5] which grant similar treatment for counterfactual samples, where a counterfactual sample of an individual x is defined to be a "synthetic" individual who is similar to x *except for* the protected attribute. However, counterfactual fairness cannot guarantee general and stronger individual-level fairness for *any* two similar individuals.

We mainly focus on reconciling accuracy and two types of fairness (i.e., group fairness and individual fairness). Due to the conflict between group and individual fairness [11], we aim to achieve a better *trade-off* between them. To solve this problem, we propose an approach called **GIFair** (for group and individual fair representations), which transforms the original dataset into a *fair representation*. To reconcile group and individual fairness in the learned representation, we use two adversaries, one for group fairness and the other for individual fairness, instead of using only one adversary in the related studies. For group (fairness) adversary, we apply an effective formation of target function, which better guarantees group fairness. For individual (fairness) adversary, we form its target function with a metric called *yNN* based on k-nearest neighbors, which addresses the explicit individual fairness of treating any similar individuals equally. We propose a well-designed training algorithm to reconcile all concepts in our structure. Compared to the existing adversarial learning studies that only consider accuracy and group fairness, we handle a more complicated problem with a better performance, e.g., we achieve a 3% improvement in accuracy and 40% improvement in group fairness on dataset COMPAS compared with baselines.

To further optimize GIFair, we propose a focal loss function so that the three targets receive more balanced attention. GIFair with focal loss function obtains even better trade-off performance (e.g., 30% improvement of group fairness under the same level of individual fairness) compared with the original GIFair.

We conduct extensive experiments on three real datasets to study the trade-off among accuracy, group fairness and individual fairness. The results show that compared with many baseline algorithms, GIFair can achieve better performance, e.g., GIFair can achieve up to 2% improvement in accuracy under the same individual fairness performance on dataset Adult.

The contributions of our work are as follows. (1) We design a novel structure of adversarial representation learning with two adversaries for group fairness and individual fairness, respectively. (2) We design a training algorithm that can well reconcile the two adversaries in our structure. Ablation analysis is conducted to show its superiority. (3) We propose a focal loss function to ensure balanced attention of two types of fairness and accuracy. (4) The experiments conducted on 3 real datasets show that GIFair can reconcile good fairness with high accuracy.

The rest of this paper is organized as follows. Section 2 reviews related work. Section 3 presents the preliminaries. Section 4 describes our solution to the fair classification problem. Then, Sect. 5 reports experimental results and our analysis. Finally, Sect. 6 concludes this paper.

2   Related Work

Most machine learning studies about fairness can be classified into *pre-processing*, *in-processing* and *post-processing*. *Pre-processing* approaches directly modify data-sets to remove discrimination [6]. *In-processing* approaches modify the classifier to improve its fairness performance [7,12]. *Post-processing* approaches directly change the predicted outcomes of the learned predictors [2].

Learning Fair Representations. Recently, *fair representation learning* [7] attracts great attention in fair machine learning, which is to learn a debiased representation so that the downstream tasks could satisfy fairness requirements. In this branch, iFair [12] considers a probabilistic mapping to the representation space to address both accuracy and individual fairness (which uses a similar fairness notion as in this paper) but fails to address group fairness as we do. DualFair [10] applies a contrastive self-supervised learning approach to obtain the representation satisfying both group fairness and counterfactual fairness. However, although LFR [7] and DualFair [10] set both group and individual fairness as targets, as mentioned in Sect. 1, they are not effective enough to address individual fairness. LFR [7] uses an *indirect* individual fairness formation that minimizes the deviation between each data point and its representation, and thus the individual fairness of the representation relies on the individual fairness of the original dataset, which is not always ensured. DualFair [10] focuses on counterfactual fairness but does not ensure individually fair results for *any* two similar samples. In comparison, we form our individual fairness notion based on an explicit target of treating any similar individuals equally.

Among those approaches, adversarial representation learning has been broadly explored. ALFR [8] provides a framework of learning representations that minimize the performance of the adversary which predicts the protected

attribute of the representation. LAFTR [13] follows this framework to explore
adversarial learning as a method of obtaining a representation to mitigate unfair
prediction outcomes. IPM [14] proposes the integral probability metric adopted
in an adversary such that a good theoretical guarantee on group fairness is
obtained. However, all these existing methods focuses on group fairness only,
while our method GIFair (following the idea of adversarial representation learning) reconciles both group and individual fairness by a novel structure of two
adversaries.

3   Preliminaries

In the fair classification problem, we are given a dataset D containing N data
points. The i-th data point in D, denoted by x_i where $i \in [1, N]$, has a list X
of d features, i.e., $x_i \in \mathbb{R}^d$. Each x_i is also associated with an outcome attribute
Y for classification and a protected attribute A representing the group membership (e.g., gender). Following [7,8,13], we assume binary outcome attribute
and binary protected attribute (i.e., $Y \in \{0,1\}$ and $A \in \{0,1\}$). We assume
that values 1 and 0 represent the protected group (e.g., females) and the unprotected group (e.g., males), respectively. We thus denote D_1 and D_0 to be the
subsets of D containing all data points in the protected and unprotected group,
respectively.

The basic goal of the fair classification problem is to obtain a classifier η
that can predict an outcome $\eta(x_i) \in \{0,1\}$ of data point x_i for $i \in [1, N]$ in the
dataset D such that some fairness criteria are satisfied.

To achieve fairness, we follow common approaches to optimize some fairness
metrics. For group fairness, we use two popular metrics, the *demographic parity
gap* [3] and *equalized odd distance* [2]. Given a classifier η and dataset D, the
demographic parity gap of η for D, denoted by $\Delta DP_D(\eta)$, is defined to be the
absolute difference between the positive rate of D_0 and the positive rate of D_1
Namely,

$$\Delta DP_D(\eta) = |\frac{1}{|D_1|}\sum_{x_i \in D_1} \eta(x_i) - \frac{1}{|D_0|}\sum_{x_j \in D_0} \eta(x_j)| \tag{1}$$

The *equalized odd distance* of η for D, denoted by $\Delta EO_D(\eta)$, is defined to be
the sum of the absolute difference between the true positive rate (TPR) of D_0
and the TPR of D_1, and the absolute difference between the false positive rate
FPR of D_0 and the FPR of D_1. In this paper, we use $\Delta DP_D(\eta)$ as our major
group fairness metric, but we also test $\Delta EO_D(\eta)$ as an alternative metric. For
both $\Delta DP_D(\eta)$ and $\Delta EO_D(\eta)$, smaller values indicate better group fairness.

Individual fairness is another perspective of fairness, which requires that two
similar individuals (i.e., data points) should be treated similarly in terms of the
predicted outcome [4]. Consider a data point x_i. Let $\mathcal{N}_D^k(x_i)$ denote the set of
k nearest neighbors of x_i in D, where k is a positive integer. Note that $\mathcal{N}_D^k(x_i)$
is computed based on the features X only (but not the protect attribute A).
This is because the similarity of two individuals should be independent to A.

To quantify the individual fairness, we adapt a commonly applied metric called *yNN* [7], which measures the consistency of the prediction results among similar data points. Specifically, given a classifier η, a positive integer k and dataset D, the yNN of η for D and k, denoted by $\Delta yNN_{D,k}(\eta)$, is defined to be

$$\Delta yNN_{D,k}(\eta) = 1 - \frac{\sum_{x_i \in D} \sum_{x_j \in \mathcal{N}_D^k(x_i)} |\eta(x_i) - \eta(x_j)|}{k \cdot N} \tag{2}$$

which captures the average difference between the predicted outcome of a data point x_i and that of a nearest neighbor x_j of x_i. This difference is 0 if x_i and x_j have the same predicted outcome and 1 otherwise. According to Eq. 2, larger $\Delta yNN_{D,k}(\eta)$ indicates better individual fairness.

Moreover, we introduce the basic concept of generative adversarial network (GAN) [15]. It has two components, namely a *generator* G and a *discriminator* C. G aims at deceiving C by constructing synthetic data $G(z)$ that could match the real data distribution P_{data}. C aims at distinguishing whether the data comes from P_{data} or $G(z)$. Both components improve their ability through learning. That is, G is trained to generate $G(z)$ that cannot be distinguished from the real data, while C is trained to identify the outcome of $G(z)$ more accurately.

4 Methodology

4.1 Problem Statement

In this work, we follow adversarial representation learning to tackle the fair classification problem, which is to learn a representation Z by re-constructing the features X in the original dataset D. The learning goal is that any classifier trained on the representation Z is accurate to predict the outcome attribute Y and is also fair in terms of both group fairness and individual fairness.

Due to the conflict of group and individual fairness [11], the two fairness goals could not be satisfied simultaneously in most cases (an extended analysis on their incompatibility is given in our supplementary material [16]). We thus set our optimization goal of classifier η such that a balanced trade-off can be obtained among accuracy, group fairness and individual fairness.

4.2 Model

First proposed by [8], plenty of existing studies follow a general framework of adversarial representation learning for fair classification. This framework uses an *encoder* as the *generator* to generate the representation Z from X which aims to obfuscate the group membership and thus ensure group fairness. To achieve that, an *adversary* as the *discriminator* is set up to identify the group of the generated representation Z. By adversarial learning [15], while the adversary improves its ability of group identification, the encoder is also well trained to generate group-obfuscated representation Z. Finally, a (group) fair representation is obtained.

However, this framework so far only addresses group fairness. It remains unsolved how to accommodate individual fairness into this framework.

With this motivation, we propose our model called **GIFair** (**G**roup **I**ndividual **Fair**). As illustrated in Fig. 1, GIFair consists of an encoder f, a classifier g and *two* adversaries, namely *group (fairness) adversary* h_1 and *individual (fairness) adversary* h_2. GIFair seeks to learn a representation Z by reconstructing the original features X of each data point in D using the encoder f. Classifier g, which predicts the outcome Y from representation Z, seeks to preserve the prediction accuracy. In addition, GIFair aims at achieving group fairness by the group adversary h_1 and individual fairness by the individual adversary h_2. Next, we introduce the details of all components and how they interact with each other.

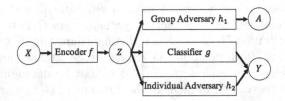

Fig. 1. Structure of GIFair

Encoder. An encoder $f\colon \mathbb{R}^d \to \mathbb{R}^{d'}$ maps a data point x_i into a d'-dimensional vector, denoted by $z_i = f(x_i)$. The representation Z of the original dataset is formed by encoding all data points in D, namely, $Z = f(X) = \{f(x_i)|x_i \in D\}$.

Classifier. We use a classifier $g\colon \mathbb{R}^{d'} \to \{0,1\}$ to predict the outcome $g(z_i)$ of each z_i in Z and form the outcome set $g(Z) = g(f(X))$. To preserve utility, we minimize a suitable classification loss function (i.e., cross-entropy) between $g(f(X))$ and Y, denoted by $L_c(g(f(X)), Y)$ (written as L_c for simplicity).

Group Adversary. To achieve group fairness of Z, the group adversary $h_1\colon \mathbb{R}^{d'} \to \{0,1\}$ is included. Given a representation $z_i = f(x_i) \in Z$, h_1 generates a value $h_1(z_i) \in \{0,1\}$, which is the predicted group of z_i. Thus, we denote the set of predicted groups of Z to be $h_1(Z) = h_1(f(X))$. The objective of h_1 is to *differentiate* representations in different groups. Note that this objective *differs* from making any $h_1(z_i)$ exactly equal to the protected attribute of x_i. Instead, h_1 is only interested in giving different group labels to two representations in different groups. It is thus interesting to observe that if any $h_1(z_i)$ is wrongly predicted, h_1 also has strong differentiation performance. Therefore, following [7], we form the group (fairness) loss function on $h_1(f(X))$ and A, denoted by $L_g(h_1(f(X)), A)$ (written as L_g for simplicity), as follows.

$$L_g = L_g(h_1(f(X)), A) = |\frac{\sum\limits_{x_i \in D_0} h_1(f(x_i))}{|D_0|} - \frac{\sum\limits_{x_j \in D_1} h_1(f(x_j))}{|D_1|}| \tag{3}$$

Here, higher L_g indicates either predicting more items in D_0 as 1 and more items in D_1 as 0 (mostly wrong), or predicting more items in D_0 as 0 and more items in D_1 as 1 (mostly correct), both leading to better differentiation of representations from different groups. Thus, h_1 is trained to *maximize* L_g.

Individual Adversary. Individual fairness requires that individuals who are similar on their features X should be *indistinguishable* in terms of the predicted outcome of their representation Z (i.e., to be given the same predicted outcome Y). To achieve individual fairness in Z, another adversary $h_2 \colon \mathbb{R}^{d'} \to \{0,1\}$ is included. Specifically, for each representation $z_i = f(x_i) \in D$, h_2 predicts an outcome $h_2(z_i) \in \{0,1\}$ (of attribute Y) such that, for another representation $z_j = f(x_j)$, if x_i and x_j are similar (e.g., x_j is a nearest neighbor of x_i), the predicted outcome of z_j should be *distinguishable* with the predicted outcome of z_i, i.e., $h_2(z_j) \neq h_2(z_i)$. We formalize the individual (fairness) loss function on $h_2(f(X))$, denoted by $L_i(h_2(f(X)))$ (written as L_i for simplicity), as follows to capture the above objective, where a conceptual notation $h_2(Z) = h_2(f(X))$ is also used here to denote the process of generating all $h_2(z_i)$ for $z_i \in Z$.

$$L_i = L_i(h_2(f(X))) = \frac{\displaystyle\sum_{x_i \in D}\sum_{x_j \in \mathcal{N}_D^k(x_i)} |h_2(f(x_i)) - h_2(f(x_j))|}{k \cdot N} \tag{4}$$

When L_i is larger, $h_2(f(x_i)) \neq h_2(f(x_j))$ holds for more pairs of similar data points x_i and x_j in D. Thus, the goal of adversary h_2 is to *maximize* L_i so that h_2 is more capable of distinguishing similar data points.

To find the k nearest neighbors of a data point in D, a suitable similarity metric is needed. In this work, we choose the Euclidean distance (a commonly applied metric) on all features X as the similarity metric, but not the representations $f(X)$ for distance computation. This is to ensure that we find the data points that are "really" similar to their original features. Note that another similarity metric (that could be more suitable for a specific dataset) also works, which only influences the result of finding the nearest neighbors.

Total Loss. The total loss function $L(f, g, h_1, h_2)$ is formalized to be the weighted sum of the classification loss function, group loss function and individual loss function based on three coefficients α, β and δ, respectively.

$$L(f, g, h_1, h_2) = \alpha \cdot L_c + \beta \cdot L_g + \delta \cdot L_i \tag{5}$$

The coefficients α, β and δ provide a trade-off among accuracy, group fairness and individual fairness. We train our model with a min-max optimization: $\min_{f,g} \max_{h_1,h_2} \mathbb{E}_{X,A,Y}[L(f, g, h_1, h_2)]$ following adversarial learning [15].

Training Algorithm. We train our model in a number of epochs. In each epoch, we first sample a mini-batch D' from the dataset D. Next, we do the training for this epoch in 3 steps. In Step 1 and Step 2, we freeze the parameters of f and g, and then, we train the group adversary h_1 and individual adversary h_2, respectively, such that their objective functions are maximized. Finally, in Step 3, f and g are trained such that the total loss function $L(f, g, h_1, h_2)$ on D'

is minimized. In this way, the group fairness and individual fairness can both be improved in the generated representation Z, and meanwhile the accuracy of classifier g, which is encoded in the total loss function, is also improved.

Although it is not theoretical guaranteed that the adversarial learning will always converge, several heuristics that we apply could encourage its convergence practically including training sufficient epochs and using mini-batches [17,18]. In our algorithm, we aim at optimizing the group fairness and the individual fairness, and finally, our results in Sect. 5 show the balanced trade-off between the two targets (e.g., 30% improvement of group fairness under the same level of individual fairness). This verifies the practical convergence of our algorithm.

4.3   Theoretical Properties of Loss Functions

We give the theoretical properties to show the effectiveness of using our loss functions to ensure fairness. First, we show that the optimal value of L_g can upper-bound the demographic parity gap of any classifier trained on representation Z. In the supplementary material [16], we provide the proofs.

Lemma 1. *For a group adversary h_1, the optimal value of $L_g(h_1(Z), A)$ (denoted by $L_g(h_1^*(Z), A)$) is at least the demographic parity gap of any classifier η on representation Z, i.e., $L_g(h_1^*(Z), A) \geq \Delta DP_Z(\eta)$.*

In Lemma 1, we connect $L_g(h_1(Z), A)$ with $\Delta DP_Z(\eta)$ (i.e., the performance of Z), and thus we can obtain the worst $\Delta DP_Z(\eta)$ performance of any classifier trained on Z given the optimal group adversary h_1^*. This shows the effectiveness of using $L_g(h_1(f(X)), A)$ as the group loss function.

Analogously, we want to show the effectiveness of the individual loss function $L_i(h_2(Z))$. We consider the yNN "variant" of a classifier η trained on representation Z, denoted by $\Delta yNN'_{Z,k}(\eta)$, which is the same as the yNN metric except that the k-NN of any sample $z_i(= f(x_i))$ for $z_i \in Z$ are defined based on the original dataset D (namely, $\mathcal{N}_Z^k(z_i) = \{f(x_j)|x_j \in \mathcal{N}_D^k(x_i)\}$). This is to ensure that the measurement is based on the "real" similarity relationships of the data points. Lemma 2 shows that, for any classifier η trained on Z, $\Delta yNN'_{Z,k}(\eta)$ is lower-bounded by a value related to the optimal value of $L_i(h_2(Z))$.

Lemma 2. *For an individual adversary h_2 and any classifier η on representation Z, $\Delta yNN'_{Z,k}(\eta) \geq 1 - L_i(h_2^*(Z))$, where $L_i(h_2^*(Z))$ denotes the optimal value of $L_i(h_2(Z))$.*

In Lemma 2, we can also obtain the worst $\Delta yNN'_{Z,k}(\eta)$ performance given the optimal individual adversary h_2^*, showing that our individual loss L_i is effective.

4.4   Optimization with Focal Loss

To this end, we have formed our GIFair structure. However, we notice that the ranges of the three losses in Eq. 5 have large differences (e.g., the value of L_i

is much smaller than the other two losses). Since our target is to minimize the total loss, the loss with a smaller value receives less attention.

To solve this issue, we exploit the focal loss function [19] to alleviate the imbalance among the three losses. Consider an item with two possible outcomes 1 and 0. Let p be the estimated probability with outcome 1. We define a variable p_t to be p if the *true* outcome of this item is 1 and to be $1 - p$ otherwise. The formulation of *Focal Loss function* is $FL(p_t) = -(1 - p_t)^\gamma \cdot \log(p_t)$, where $\gamma \geq 0$ is a focusing parameter and $(1 - p_t)^\gamma$ is regarded as a *weight* term. We notice that if the value of p_t is high, its weight $(1 - p_t)^\gamma$ will be low. Thus, less (resp. more) weight is given to an item with higher (resp. lower) p_t value. Based on this idea, we re-design our total loss function by adjusting the weights of the three terms:

$$L(f, g, h_1, h_2) = (1 - L_c)^\gamma \cdot L_c + (1 - L_g)^\gamma \cdot L_g + (1 - L_i)^\gamma \cdot L_i \qquad (6)$$

In this equation, if the value of one loss is small (resp. large), its weight is large (resp. small). In this way, we can balance the values of the three losses with their weights. Each loss could receive similar attention during training.

Table 1. Statistics of Datasets

Dataset	Train/Test	Protected Attribute ($A = 1/0$)	$P(A = 1)$	$P(Y = 0)$
COMPAS	4,321/1,851	*race* (African-Americans/other races)	0.34	0.54
Adult	30,162/15,060	*gender* (females/males)	0.33	0.75
German	700/300	*age* (the aged/the young)	0.27	0.7

5 Experiments and Analysis

In this section, we conducted extensive experiments to evaluate the effectiveness of GIFair. We used three common real datasets: COMPAS, Adult and German. Table 1 lists the statistics. **COMPAS** [20] is used to predict whether a criminal defendant will recidivate ($Y = 1$) or not ($Y = 0$). **Adult** [21] is used to predict each person's income ($Y = 1$ if income > 50K/y, and $Y = 0$ otherwise). **German** [22] classifies each individual as good ($Y = 0$) or bad ($Y = 1$) credit risks.

We selected LAFTR [13], LFR [7], iFair [12] and DualFair [10] as baselines. We also include UNFAIR, which is a normal classification algorithm that does not consider fairness. If the original loss function (i.e., Eq. 5) is used, our algorithm is denoted as GIFair, while GIFair-focal denotes our algorithm on the focal loss function (i.e., Eq. 6). We implemented all algorithms in Python.

We focus on the classification accuracy, group fairness and individual fairness. (1) For accuracy, we use *accuracy* (denoting ACC) which is defined to be 1 minus the average difference between the outcome and the predicted outcome of all data points, and *F-1 score* (denoting $F1$) which is defined to be the harmonic mean of the precision and the recall of a classifier. (2) For group fairness, we adopt the two metrics as introduced in Sect. 3, namely *demographic parity gap* (denoting ΔDP) and *equalized odds distance* (denoting ΔEO). (3) For individual fairness, we use *yNN*, denoted by ΔyNN (introduced in Eq. 2).

We varied β and δ in GIFair from 0.1 to 20, while α is fixed to 1. For baselines, we also changed their coefficients from 0.1 to 20. For GIFair-focal, we varied γ from 0.05 to 5. By default, we set k to 10 when computing the k-nearest neighbors for yNN according to [12]. For each coefficient setting and each model, we trained it 5 times (using different random seeds) and obtained the mean performance on the test datasets. The implementation details of algorithms are included into the supplementary materials [16]. In the following, we show the experimental results.

Overall Comparison. Due to lack of space, we show the overall comparison of our GIFair algorithm with all baselines for the best value achieved for each measurement in [16]. GIFair outperforms all the baselines on most metrics.

Trade-off Studies. We studied the trade-off between any two terms from accuracy, group fairness and individual fairness. We compared with the baselines that also study the trade-off. To show which algorithm performs better under multi-metrics, we plotted the Pareto front curves (widely used in existing trade-off studies [12,13], which only shows the dominating points of multi-metrics for better illustration). We also include baseline UNFAIR without weights for trading-off (thus shown as a star mark). Since the group fairness metrics are favored with smaller values, we plot 1 minus the group fairness metric, so that for each figure, the right-top points (high values along each axis) are preferable. We show the results on dataset German, while we obtain similar results for the other two datasets, which are reported in our supplementary material [16].

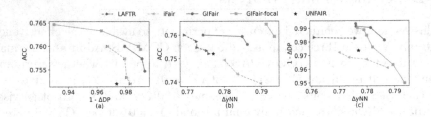

Fig. 2. Trade-off Curves on Dataset German

Accuracy and Group Fairness. Figure 2(a) shows the trade-off between accuracy and group fairness, with the default metric ACC and ΔDP, respectively. Compared with baselines, both GIFair and GIFair-focal have superior trading-off ability by reaching the most upper-right location. More closely, at the same level of accuracy ($ACC \approx 0.76$), the best ΔDP that baselines could achieve is at least

0.03 (i.e., $1 - \Delta DP < 0.97$), while the ΔDP values of our GIFair and GIFair-focal are around 0.02 and 0.01, improving the best baseline by 33% and 67%, respectively. For the same level of group fairness achieved (e.g., $\Delta DP \approx 0.02$), our GIFair and GIFair-focal obtain slightly better accuracy. The above indicates our better reconciliation between group fairness and accuracy compared with baselines, because we use an effective group fairness target, which ensures group fairness more easily without sacrificing accuracy too much. GIFair-focal could reach the highest ACC of around 0.765 but at a cost of sacrificing group fairness.

Accuracy and Individual Fairness. Figures 2(b) shows the trade-off between accuracy and individual fairness. GIFair and GIFair-focal still obtain the best trade-off. When ACC is fixed to around 0.76, the baseline with the best individual fairness has around 0.772 ΔyNN, while the ΔyNN of GIFair-focal reaches 0.792 with 2.6% improvement. Moreover, the baseline iFair could also obtain high ΔyNN of around 0.79 but with its ACC below 0.74, while our GIFair-focal keeps ACC above 0.76, which improves iFair by more than 3%. This similarly indicates that our algorithms better reconcile individual fairness and accuracy than iFair even though iFair has the same individual fairness target, because using adversarial learning could achieve the reconciliation more effectively.

Group Fairness and Individual Fairness. Our algorithms also obtain superior trade-off between the two types of fairness as shown in Fig. 2(c). GIFair-focal achieves the highest ΔyNN (0.794), since it uses the focal loss function to effectively give larger weight to individual fairness while down-weigh group fairness. GIFair could also obtain good individual fairness (e.g., $\Delta yNN = 0.786$), while its group fairness is only slightly downgraded (with $\Delta DP = 0.02$).

Ablation Studies. We conducted ablation studies for the two adversaries in GIFair with the following variants. (1) GIFair without group adversary h_1 (i.e., GIFair-w/o-h_1), by skipping Step 1 of training h_1. (2) GIFair without individual adversary h_2 (i.e., GIFair-w/o-h_2), by skipping Step 2 of training h_2. (3) GIFair without h_1 and h_2 (i.e., GIFair-w/o-h_1-h_2), by skipping both Step 1 and Step 2.

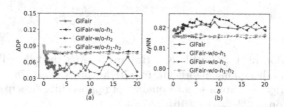

Fig. 3. Ablation Studies of GIFair on Dataset German

Figure 3(a) and (b) illustrate the ablation study results on dataset German. Without group adversary h_1, GIFair-w/o-h_1 has much larger ΔDP (i.e., worse group fairness) than the original GIFair. This verifies the effectiveness of improving group fairness using the group adversary. Similarly, GIFair has larger yNN

than GIFair-w/o-h_2, indicating that the individual adversary h_2 could effectively improve individual fairness. Without both adversaries, GIFair-w/o-h_1-h_2 obtains bad performance for both group and individual fairness.

Case Studies. We conducted case studies for the classification results regarding group and individual fairness. When only individual fairness is optimized (i.e., setting β to 0) for dataset COMPAS, we observe a representative result where 47% of the African-American group will recidivate, while this proportion for the other races is only 29%. When both group and individual fairness are optimized (i.e., setting all parameters to 1), the recidivation proportions among African-Americans and other races are predicted to be 40% and 38%, respectively, which is much fairer. Moreover, there exist some pairs of similar defendants who only have 1 day difference on the days between screening and arrest and have the same value for all other attributes. When only group fairness is optimized (i.e., setting δ to 0), we found that the number of these pairs of similar defendants that obtain different prediction results is 14. This number improves to only 1 when both group and individual fairness are optimized.

6   Conclusion

In this paper, we propose an adversarial learning structure, GIFair, with two adversaries for group fairness and individual fairness, respectively. With a designed training algorithm, GIFair can reconcile utility with group and individual fairness during generating a representation on the original dataset. We also propose a focal loss function that can better balance all the goals in GIFair. In our experiments on 3 real datasets, GIFair outperforms baselines with better fairness and higher accuracy. For future work, we would like to achieve a holistic optimization for utility and multiple fairness goals at the same time, and explore the problem on intersectional or unknown group.

Acknowledgements. We greatly thank Zheng Zhang for his contribution on this paper.

References

1. Zehlike, M., Bonchi, F., Castillo, C., Hajian, S., Megahed, M., Baeza-Yates, R.: Fa*ir: a fair top-k ranking algorithm. In: CIKM, pp. 1569–1578 (2017)
2. Hardt, M., Price, E., Srebro, N.: Equality of opportunity in supervised learning. In: NeurIPS, pp. 3323–3331 (2016)
3. Kamiran, F., Calders, T.: Data preprocessing techniques for classification without discrimination. In: KAIS, vol. 33, pp. 1–33 (2011)
4. Dwork, C., Hardt, M., Pitassi, T., Reingold, O., Zemel, R.: Fairness through awareness. In: ITCS, pp. 214–226 (2012)
5. Kusner, M.J., Loftus, J., Russell, C., Silva, R.: Counterfactual fairness. In: Advances in Neural Information Processing Systems, vol. 30 (2017)
6. Salimi, B., Rodriguez, L., Howe, B., Suciu, D.: Interventional fairness: causal database repair for algorithmic fairness. In: SIGMOD, pp. 793–810 (2019)

7. Zemel, R., Wu, Y., Swersky, K., Pitassi, T., Dwork, C.: Learning fair representations. In: ICML, vol. 28, no. 3, pp. 325–333 (2013)
8. Edwards, H., Storkey, A.: Censoring representations with an adversary. In: ICLR (2016)
9. Zhao, H., Coston, A., Adel, T., Gordon, G.J.: Conditional learning of fair representations. In: ICLR (2020)
10. Han, S., et al.: Dualfair: fair representation learning at both group and individual levels via contrastive self-supervision, arXiv preprint arXiv:2303.08403 (2023)
11. Binns, R.: On the apparent conflict between individual and group fairness. In: Proceedings of the Conference on Fairness, Accountability, and Transparency (2020)
12. Lahoti, P., Gummadi, K.P., Weikum, G.: ifair: learning individually fair data representations for algorithmic decision making. In: ICDE, pp. 1334–1345 (2019)
13. Madras, D., Creager, E., Pitassi, T., Zemel, R.: Learning adversarially fair and transferable representations. In: ICML, vol. 80, pp. 3384–3393 (2018)
14. Kim, D., Kim, K., Kong, I., Ohn, I., Kim, Y.: Learning fair representation with a parametric integral probability metric, arXiv preprint arXiv:2202.02943 (2022)
15. Goodfellow, I.J., et al.: Generative adversarial nets. In: NeurIPS (2014)
16. Liu, H., Wong, R.C.-W.: Adversarial learning of group and individual fair representations (supplementary material) (2024). https://github.com/satansin/GIFair
17. Saxena, D., Cao, J.: Generative adversarial networks (GANs) challenges, solutions, and future directions. ACM Comput. Surv. **54**(3), 1–42 (2021)
18. Salimans, T., Goodfellow, I., Zaremba, W., Cheung, V., Radford, A., Chen, X.: Improved techniques for training GANs. In: Advances in Neural Information Processing Systems, vol. 29 (2016)
19. Lin, T.-Y., Goyal, P., Girshick, R.B., He, K., Dollár, P.: Focal loss for dense object detection. In: ICCV, pp. 2999–3007 (2017)
20. Angwin, J., Larson, J., Mattu, S., Kirchner, L.: Machine bias: risk assessments in criminal sentencing. ProPublica (2016). https://www.propublica.org/article/machine-bias-risk-assessments-in-criminal-sentencing
21. Becker, B., Kohavi, R.: Adult. UCI Machine Learning Repository (1996). https://doi.org/10.24432/C5XW20
22. Hofmann, H.: Statlog (German Credit Data). UCI Machine Learning Repository (1994). https://doi.org/10.24432/C5NC77

Class Ratio and Its Implications for Reproducibility and Performance in Record Linkage

Jeremy Foxcroft[1]([✉]) [iD], Peter Christen[2] [iD], and Luiza Antonie[1] [iD]

[1] School of Computer Science, Reynolds Building, University of Guelph, 474 Gordon Street, Guelph, ON N1G 1Y4, Canada
{jfoxcrof,lantonie}@uoguelph.ca
[2] Scottish Centre for Administrative Data Research, University of Edinburgh, Bayes Centre, 47 Potterrow, Edinburgh EH8 9BT, UK
peter.christen@ed.ac.uk

Abstract. Record linkage is the process of identifying and matching records from different datasets that refer to the same entity. This process can be framed as a pairwise binary classification problem, where a classification model predicts if a pair of records match (i.e., refer to the same entity) or not. Even though training data is paramount in model building and the subsequent predictions, there is a lack of reporting in the literature on training data details, especially the ratio of matching to non-matching examples. The absence of adequate reporting has a significant impact on both the model building and reproducibility of research studies. In this paper we demonstrate how the performance measures commonly used in record linkage (precision, recall, and F_1-measure) vary with respect to this ratio. Specifically, we show that different class imbalance ratios in training data have a substantial impact in classifier performance, with more imbalanced training data resulting in lower performance. Furthermore, we examine the impact on performance when the class ratio between the test data and the training data is changed. Our extensive experimental study allows us to offer practical advice for constructing training data, building record linkage models, measuring performance, and reporting on the training data details.

Keywords: training data · entity resolution · reproducibility · class imbalance · evaluation · precision · recall · F_1-measure

1 Introduction

Record linkage, also known as deduplication and entity resolution, is the process of finding records or entities which refer to the same underlying entity across a

We acknowledge the support of the Natural Sciences and Engineering Research Council of Canada (NSERC), [Discovery Grant: Bias and Representativeness in Linked Data]. This work was partially supported by the *Scottish Historic Population Platform* (SHiPP), and the support of the UK Economic and Social Research Council (ESRC) through project ES/W010321/1, is gratefully acknowledged.

© The Author(s), under exclusive license to Springer Nature Singapore Pte Ltd. 2024
D.-N. Yang et al. (Eds.): PAKDD 2024, LNAI 14645, pp. 194–205, 2024.
https://doi.org/10.1007/978-981-97-2242-6_16

single or multiple datasets [7]. Record linkage is a challenging problem particularly for datasets that are heterogeneous and contain records with poor data quality [15,17], which is the norm with most real-world applications. Considering the complexity of the record linkage problem and its associated data challenges, research continues in this field with new models being proposed and investigated across diverse research and application domains [1,2,5,6,9,21].

The record linkage problem is often framed as a binary classification problem [20], where every pair of records in the input space is assessed and predicted to either be a match (two records referring to the same entity) or a non-match (two records referring to two different entities). The supervised machine learning models that perform this task are trained on pairs of records that have been labeled as either matching or non-matching, and it is from these labeled record pairs that the classification models are able to learn patterns and form decision boundaries to make predictions about previously unseen pairs of records. Thus, it is this training data that is the pivotal factor in model building and subsequent making of predictions. Yet there is a lack of reporting on training data details in the literature [11,19,23,25], even though solutions have been proposed for data and model reporting [12,13,24].

This gap has significant implications for understanding the true performance of models, as well as for ensuring reproducibility and transparency. More recently, conferences and journals have started to recommend or require checklists for reproducibility [27]. This is a step in the right direction, but to our knowledge there has not been a checklist or a requirement to report class ratios. A large study on reproducibility [18] in a variety of domains where supervised classification is employed reports many articles that do not report on the training/testing split(s) used.

Training data is costly to collect, particularly in the context of record linkage. In addition, for record linkage, most of the available training data consists of positive examples and that is what is usually described in publications; the number and selection process for negative examples is not so commonly detailed. This is significant, as in record linkage it is natural for the number of negative examples to vastly exceed the number of positive examples [7].

In this paper we investigate how the training data class ratio affects the performance of classification models. Is it possible to reliably report higher performance on domain standard performance measures (such as the F_1-measure [8,29]) by varying the number of labeled non-matching examples relative to the number of labeled matching examples during training and/or testing? Should the impact of varying these class ratios be negligible or random, this would not be an issue. If, on the other hand, varying these ratios biases the reported performance in a predictable fashion, then it is problematic to not report these numbers.

We demonstrate through our extensive experimental study that both the training and testing class ratios matter, and we investigate the effect of these ratios on different deployment scenarios. Our study allows us to offer practical advice for constructing training data, building record linkage models, and reporting on the training data details.

2   Methodology

In this section we provide a description of the methodology used to assess the impact of varying the number of matching pairs to non-matching pairs in the training and testing data on the performance of a record linkage model. Our methodology involves performing supervised record linkage on multiple benchmark datasets. Our focus is on building and measuring the performance for the classification models. We assume that cleaning, standardization, blocking and feature engineering have already been performed [7].

2.1   Data Partitioning

We start by separating a dataset of labeled pairs into the sets M (matches, where the two records in each pair refer to the same entity) and N (non-matches, where the records in each pair refer to different entities). We randomly sample record pairs from N such that the number of sampled pairs does not exceed $5 \times |M|$ (this is an upper bound dictated by the datasets we explore in our experimental study, shown in Table 1). From these sampled non-matching record pairs and the full set of matching record pairs M, we form 10 stratified folds for 10 fold cross validation [16]. We use a 8:1:1 ratio, where in each of the 10 runs 8 folds are used for training, 1 fold is used for validation, and 1 fold is used for testing.

The classification problem in record linkage is generally framed as a single label binary class prediction. During the training stage we build our classification model on different class ratios. We subset the non-matching examples within each fold to achieve the desired ratio of matching to non-matching pairs, considering five ratios of matching to non-matching pairs: 1:1, 1:2, 1:3, 1:4, and 1:5.

We consider the ratio of matching to non-matching pairs separately in both training and testing phases, performing a 5×5 grid search. In most research applications, models are evaluated on a test set that contains the same class imbalance as the training data. In more practical situations where record linkage models are deployed on previously unseen real-world data, it is often not possible to know the class ratio of the deployment class while the model is being trained. As such, it is also interesting to look at the impact that a relative abundance or deficit of non-matching pairs during training has on model performance when deployed in a different scenario than encountered in the training phase. The partitioning of the data across the ten folds and five ratios is shown in Fig. 1.

2.2   Classification and Evaluation

To perform the pairwise classification, we train models on the training folds described in Sect. 2.1. Given a pair of records, a model returns the probability that they refer to the same entity. During this stage, we evaluate the

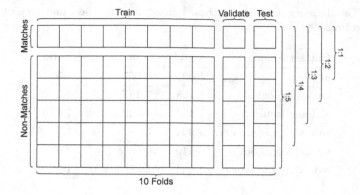

Fig. 1. Match and non-match allocation across folds and ratios, as described in Sect. 2.1

performance of our classification model on the test set. The performance is measured through precision (the fraction of all positive predictions that are actual positives), recall (the fraction of all actual positives that are predicted to be positive) and F_1-measure (the harmonic mean of precision and recall). Expanding these definitions, F_1-measure can be defined in terms of the number of true positives (TP), false positives (FP), and false negatives (FN) as $F_1 = 2 \times TP/(2 \times TP + FP + FN)$ [8].

To calculate F_1-measure, the number of TPs, FPs, and FNs within the test set must first be identified. A classification model returns the conditional probability p that an individual pair is a match. A decision rule in the form of a numeric threshold $t \in [0, 1]$ is required to convert these probabilities into binary predictions (i.e., only pairs where $p > t$ are considered to match). Once a pair has been assigned a binary prediction, it can be identified as a TP, FP, or FN.

A threshold $t = 0.5$ may seem a natural starting point, but using a threshold other than 0.5 to assign binary predictions can often increase the performance as measured by the F_1-measure [22]. We select a threshold for each model that maximizes the F_1-measure on the validation set, and use this threshold when computing precision, recall, and the F_1-measure for the test set.

3 Experimental Study

We now introduce the datasets we use in our experimental study and present our results. All results and source code to reproduce these experiments are available in a public GitHub repository.[1] In this repository, we also report results on additional performance measures.

---

[1] https://github.com/foxcroftjn/PAKDD-Class-Ratio.

Table 1. The datasets, labeled matches, labeled non-matches, and (rounded) class ratios taken from [28].

Dataset Name	Matches	Non-Matches	Ratio
abt-buy	1 095	6 067	1:5
amazon-google	1 298	7 142	1:5
walmart-amazon	1 154	14 425	1:12
wdc_xlarge_computers	9 991	59 571	1:5
wdc_xlarge_shoes	4 440	39 088	1:8
wdc_xlarge_watches	9 564	53 105	1:5

3.1   Datasets

In our study, we use a variety of record linkage datasets that were prepared by Primpeli and Bizer [28] and available online.[2] As part of their work, they published labeled pairs of records and similarity vectors suitable for supervised machine learning models (e.g., Random Forests [3], Support Vector Machines [10] classifiers) for 21 publicly available labeled datasets for record linkage.

We report results only for the datasets with at least 1 000 labeled matching pairs. This filters out a number of the less comprehensively labeled datasets. In addition, we require each dataset to contain at least five labeled non-matching pairs for every matching pair, to ensure we can train and test a model on a 1:5 class imbalance without compromising the set of record pairs which characterize the matching class. Finally, we do not use datasets where the Random Forests model in [28] achieved an F_1-measure ≥ 0.99, as these linkage tasks are too easy [26] to draw meaningful conclusions from in our work. After applying this filtering criteria we are left with the six datasets summarized in Table 1.

3.2   Results

We consider three different architectures of classification model: Random Forest (RF) [3], Support Vector Machine (SVM) [10], and Entity Matching Transformer (EMT) [4]. The first two model architectures are traditional machine learning techniques; the third uses the roBERTa attention-based transformer architecture to achieve near state-of-the-art results. We rely on existing implementations for each of these architectures [4,28]; the specific details of how the models are implemented is outside the scope of this work. By performing this experiment using both traditional machine learning and deep learning, we demonstrate that our results are not an artifact of a specific classification method, but rather that they generalize across a variety of commonly used classification approaches.

Figure 2 shows the F_1-measure results for all the architecture/dataset combinations. It is interesting to observe that a consistent gradient has emerged.

---

[2] https://data.dws.informatik.uni-mannheim.de/benchmarkmatchingtasks.

RF — **SVM** — **EMT** (columns); Test Ratio (1:1, 1:2, 1:3, 1:4, 1:5) across bottom; Train/Validation Ratio (1:1–1:5) on right.

abt-buy

	RF					SVM					EMT				T/V
	.90	.85	.81	.77	.74	.88	.82	.77	.72	.69	.96	.94	.91	.88	.86 — 1:1
	.90	.87	.85	.82	.80	.87	.85	.83	.80	.78	.96	.95	.94	.93	.92 — 1:2
	.90	.88	.85	.83	.81	.87	.85	.83	.81	.80	.96	.95	.94	.93	.92 — 1:3
	.88	.87	.86	.84	.83	.88	.86	.84	.82	.80	.96	.95	.94	.94	.93 — 1:4
	.89	.88	.86	.85	.84	.86	.84	.83	.82	.81	.96	.95	.94	.94	.93 — 1:5

amazon-google

.89	.84	.80	.76	.73	.87	.81	.76	.71	.66	.93	.89	.86	.83	.80 — 1:1
.89	.85	.82	.79	.77	.87	.82	.78	.74	.70	.93	.91	.88	.86	.83 — 1:2
.88	.85	.83	.80	.78	.84	.80	.77	.74	.71	.92	.90	.88	.86	.84 — 1:3
.87	.85	.83	.81	.79	.83	.80	.77	.75	.72	.91	.90	.89	.87	.85 — 1:4
.86	.84	.82	.81	.79	.84	.80	.78	.75	.72	.92	.90	.89	.87	.85 — 1:5

walmart-amazon

.98	.96	.95	.93	.92	.98	.96	.95	.94	.92	.99	.98	.97	.96	.95 — 1:1
.98	.97	.96	.94	.93	.97	.97	.96	.95	.94	.98	.98	.97	.97	.96 — 1:2
.98	.97	.96	.96	.95	.97	.96	.96	.95	.94	.99	.98	.98	.97	.97 — 1:3
.97	.97	.97	.96	.95	.97	.96	.96	.95	.94	.98	.98	.98	.97	.97 — 1:4
.98	.97	.97	.96	.96	.97	.96	.96	.95	.94	.98	.98	.98	.97	.97 — 1:5

wdc_xlarge_computers

.87	.81	.75	.71	.66	.84	.76	.69	.64	.59	.98	.98	.97	.96	.95 — 1:1
.87	.83	.80	.76	.73	.83	.78	.74	.70	.66	.98	.98	.97	.97	.96 — 1:2
.85	.83	.81	.78	.76	.81	.77	.74	.71	.68	.98	.98	.97	.97	.96 — 1:3
.85	.83	.81	.79	.77	.80	.77	.74	.72	.69	.98	.98	.97	.97	.97 — 1:4
.85	.83	.81	.79	.77	.80	.77	.74	.71	.69	.97	.97	.97	.97	.96 — 1:5

wdc_xlarge_shoes

.84	.76	.69	.63	.59	.80	.69	.61	.54	.50	.96	.94	.91	.89	.87 — 1:1
.83	.77	.72	.67	.63	.80	.70	.63	.57	.53	.97	.95	.93	.92	.90 — 1:2
.81	.77	.73	.69	.66	.77	.70	.64	.59	.55	.96	.95	.94	.92	.91 — 1:3
.81	.77	.73	.70	.67	.75	.69	.64	.60	.56	.96	.95	.94	.93	.92 — 1:4
.78	.75	.72	.70	.67	.72	.68	.64	.60	.57	.96	.95	.94	.92	.92 — 1:5

wdc_xlarge_watches

.91	.88	.85	.82	.79	.88	.83	.79	.75	.72	.98	.96	.95	.94	.93 — 1:1
.91	.89	.87	.85	.83	.87	.84	.80	.77	.75	.97	.97	.96	.95	.95 — 1:2
.91	.89	.87	.86	.84	.85	.83	.81	.79	.77	.97	.97	.96	.95	.95 — 1:3
.90	.89	.87	.86	.85	.83	.82	.80	.79	.77	.97	.96	.96	.95	.95 — 1:4
.89	.88	.87	.86	.85	.82	.80	.79	.78	.77	.97	.96	.96	.96	.95 — 1:5

Test Ratio: 1:1 1:2 1:3 1:4 1:5   (for each of RF, SVM, EMT)

Right axis label: Train/Validation Ratio

Fig. 2. F_1-measure for all model architectures, train/test ratio combinations, and datasets.

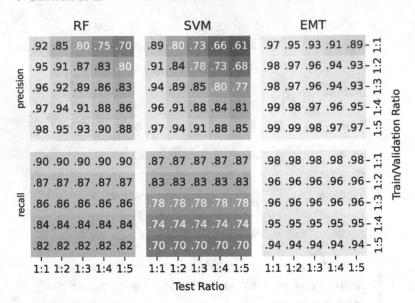

Fig. 3. Precision and recall for all model architectures on the amazon-google dataset.

The left column for each architecture/dataset, where the test set contains the fewest non-matching pairs, consistently contains the highest reported performance. Moving from the top left to the bottom right consistently causes F_1-measure performance to monotonically decrease. This tells us that we can worsen the reported performance of a methodology just by increasing the number of non-matching pairs in the training and testing data (or conversely, we can bolster the reported performance by reducing the number of non-matching pairs). Most work published in record linkage lies somewhere on this diagonal, as it is normal to train and test on data that have a single fixed class ratio in a controlled environment. The problematic part is that when class ratios are not reported, it is unknown where on this diagonal reported results lie. Class ratios higher than 1:5 are regularly used when performing record linkage [28], which in turn leads to the reported F_1-measure results skewing even lower than in this experiment.

Another commonality across all the architecture/dataset combinations is that the top right corner consistently contained the lowest reported F_1-measure performance, whereas the performance in the bottom left was in some cases the highest we obtained in our evaluation. We believe this trend can be used to inform the training data selection for real-world models, where the class ratio of the deployment data is not known when training a model. The low number in the top right corner reflects the consequences of training with a lower class ratio than testing. In contrast, the bottom left number reflects the consequences of a relative abundance of non-matching pairs during training. As such, we conclude that it is better to overestimate than underestimate the class imbalance of the deployment environment when choosing the class imbalance of a training set.

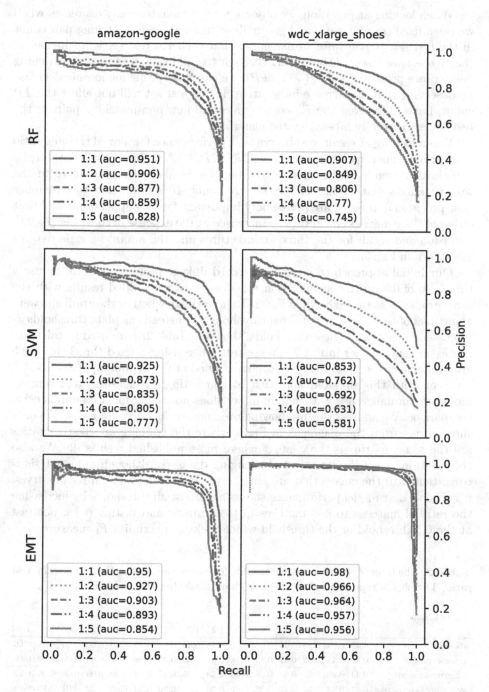

Fig. 4. Precision-recall curves and area under the curve (auc) for the two most challenging datasets. Only experiments with the same train/test ratio are shown.

When looking at precision, we observe that it monotonically decreases when we use a fixed training ratio and gradually increase the non-matching pair count in the test set. It is helpful to remember that each $1{:}n$ test set is a superset of the $1{:}(n-1)$ test set, differing only through the addition of more non-matching pairs. Since precision is $TP/(TP+FP)$, we can justify this monotonic decrease by observing that an increase in negatives in the test set will not affect the TP count, but will increase the FP count (unless all new non-matching pairs in the test set are correctly labeled by the classifier).

When looking at recall, we observe that performance for a fixed training ratio is invariant. Since recall is defined as $TP/(TP+FN)$, this consistency can be explained by remembering that increasing the number of non-matches in the test set affects neither the TP or the FN count. It is important to remember that precision and recall are not equally important for all applications, and that either of these metrics can always be increased at the expense of the other [14,22]. Precision and recall for the three architectures and the amazon-google dataset are shown in Fig. 3.

Our initial approach to measuring record linkage performance was to use a threshold of 0.5 on the classification function C, which yielded results with the same gradients as those shown in Fig. 2. To investigate whether the gradients were an artifact of this likely sub-optimal threshold, we instead calculate thresholds as discussed in Sect. 2.2. These thresholds, shown in Table 2, were used to calculate the F_1-measure values in Fig. 2. The results when using a fixed threshold of 0.5 are available in the GitHub repository mentioned at the beginning of Sect. 3.

Even with this updated approach to computing F_1-measure, representing model performance using a single number does not make for a comprehensive comparison. We address this by showing precision-recall curves in Fig. 4. We only show curves from the diagonals in Fig. 2 where the training and testing ratios are the same, as this is the context where most published results lie. We also choose to focus on the two most challenging datasets, although our analysis is consistent with the curves that are not shown. From the precision-recall curves, it is visible that model performance suffers at almost all thresholds by increasing the ratio of matches to non-matches in the training and testing data, not just at the 0.5 threshold or the threshold which seeks to maximize F_1-measure.

Table 2. The binary classification thresholds for experiments with the same train/test ratio. Thresholds reported as 0.00 are less than 0.005 (but still greater than 0).

Dataset Name	RF					SVM					EMT				
	1:1	1:2	1:3	1:4	1:5	1:1	1:2	1:3	1:4	1:5	1:1	1:2	1:3	1:4	1:5
abt-buy	0.45	0.43	0.36	0.44	0.39	0.41	0.48	0.45	0.34	0.40	0.00	0.00	0.05	0.01	0.00
amazon-google	0.49	0.45	0.46	0.42	0.43	0.46	0.37	0.35	0.36	0.28	0.00	0.00	0.00	0.00	0.00
walmart-amazon	0.44	0.40	0.45	0.48	0.46	0.46	0.53	0.45	0.47	0.43	0.00	0.99	0.06	0.96	0.07
wdc_xlarge_computers	0.44	0.42	0.44	0.40	0.38	0.41	0.39	0.37	0.30	0.23	0.02	0.02	0.01	0.10	0.04
wdc_xlarge_shoes	0.43	0.39	0.38	0.34	0.38	0.35	0.26	0.24	0.20	0.17	0.00	0.00	0.00	0.02	0.01
wdc_xlarge_watches	0.50	0.48	0.43	0.44	0.43	0.49	0.39	0.37	0.34	0.33	0.97	0.79	0.05	0.33	0.31

4   Discussion and Recommendations

The two key findings of our work are:

1. The F_1-measure can be artificially lowered or raised in a predictable direction by increasing the number of non-matching pairs in the training and/or testing sets of a record linkage problem. The impact of varying this ratio means that it is important to report this ratio when reporting the results of a record linkage methodology. Stating the number of labeled matching pairs is not sufficient.
2. When the deployment environment class imbalance is unknown, it is safest to err on the side of including more non-matching pairs during training. There is a consistently larger performance penalty to underestimating this ratio as compared to overestimating it.

Following from these findings, our recommendations for building training data and reporting are as follows:

1. Document [12] and report on the construction of the training data.
2. Report the class ratio or both the number of matching and non-matching pairs that are used to build the classification model used in record linkage.
3. Add more non-matching pairs to the training data when the class imbalance in the deployment environment is unknown.

Finally, it is worth commenting on the training time required for each of the model architectures. Each Random Forest took only seconds to train, as the training process parallelizes effectively across a multi-core CPU. Training the SVM models does not naturally parallelize, so models sometimes took a couple minutes to train (also using only a CPU). The EMT models were trained for 3 h each using a GPU. It is worth considering if the higher performance of a deep learning approach is always worth the significantly higher demand for specialized hardware and training time.

5   Conclusions and Future Work

There are many factors that come into play when preparing training data for binary classification of record linkage problems. The focus of this paper is on how class ratio affects F_1-measure, one of the most common performance measures used for classification and record linkage problems.

The impact of other aspects of training data creation remain the potential subject of future work. This includes how labeled non-matching pairs are sourced (i.e., random sampling? hard negative mining [11,28]?) and strategies for reducing the size of the training data to accelerate model training without compromising model quality. Other future research directions are to investigate if the findings discovered in this study hold for multi class settings and other application domains where imbalanced classes are common.

References

1. Akgün, Ö., Dearle, A., Kirby, G.N.C., Christen, P.: Using metric space indexing for complete and efficient record linkage. In: Phung, D., Tseng, V., Webb, G., Ho, B., Ganji, M., Rashidi, L. (eds.) PAKDD 2018. LNCS, vol. 10939, pp. 89–101. Springer, Cham (2018). https://doi.org/10.1007/978-3-319-93040-4_8
2. Anindya, I.C., Kantarcioglu, M., Malin, B.: Determining the impact of missing values on blocking in record linkage. In: Yang, Q., Zhou, Z.H., Gong, Z., Zhang, M.L., Huang, S.J. (eds.) PAKDD 2019. LNCS, vol. 11441, pp. 262–274. Springer, Heidelberg (2019). https://doi.org/10.1007/978-3-030-16142-2_21
3. Breiman, L.: Random forests. Mach. Learn. **45**(1), 5–32 (2001). https://doi.org/10.1023/A:1010933404324
4. Brunner, U., Stockinger, K.: Entity matching with transformer architectures - a step forward in data integration. In: Proceedings of the 23rd EDBT (2020). https://doi.org/10.21256/ZHAW-19637
5. Cao, X., Zheng, Y., Shi, C., Li, J., Wu, B.: Link prediction in schema-rich heterogeneous information network. In: Bailey, J., Khan, L., Washio, T., Dobbie, G., Huang, J., Wang, R. (eds.) PAKDD 2016. LNCS, vol. 9651, pp. 449–460. Springer, Cham (2016). https://doi.org/10.1007/978-3-319-31753-3_36
6. Cao, Y., Peng, H., Yu, P.S.: Multi-information source HIN for medical concept embedding. In: Lauw, H., Wong, R.W., Ntoulas, A., Lim, E.P., Ng, S.K., Pan, S. (eds.) PAKDD 2020. LNCS, vol. 12085, pp. 396–408. Springer, Heidelberg (2020). https://doi.org/10.1007/978-3-030-47436-2_30
7. Christen, P.: Data Matching - Concepts and Techniques for Record Linkage, Entity Resolution, and Duplicate Detection. Data-Centric Systems and Applications, Springer, Heidelberg (2012). https://doi.org/10.1007/978-3-642-31164-2
8. Christen, P., Hand, D.J., Kirielle, N.: A review of the F-measure: its history, properties, criticism, and alternatives. ACM Comput. Surv. **56**(3), 1–24 (2023)
9. Christen, P., Ranbaduge, T., Schnell, R.: Linking Sensitive Data. Springer, Heidelberg (2020). https://doi.org/10.1007/978-3-030-59706-1
10. Cortes, C., Vapnik, V.: Support-vector networks. Mach. Learn. **20**(3), 273–297 (1995). https://doi.org/10.1007/BF00994018
11. Fakhraei, S., Mathew, J., Ambite, J.L.: NSEEN: neural semantic embedding for entity normalization. In: Brefeld, U., Fromont, E., Hotho, A., Knobbe, A., Maathuis, M., Robardet, C. (eds.) ECML PKDD 2019. LNCS, vol. 11907, pp. 665–680. Springer, Heidelberg (2019). https://doi.org/10.1007/978-3-030-46147-8_40
12. Gebru, T., et al.: Datasheets for datasets. Commun. ACM **64**(12), 86–92 (2021). https://doi.org/10.1145/3458723
13. Gilbert, R., et al.: Guild: guidance for information about linking data sets. J. Public Health **40**, 191–198 (2017)
14. Hand, D.J., Christen, P.: A note on using the F-measure for evaluating record linkage algorithms. Stat. Comput. **28**(3), 539–547 (2018)
15. Harron, K., et al.: Challenges in administrative data linkage for research. Big Data Soc. **4**(2) (2017). https://doi.org/10.1177/2053951717745678. pMID: 30381794
16. Hastie, T., Tibshirani, R., Friedman, J.: The Elements of Statistical Learning. Springer Series in Statistics, Springer, New York (2001). https://doi.org/10.1007/978-0-387-21606-5
17. Herzog, T., Scheuren, F., Winkler, W.: Data Quality and Record Linkage. Springer, New York (2007). https://doi.org/10.1007/0-387-69505-2

18. Kapoor, S., Narayanan, A.: Leakage and the reproducibility crisis in ML-based science (2022). https://doi.org/10.48550/ARXIV.2207.07048
19. Kooli, N., Allesiardo, R., Pigneul, E.: Deep learning based approach for entity resolution in databases. In: Nguyen, N.T., Hoang, D.H., Hong, T., Pham, H., Trawinski, B. (eds.) ACIIDS 2018. LNCS, vol. 10752, pp. 3–12. Springer, Cham (2018). https://doi.org/10.1007/978-3-319-75420-8_1
20. Köpcke, H., Rahm, E.: Frameworks for entity matching: a comparison. Data Knowl. Eng. **69**(2), 197–210 (2010). https://doi.org/10.1016/j.datak.2009.10.003
21. Koumarelas, I., Papenbrock, T., Naumann, F.: Mdedup: duplicate detection with matching dependencies. Proc. VLDB Endow. **13**(5), 712–725 (2020). https://doi.org/10.14778/3377369.3377379
22. Lipton, Z.C., Elkan, C., Naryanaswamy, B.: Optimal thresholding of classifiers to maximize F1 measure. In: Calders, T., Esposito, F., Hüllermeier, E., Meo, R. (eds.) ECML PKDD 2014. LNCS, vol. 8725, pp. 225–239. Springer, Heidelberg (2014). https://doi.org/10.1007/978-3-662-44851-9_15
23. Makri, C., Karakasidis, A., Pitoura, E.: Towards a more accurate and fair SVM-based record linkage. In: Tsumoto, S., et al. (eds.) International Conference on Big Data, Osaka, pp. 4691–4699. IEEE (2022). https://doi.org/10.1109/BigData55660.2022.10020514
24. Mitchell, M., et al.: Model cards for model reporting. In: Proceedings of the Conference on Fairness, Accountability, and Transparency, FAT 2019, pp. 220–229. Association for Computing Machinery, New York (2019). https://doi.org/10.1145/3287560.3287596
25. Mudgal, S., et al.: Deep learning for entity matching: a design space exploration. In: Proceedings of the 2018 International Conference on Management of Data, SIGMOD 2018, pp. 19–34. Association for Computing Machinery, New York (2018). https://doi.org/10.1145/3183713.3196926
26. Papadakis, G., Kirielle, N., Christen, P., Palpanas, T.: A critical re-evaluation of benchmark datasets for (deep) learning-based matching algorithms. In: IEEE International Conference on Data Engineering (ICDE), Utrecht (2024)
27. Pineau, J., et al.: Improving reproducibility in machine learning research (a report from the neurips 2019 reproducibility program). J. Mach. Learn. Res. **22**(1), 1–20 (2021)
28. Primpeli, A., Bizer, C.: Profiling entity matching benchmark tasks. In: Proceedings of the 29th ACM International Conference on Information and Knowledge Management, CIKM 2020, pp. 3101–3108. Association for Computing Machinery, New York (2020). https://doi.org/10.1145/3340531.3412781
29. Shaw, W., Burgin, R., Howell, P.: Performance standards and evaluations in IR test collections: cluster-based retrieval models. Inf. Process. Manag. **33**(1), 1–14 (1997). https://doi.org/10.1016/S0306-4573(96)00043-X

Clustering

Clustering-Friendly Representation Learning for Enhancing Salient Features

Toshiyuki Oshima$^{(\boxtimes)}$, Kentaro Takagi, and Kouta Nakata

Corporate R&D Center, Toshiba Corporation, 1, Komukai Toshiba-cho, Saiwai-ku, Kawasaki, Kanagawa, Japan
{toshiyuki1.oshima,kentaro1.takagi,kouta.nakata}@toshiba.co.jp

Abstract. Recently, representation learning with contrastive learning algorithms has been successfully applied to challenging unlabeled datasets. However, these methods are unable to distinguish important features from unimportant ones under simply unsupervised settings, and definitions of importance vary according to the type of downstream task or analysis goal, such as the identification of objects or backgrounds. In this paper, we focus on unsupervised image clustering as the downstream task and propose a representation learning method that enhances features critical to the clustering task. We extend a clustering-friendly contrastive learning method and incorporate a *contrastive analysis* approach, which utilizes a reference dataset to separate important features from unimportant ones, into the design of loss functions. Conducting an experimental evaluation of image clustering for three datasets with characteristic backgrounds, we show that for all datasets, our method achieves higher clustering scores compared with conventional contrastive analysis and deep clustering methods.

1 Introduction

Clustering is one of the most fundamental methods for unsupervised machine learning and it aims to classify objects based on measures of similarity between unlabeled samples. Advancements in deep learning techniques have produced *deep clustering* methods, in which feature extraction with deep neural networks (DNNs) and a clustering process are integrated at various levels [23]. Self-supervised learning (SSL) has also recently been attracting attention in the representation learning field. SSL methods learn representations by solving user-defined pretext tasks, such as instance discrimination tasks and contrastive learning, and their impressive performance in capturing visual features has been demonstrated even for complex real-world datasets [3–5, 21]. *High-resolution* representations extracted by SSL are naturally expected to be applied to clustering. Several approaches to simultaneously performing SSL and clustering tasks have been proposed and have achieved state-of-the-art performance [10, 13, 18, 20].

T. Oshima and K. Takagi—First two authors have equal contribution.

© The Author(s), under exclusive license to Springer Nature Singapore Pte Ltd. 2024
D.-N. Yang et al. (Eds.): PAKDD 2024, LNAI 14645, pp. 209–220, 2024.
https://doi.org/10.1007/978-981-97-2242-6_17

Fig. 1. Distribution of feature representations on ImageNet-10 learned by IDFD (Our reproduction of Fig. 6 in [18]). Some samples are grouped according to the features of the mesh structure.

However, simply capturing features at high resolution or in ways that are easy for machines to understand does not necessarily improve clustering scores, and can even worsen them, due to differences in human and algorithmic criteria for feature importance. As a concrete example, Fig. 1 shows instance discrimination and feature decorrelation (IDFD), an SSL method, forming a cluster depending on features of a prominent mesh structure (foreground component) rather than objects such as a leopard or orange. Even though each feature is correctly extracted, this is not a desirable situation in the context of clustering that focuses on objects. Similar problems can arise when the methods are applied to real-world problems. In the case of inspection images taken at a factory, for example, it is not uncommon for complex product patterns to become more prominent than the object defects targeted for classification [17]. For this reason, models require proper guidance so that only features that are important for the targeted clustering are picked up.

Contrastive analysis (CA) [9] is one technique for providing inductive bias with unsupervised methods to distinguish important from unimportant components. In CA settings, two datasets are prepared: a target dataset that includes salient features of interest for clustering and a background dataset that contains only information that is ineffectual for clustering and should be discarded. Several unsupervised methods making use of CA settings have recently been proposed [1,2,8]. These architectures have successfully isolated and extracted variables of interest from backgrounds, but they use classical architectures and learning schemes such as matrix decomposition and simple encoder-decoder structures, and they mainly minimize reconstruction loss functions. There is thus room for improvement in the ability to learn representations for more complex real-world datasets.

In this paper, we propose contrastive IDFD (cIDFD), a method that combines a CA setting with the latest SSL scheme based on instance discrimination. cIDFD learns unimportant features at high resolutions from the background dataset via normal instance discrimination and feature decorrelation. Similarities among unimportant features in target samples are also obtained and utilized to guide networks toward extracting only features that are important for desirable clustering. This feature extraction is performed by minimizing a newly defined loss function, the contrastive instance discriminative loss function. We adapt our method to various datasets with characteristic background patterns. Our method shows improvements over conventional methods in terms of evaluating clustering tasks.

Our main contributions are as follows:

- We propose a new clustering method based on instance discrimination with a CA setting that selectively extracts features meaningful for the user-defined clustering objective.
- We conduct experimental evaluations of clustering for three challenging datasets with characteristic backgrounds. The results show that cIDFD achieves higher clustering scores than conventional contrastive analysis models and state-of-the-art SSL methods.

2   Related Works

There have recently been proposals of contrastive learning as pretext tasks for SSL, in which positive pairs generated by data augmentation from the same data are brought closer together, and negative samples generated from different data are kept apart [4,21]. Features learned by the pretext tasks have high generalization and achieve excellent performance on a variety of downstream tasks. Although negative pairs played an important role in avoiding collapsing solutions, there are some drawbacks, such as the need to increase batch size, and thus several methods that do not use negative pairs have been proposed [3,5].

Deep clustering has shown superior performance in a variety of areas. End-to-end learning was proposed to simultaneously perform representation learning and clustering [23]. To improve representation learning ability, recent works integrate clustering and SSL, in particular contrastive learning [6,10,13,14,18–20,22]. IDFD [18] focuses on the representation learning phase and proposes a clustering-friendly representation learning method that uses instance discrimination loss and a proposed feature decorrelation loss motivated by the properties of classical spectral clustering. They achieved high performance, despite using k-means in the clustering phase. MiCE [19] and CC [10] focused on end-to-end learning by integrating contrastive learning and clustering. SCAN [20], RUC [14], NNM [6], TCL [22], and SPICE [13] focused on the clustering phase and are multi-stage deep clustering methods.

The principles of contrastive analysis [9] were proposed as a method to separate features to be emphasized from irrelevant features that should be suppressed. cPCA [1] utilizes that principle to separate the principal components to be emphasized from those that should be suppressed, introducing a dataset without features to be emphasized. Contrastive singular spectrum analysis (cSSA) [8] extends the cPCA concept to analyze time-series datasets. Contrastive VAE (cVAE) [2] was also developed to understand complex, nonlinear relations between latent variables and inputs.

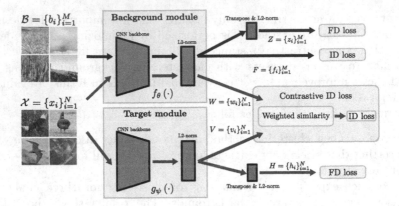

Fig. 2. cIDFD framework.

3   Proposed Method

Our goal is to learn only those representations that are appropriate to clustering for a target dataset $\mathcal{X} = \{x_i\}_{i=1}^{N}$ and cluster these samples into meaningful groups under conditions where a background dataset $\mathcal{B} = \{b_i\}_{i=1}^{M}$ can be prepared. cIDFD utilizes a dataset $\mathcal{B}$ to reject the influences of background features that are unimportant with respect to clustering for the target dataset $\mathcal{X}$. In this section, we describe the model architecture, loss computation, and actual training process of cIDFD.

3.1   The Framework of cIDFD

As Fig. 2 shows, we use two embedding functions, f_θ and g_ψ, which map images to feature vectors distributed on a d-dimensional unit sphere. These functions are modeled as deep neural networks with parameters θ and ψ, which are typically a CNN backbone and an L2-normalization layer. f_θ is learned from background dataset $\mathcal{B}$ and assigned the role of extracting background features $W = \{w_i\}_{i=1}^{N}$ and $Z = \{z_i\}_{i=1}^{M}$ for image samples from $\mathcal{X}$ and $\mathcal{B}$, respectively. The other embedding function g_ψ is learned to extract target features $V = \{v_i\}_{i=1}^{N}$ from $\mathcal{X}$. To train the background branch f_θ, we conduct minimizing instance discrimination and feature decorrelation loss, following the same process as in a previous work [18]. For target branch g_ψ, parameters are optimized by a newly defined loss function, contrastive instance discrimination loss, and feature decorrelation loss. In the computation of contrastive instance discrimination loss, weighted similarity between negative samples, which includes influences from W and V, is used as input to a non-parametric softmax classifier.

3.2   Loss for Background Feature Extraction

We apply the instance discrimination proposed in [21] to learn representations from a background dataset. For given samples $\{b_i\}_{i=1}^{M}$, the corresponding repre-

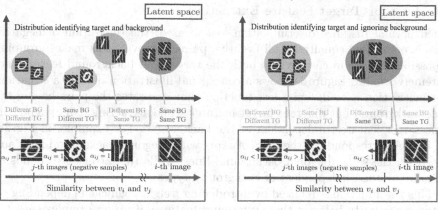

(a) Normal instance discrimination   (b) Contrastive instance discrimination

Fig. 3. Illustration of the problem in normal instance discrimination (a) and the contrastive instance discrimination strategy (b) for datasets with characteristic background patterns. Here, the number is called target (TG), the pattern of lines is called background (BG) for Stripe MNIST. The red arrows denote the strength of the repulsive force according to $\alpha_{ij}(Color figure online)$ for the i-th and j-th images.

sentations are $\{z_i\}_{i=1}^M$ with $z_i = f_\theta(b_i)$, where z_i is normalized to $\|z_i\| = 1$. The probability of representation z being assigned to the ith class is given by the non-parametric softmax formulation

$$P(i|z) = \frac{\exp(z_i^T z/\tau_b)}{\sum_{j=1}^M \exp(z_j^T z/\tau_b)},$$ (1)

where the dot product $z_i^T z$ is how well z matches the ith class and τ_b is a temperature parameter that determines the concentration of distribution. The learning objective is to maximize the joint probability $\prod_{i=1}^M P(i|f_\theta(b_i))$ as

$$\mathcal{L}_{I,b} = -\sum_{i=1}^M \log\left(\frac{\exp(z_i^T z_i/\tau_b)}{\sum_{j=1}^m \exp(z_j^T z_i/\tau_b)}\right).$$ (2)

We use a constraint for orthogonal features proposed in [18]. The objective is to minimize

$$\mathcal{L}_F(F) = \mathcal{L}_{F,b} = -\sum_{l=1}^d \log\left(\frac{\exp(f_l^T f/\tau_2)}{\sum_{j=1}^d \exp(f_j^T f/\tau_2)}\right),$$ (3)

where $F = \{f_l\}_{l=1}^d$ are latent feature vectors defined by the transpose of the latent vectors Z, d is the dimensionality of the representations, and τ_2 is the temperature parameter.

3.3 Loss for Target Feature Extraction

When normal instance discrimination loss is used, the separation of negative pairs is performed equally for all negative pairs. However, this makes grouping impossible, except in cases where both the target and background features are extremely similar. Figure 3a shows a conceptual illustration of such a situation. Our motivation is to design a loss function that separates the pairs having the same background but different target features while attracting pairs having the same target features, independent of the background. For example, in Fig. 3a, the former corresponds to the pairs of *zeros* with diagonal background lines and *ones* with the same diagonal background lines, while the latter corresponds to the pairs of *ones*, independent of background types.

This loss function is realized by introducing weight coefficients depending on pairwise similarity between the background features of target samples $\{w_i\}_{i=1}^N$. Our new loss function, namely, contrastive instance discrimination loss, is defined as

$$\mathcal{L}_{CI} = -\sum_{i=1}^{N} \log \left(\frac{\alpha_{ii} \exp(v_i^T v_i / \tau_x)}{\sum_{j=1}^{N} \alpha_{ij} \exp(v_i^T v_j / \tau_x)} \right), \tag{4}$$

where α_{ij} is the weight coefficient between the i-th and j-th samples that determine how strongly target features v_i and v_j are pulled apart in the learning process. This coefficient is formulated simply as

$$\alpha_{ij} = \exp(w_i^T w_j / \tau_{xb}). \tag{5}$$

When the background features of the i-th and j-th images are similar, α_{ij} becomes large, causing their repulsive force in the target feature space to increase. This effect, shown in Fig. 3b, reduces the similarities between the feature vectors of *zeros* with diagonal lines and *ones* with diagonal lines. α_{ij} also becomes large for pairs having the same background and target, but there is also the usual contrastive learning effect, and their repulsion becomes weakened. For pairs of small α_{ij}, the repulsive force is relatively weak, and thus samples with the same target and different backgrounds are weakly separated. We experimentally demonstrated that our method works according to the perspectives described above; the details are given in Sect. 4.4.

Temperature parameters τ_x control the distribution of feature vectors, and τ_{xb} controls the magnitude of the weight coefficient α_{ij}. The differences from IDFD are the background module f_θ and α_{ij}. In the case of $\lim_{\tau_{xb} \to \infty} \alpha_{ij} = 1$, cIDFD is consistent with IDFD because the contribution from the background module f_θ to the target module g_ϕ disappears.

3.4 Two-Stage Learning

We consider separately learning the embedding functions f_θ and g_ψ. In the first step, the f_θ branch learns a background dataset $\mathcal{B}$ by minimizing the objective function

$$\mathcal{L}_{bg} = \mathcal{L}_{I,b} + \mathcal{L}_F(F). \tag{6}$$

Algorithm 1. Two-stage learning in cIDFD.

Input:

    dataset $\mathcal{X} = \{x_i\}_{i=1}^{N}$ and $\mathcal{B} = \{b_i\}_{i=1}^{M}$; structure of embedding function f_θ and g_ψ; memory banks $\bar{V} = \{\bar{v}_i\}_{i=1}^{N}$, $\bar{W} = \{\bar{w}_i\}_{i=1}^{N}$, $\bar{Z} = \{\bar{z}_i\}_{i=1}^{M}$; training epochs E_f and E_g;.

1: Initialize parameters θ, ψ, $\bar{V}$, $\bar{W}$, $\bar{Z}$
2: **while** $epoch < E_f$ **do**   ▷ training for f_θ
3:     **for** each minibatch $\mathcal{B}_b$ **do**
4:         compute $z_i = f_\theta(b_i)$ for b_i in minibatch $\mathcal{B}_b$;
5:         compute the IDFD loss $\mathcal{L}_{bg}$ through Eq.(6);
6:         update θ by the optimizer;
7:         update $\bar{z}_i$ by the moving average;
8:     **end for**
9:     $epoch \leftarrow epoch + 1$
10: **end while**
11: **while** $epoch < E_g$ **do**   ▷ training for g_ψ
12:     **for** each minibatch $\mathcal{X}_b$ **do**
13:         compute $v_i = g_\psi(x_i)$ and $w_i = f_\theta(x_i)$ for x_i in minibatch $\mathcal{X}_b$;
14:         compute the cIDFD loss $\mathcal{L}_{tg}$ through Eq.(7);
15:         update ψ by the optimizer;
16:         update $\bar{v}_i$ and $\bar{w}_i$ by the moving average;
17:     **end for**
18:     $epoch \leftarrow epoch + 1$
19: **end while**

During this step, the input samples are only from $\mathcal{B}$, and the network parameters of g_ψ are not updated. After learning, f_θ works as an extractor of features to be discarded. In the second step, we freeze the parameters of f_θ and train g_ψ by dataset $\mathcal{X}$. The objective function is a composition of the contrastive instance discriminative loss (4) and feature decorrelation loss for the target features,

$$\mathcal{L}_{tg} = \mathcal{L}_{CI} + \mathcal{L}_F(H), \tag{7}$$

where $H = \{h_l\}_{l=1}^{d}$ are the vectors defined by the transpose of V. The above learning process is summarized as Algorithm 1.

4 Experiments

4.1 Datasets

We evaluated the performance of cIDFD on three datasets created from commonly used public datasets. Each dataset contains both target and background datasets. Table 1 summarizes the key details. Figure 4 shows sample images from each dataset.

Table 1. Image datasets used in the experiments.

Dataset	Image size	Samples	classes
Stripe MNIST	$28 \times 28 \times 1$	60000	10
Stripes	$28 \times 28 \times 1$	10000	4
CelebA-ROH	$178 \times 218 \times 3$	21065	2
CelebA-RNH	$178 \times 218 \times 3$	20000	–
Birds400-ABC	$224 \times 224 \times 3$	20822	144
Landscape Pictures	$224 \times 224 \times 3$	41733	–

Stripe MNIST. We created a synthetic image dataset using handwritten digits from the Modified National Institute of Standards and Technology (MNIST) database [7] and randomly generated artificial stripe patterns, which are shown in Fig. 4a. Our goal was to cluster the ten handwritten digits independently of the background patterns. We also prepared a background dataset of stripe-pattern images that are almost the same as those in the target dataset, but not used in its creation (Fig. 4b).

CelebA-ROH. We assume a clustering task that focuses on certain features of facial images. We made datasets from the popular celebrity facial images dataset CelebA [11]. As a target dataset, we collected images with the target attributes "receding hairlines" or "wearing hats" (Fig. 4c). We used the remaining celebrities as the background dataset, which we call CelebA-RNH (Fig. 4d). Our goal was to cluster the two target attributes independently of other attributes such as eyeglasses, hair color, or gender.

Birds400-ABC. As a target dataset, we collected images from Birds400 [15], which are derived from the Kaggle datasets. Birds400 contains images of 400 types of birds with various backgrounds. From the training split of the original dataset, we utilized 144 bird species with names starting with A, B, or C (Fig. 4e). To realize *bird-oriented* clustering by cIDFD, we used the Landscape Pictures dataset [16], which includes high-quality images of natural landscapes. By randomly cropping and resizing those images, we generated 41,733 samples with 224×224 pixels for the background dataset (Fig. 4f).

4.2 Comparison with Conventional Methods

We compared cIDFD with VAE, cVAE and eight other competitive deep clustering methods: CC, MiCE, SCAN, RUC, NNM, TCL, SPICE, IDFD. Given that VAE and the eight competitive deep clustering methods have no way to handle the background dataset, only the target dataset was used. In the clustering phase, we applied simple k-means to representations for VAE, cVAE, IDFD, and cIDFD. The number of clusters k is set equal to the number of classes in each

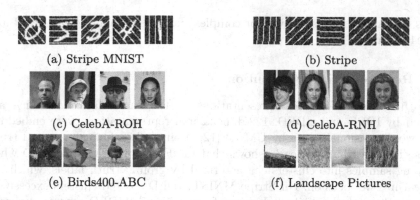

(a) Stripe MNIST

(b) Stripe

(c) CelebA-ROH

(d) CelebA-RNH

(e) Birds400-ABC

(f) Landscape Pictures

Fig. 4. Sample images from each dataset. The images on the left are from target datasets, while images on the right are from background datasets.

dataset as shown in Table 1. Clustering performance was evaluated by three popular metrics: clustering accuracy (ACC), normalized mutual information (NMI), and the adjusted rand index (ARI). These metrics give values in the range $[0, 1]$, with higher scores indicating more accurate clustering assignments.

Table 2 lists performances for each method. These results show that cIDFD clearly outperform the conventional methods. In terms of ACC, the cIDFD scores were improved by approximately 24% for Stripe MNIST, by 7% for CelebA-ROH, and by 25% for Birds400-ABC. In particular, performance under our method is better than performance under cVAE, which is the most similar method in

Table 2. Clustering results of various methods on three datasets. Results of eight deep clustering methods obtained from our experiments with official code.

Dataset	Stripe MNIST			CelebA-ROH			Birds400-ABC		
Metric	ACC	ARI	NMI	ACC	ARI	NMI	ACC	ARI	NMI
VAE	0.177	0.059	0.085	0.500	0.038	0.034	0.067	0.016	0.218
cVAE	0.578	0.421	0.563	0.776	0.299	0.208	0.070	0.016	0.200
CC	0.377	0.248	0.439	0.893	0.619	0.579	0.331	0.204	0.553
MiCE	0.349	0.233	0.420	0.714	0.185	0.149	0.265	0.202	0.562
SCAN	0.594	0.505	0.696	0.577	0.023	0.021	0.257	0.152	0.479
RUC	0.587	0.504	0.706	0.482	0.019	0.016	0.005	0.031	0.310
NNM	0.402	0.321	0.548	0.632	0.070	0.055	0.226	0.140	0.470
TCL	0.376	0.233	0.421	0.820	0.409	0.456	0.332	0.193	0.547
SPICE	0.492	0.411	0.588	0.578	0.021	0.037	0.321	0.203	0.526
IDFD	0.276	0.156	0.369	0.648	0.086	0.067	0.486	0.361	0.658
cIDFD	**0.830**	**0.809**	**0.908**	**0.969**	**0.879**	**0.796**	**0.738**	**0.664**	**0.848**

terms of CA setting usage, even for complex datasets such as CelebA-ROH and Birds400-ABC.

4.3 Representation Distribution

Figure 5 visualizes feature representations of the three datasets, which are learned by IDFD and cIDFD. 128-dimensional representations were embedded into two-dimensional space by UMAP [12]. Point colors indicate ground truth labels. The distribution clearly shows that cIDFD is preferable to IDFD when grouping samples into clusters characterized by ground truth labels, which are the features we focus on. For Stripe MNIST, IDFD created clusters excessively according to both the digits and stripes features, while cIDFD generated almost ten clusters. For CelebA-ROH, IDFD generated one distribution mixing two classes, but cIDFD generated distinct distributions according to the ground truth labels. cIDFD also correctly distinguished an extremely large number of bird species; however, in the IDFD results, samples of many classes were degenerated to several large clusters.

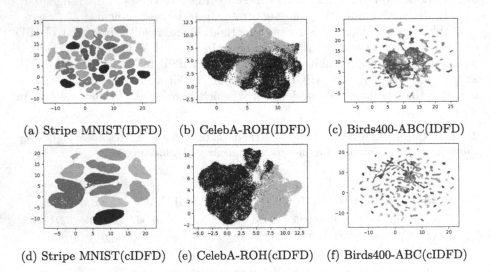

(a) Stripe MNIST(IDFD)   (b) CelebA-ROH(IDFD)   (c) Birds400-ABC(IDFD)

(d) Stripe MNIST(cIDFD)   (e) CelebA-ROH(cIDFD)   (f) Birds400-ABC(cIDFD)

Fig. 5. Visualizations of feature representations on the Stripe MNIST(10 classes), CelebA-ROH(2 classes), and Birds400-ABC(144 classes) datasets.

4.4 Similarity Distribution

To clearly understand how our method works, we conducted experiments on similarity distribution for both IDFD and cIDFD with the synthetic image dataset Stripe MNIST. Figure 6 shows the resulting histograms on four types of average similarity, which were calculated for each instance: the first type is similarity to instances of the same background and a different target; the second type is to instances of a different background and the same target; the third type is to instances of the same background and the same target; and the fourth type is to instances of a different target and a different background. In the case of IDFD (Fig. 6a), the peaks of the first and second type of similarity are located in almost same lower region. As mentioned in Sect. 3.3, this situation is problematic. In contrast, cIDFD successfully moved the peak of the second type to the same position of the third type distribution in higher region (Fig. 6b). On the other hand, the first and fourth type distributions are located in the same lower region. Consequently, instances with the same target features were clustered independently of the background features.

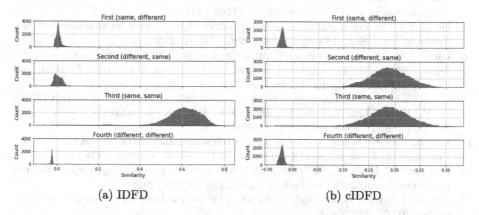

(a) IDFD   (b) cIDFD

Fig. 6. Histograms of similarity between samples in four different types of pairs. The pair types are indicated at the top of each figure. For example, the distributions of the first pairs are titled as (same, different), meaning that similarity is calculated for instances of the same background and different target features.

5   Conclusion

We presented cIDFD as a new self-supervised clustering method combining instance discrimination with feature decorrelation and contrastive analysis. Our method is designed to extract unimportant features from a background dataset and reject them in the learning process for the target dataset, resulting in clustering according to only the important features. The experimental results on the Stripe MNIST, CelebA-ROH, and Birds400-ABC datasets showed that cIDFD outperforms the state-of-the-art SSL method and similar conventional methods with contrastive analysis. Problem settings allowing utilization of background datasets appear in various fields, so we expect there to be many situations in which cIDFD can be applied.

References

1. Abid, A., Zhang, M.J., Bagaria, V.K., Zou, J.: Exploring patterns enriched in a dataset with contrastive principal component analysis. Nat. Commun. **9**, 2134 (2018)
2. Abid, A., Zou, J.: Contrastive variational autoencoder enhances salient features. arXiv:1902.04601 (2019)
3. Caron, M., Misra, I., Mairal, J., Goyal, P., Bojanowski, P., Joulin, A.: Unsupervised learning of visual features by contrasting cluster assignments. In: NeurIPS, pp. 9912–9924. Curran Associates, Inc. (2020)
4. Chen, T., Kornblith, S., Norouzi, M., Hinton, G.: A simple framework for contrastive learning of visual representations. In: ICML, pp. 1597–1607. PMLR (2020)
5. Chen, X., He, K.: Exploring simple siamese representation learning. In: CVPR, pp. 15750–15758. IEEE (2021)
6. Dang, Z., Deng, C., Yang, X., Wei, K., Huang, H.: Nearest neighbor matching for deep clustering. In: CVPR, pp. 13693–13702 (2021)
7. Deng, L.: The mnist database of handwritten digit images for machine learning research. IEEE Signal Process. Mag. **29**(6), 141–142 (2012)
8. Dirie, A.H., Abid, A., Zou, J.: Contrastive multivariate singular spectrum analysis. In: 2019 57th Annual Allerton Conference on Communication, Control, and Computing (Allerton), pp. 1122–1127 (2019)
9. Ge, R., Zou, J.: Rich component analysis. In: ICML, pp. 1502–1510. PMLR (2016)
10. Li, Y., Hu, P., Liu, Z., Peng, D., Zhou, J.T., Peng, X.: Contrastive clustering. In: AAAI, pp. 8547–8555. AAAI Press (2021)
11. Liu, Z., Luo, P., Wang, X., Tang, X.: Deep learning face attributes in the wild. In: Proceedings of International Conference on Computer Vision (ICCV) (2015)
12. McInnes, L., Healy, J., Melville, J.: UMAP: uniform manifold approximation and projection for dimension reduction. arXiv:1802.03426 (2018)
13. Niu, C., Shan, H., Wang, G.: SPICE: semantic pseudo-labeling for image clustering. arXiv:2103.09382 (2022)
14. Park, S., et al.: Improving unsupervised image clustering with robust learning. In: CVPR (2021)
15. Piosenka, G.: Birds 400 - species image classification (2022)
16. Rougetet, A.: Landscape pictures (2020)
17. Shota, M., Yukako, T.: Application of contrastive representation learning to unsupervised defect classification in semiconductor manufacturing. In: AEC/APC Symposium Asia 2021 (2021)
18. Tao, Y., Takagi, K., Nakata, K.: Clustering-friendly representation learning via instance discrimination and feature decorrelation. In: ICLR (2021)
19. Tsai, T.W., Li, C., Zhu, J.: Mice: mixture of contrastive experts for unsupervised image clustering. In: ICLR (2021)
20. Van Gansbeke, W., Vandenhende, S., Georgoulis, S., Proesmans, M., Van Gool, L.: SCAN: learning to classify images without labels. In: Vedaldi, A., Bischof, H., Brox, T., Frahm, J.-M. (eds.) ECCV 2020. LNCS, vol. 12355, pp. 268–285. Springer, Cham (2020). https://doi.org/10.1007/978-3-030-58607-2_16
21. Wu, Z., Xiong, Y., Yu, S.X., Lin, D.: Unsupervised feature learning via nonparametric instance discrimination. In: CVPR, pp. 3733–3742. IEEE (2018)
22. Li, Y., Yang, M., Peng, D., Li, T., Huang, J., Peng, X.: Twin contrastive learning for online clustering. Int. J. Comput. Vision **130**(9), 2205–2221 (2022)
23. Zhou, S., et al.: A comprehensive survey on deep clustering: taxonomy, challenges, and future directions (2022)

ImMC-CSFL: Imbalanced Multi-view Clustering Algorithm Based on Common-Specific Feature Learning

Xiaocui Li[1,2], Yu Xiao[1], Xinyu Zhang[1,2(✉)], Qingyu Shi[1,2], and Xiance Tang[1]

[1] School of Computer Science, Hunan University of Technology and Business, Changsha, China
zhangxinyu247@163.com
[2] Xiangjiang Laboratory, Changsha, China

Abstract. Clustering as one of the main research methods in data mining, with the generation of multi-view data, multi-view clustering has become the research hotspot at present. Many excellent multi-view clustering algorithms have been proposed to solve various practical problems. These algorithms mainly achieve multi-view feature fusion by maximizing the consistency between views. However, in practical applications, multi-view data' initial feature is often imbalanced, resulting in poor performance of existing multi-view clustering algorithms. Additionally, imbalanced multi-view data exhibits significant differences in feature across different views, which better reflects the complementarity of multi-view data. Therefore, it is important to fully extract feature from different views of imbalanced multi-view data. This paper proposes an imbalanced multi-view clustering algorithm based on common specific feature learning, ImMC-CSFL. Two deep networks are used to extract common and specific feature on each view, the GAN network is introduced to maximize the extraction of common feature from multi-view data, and orthogonal constraints are used to maximize the extraction of specific feature from different views. Finally, the learned imbalanced multi-view feature is input for clustering. The experiment result on three different multi-view datasets UCI Digits, BDGP, and CCV showed that our proposed algorithm had better clustering performance, and the effectiveness and robustness were verified through experiment analysis of different modules.

Keywords: imbalanced multi-view data · common-specific information network · deep feature learning network · multi-view clustering

1 Introduction

As one of the main research contents of machine learning, clustering technique has always been a research hotspot. With the rise of multi-view data, originating from different sources or modalities, multi-view clustering has been attracted significant attention from researchers, with a focus on multi-view feature learning. There is not only consistency but also differences between different views, that each view contains the common information or specific information. The key of multi-view feature learning is to comprehensively extract clustering-friendly feature from multi-view data. Multi-view data both

© The Author(s), under exclusive license to Springer Nature Singapore Pte Ltd. 2024
D.-N. Yang et al. (Eds.): PAKDD 2024, LNAI 14645, pp. 221–232, 2024.
https://doi.org/10.1007/978-981-97-2242-6_18

have similarities and disparities among different views. There are numerous excellent multi-view clustering algorithms, mainly including: multi-view spectral clustering algorithm, multi-view subspace clustering algorithm, multi-view clustering methods based on non-negative matrix decomposition, and multi-kernel based multi-view clustering algorithms.

1.1 Motivation

While multi-view clustering algorithms have received considerable attention and displayed good performance, they are exclusively suitable for multi-view data with relatively balanced initial features and perform inadequately for multi-view data with varied quality and imbalanced initial features. The imbalanced multi-view clustering is an urgent problem, which necessitate addressing. The main reason are as follows:

1) For multi-view data with imbalanced initial features, current methods fail to consider the specific information of the different view. It is worth noting that the specific information of the imbalanced multi-view data is particularly rich, which is more beneficial for extracting complementary features;
2) The extraction of the common and specific information of imbalanced multi-view data presents a significant challenge for feature fusion. Feature fusion strategies and clustering methods for the initial feature-imbalanced multi-view dataset are still relatively rare.

1.2 Contribution

To address the aforementioned problem, we propose an imbalanced multi-view clustering algorithm based on common and specific feature learning ImMC-CSFL, a novel approach that effectively integrates the common and specific information of multi-view data in a unified framework. The main contributions of our proposed approach are as follows:

(1) We design a unified framework to integrate common and specific information of imbalanced multi-view data, so that our approach can simultaneously utilize the consistency and complementarity of multi-view data.
(2) ImMC-CSFL incorporates GAN techniques and orthogonal constraints respectively to fully extract the common feature and specific feature of different views by iteratively training the common-specific information learning network and clustering network.
(3) To verify the effectiveness of ImMC-CSFL, extensive experiments were performed on three widely used clustering datasets, UCI Digits, BDGP, and CCV. Experimental results show that our common-specific multi-view feature learning model can more fully extract the feature of imbalanced multi-view data and achieve better clustering results compared with existing mainstream methods.

2 Related Work

At present, there are various excellent multi-view clustering algorithms both domestically and internationally. Based on the different mechanisms used, multi-view clustering is divided into the following categories.

Multi-view Spectral Clustering. It integrates multi-view data using graph fusion and employs spectral clustering for segmentation. Huang et al. introduced MvSCN, emphasizing intra-view invariance and inter-view consistency [1],. Zhu et al. proposed OMSC to address the limitations of two-step methods [2]. Yin et al. introduced a one-step method based on CSNE [3]. Jia et al. developed MVSC for tensor low-rank representations, focusing on intra-view and inter-view relationships [4] El Hajjar et al. presented CNESE, incorporating non-negative embedding [5].

Multi-view Subspace Clustering. It explores consistent subspaces within multi-view data to cluster similar data types. Gao et al. introduced MVSC in 2015, clustering subspace features from each view simultaneously [6]. Brbic et al. proposed low-rank sparse subspace multi-view clustering by constructing a shared affinity matrix to learn a unified subspace representation [7]. Zhang et al. addressed subspace recognition issues with flexible multi-view representation learning [8], further proposing LMSC to extract latent complementary information from multiple views [9]. Kang et al. introduced LMVSC, a large-scale multi-view subspace clustering algorithm with linear time complexity [10].

Multi-view Nonnegative Matrix Factorization Clustering. It employs non-negative matrix factorization to decompose the multi-view feature matrix into an indicator matrix and a base matrix, forming a multi-view shared indicator matrix[11]. In 2018, Zhang et al. introduced clustering analysis based on multi-view matrix decomposition, leveraging the local structure of samples [12]. Mekthanavanh et al. developed a multi-view social network video clustering model using non-negative matrix decomposition to create a shared consistent matrix from the latent feature matrix [13]. Nie et al. proposed FMVBKM for fast bilateral K-means multi-view clustering, introducing fast multi-view matrix triple decomposition [14]. Liu et al. addressed the issue of assigning equal weights to views in multi-view NMF algorithms with WM-NMF, which assigns weights to views to reduce the impact of unimportant views [15].

Multi-view Clustering Based on Multiple Kernels. It achieves clustering in a higher dimensional feature space by using a kernel function to map the sample features into this space. Liu et al. proposed a matrix-induced regularized multi-kernel k-means clustering MKKM [16], which reduces the selection of redundant kernels and enhances the diversity of kernels. Based on this, Liu et al. proposed a multicore clustering based on the subspace partitioning of nearest-neighbor kernels [17]and a missing multicore learning algorithm AMKL [18]. Sun et al. proposed MKLR-RMSC [19], a robust multi-view subspace clustering method using multi-kernel low-rank representations to extract unique and complementary view-specific information.

3   Imbalanced Multi-view Clustering Algorithm Based on Common-Specific Feature Learning (ImMC-CSFL)

The complementarity between different views of multi-view data can be reflected in the two aspects: 1) Feature consistency between different views, i.e. different views contain the consistency information; 2) Feature specificity between different views, i.e. each view contains its own differential information. To fully extract the consistency information and specificity information of multi-view data, and to leverage the complementarity of multi-view data, we proposed an imbalanced multi-view clustering algorithm based on common-specific feature learning. As shown in Fig. 1, the framework includes a deep feature extraction module, a common information learning module based on a GCN, and a specific information learning module on differential loss.

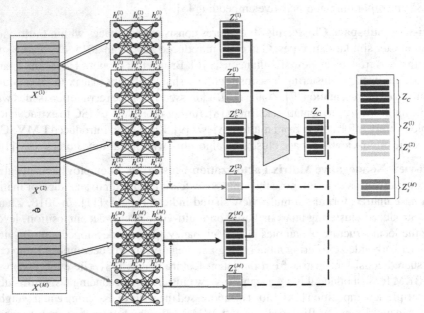

Fig. 1. The framework of imbalanced multi-view clustering algorithm based on common-specific feature learning.

3.1 Deep Feature Extraction Module

The feature extraction module based on deep convolutional networks contains two sub-networks: i.e., the common information extraction sub-network and the specific information extraction sub-network, which are used to extract the common feature across all views and the specific feature of each view. For a multi-view dataset $X = \{X^{(1)}, X^{(2)}, \ldots, X^{(M)}\}$ $X = \{X^{(1)}, X^{(2)}, \ldots, X^{(M)}\}$ $X = \{X^{(1)}, X^{(2)}, \ldots, X^{(M)}\}$ $X = \{X^{(1)}, X^{(2)}, \ldots, X^{(M)}\}$ $X = \{X^{(1)}, X^{(2)}, \ldots, X^{(M)}\}$, where M represents the number of views, $X^{(m)} \in \mathbb{R}^{d^m \times N}$, d^m is the dimensionality of samples in the m-th view, and N denotes the total number of samples. Each view is inputted into a deep learning network that connects two separate deep learning networks with various fully connected layers. The deep feature extraction module can fully extract the useful information of multi-view data, which is valuable for subsequent feature learning.

Assuming that both the common information extraction sub-network and the specific information extraction sub-network in each view consist of $n + 1$ fully connected layers, with each $k - th$ layer containing p_{sk} units, $k \in [0, n]$. The output of sample x from the

k-th layer of the common information network in the m-th view can be calculated as follows:

$$f_{ck}^m(x) = h_{ck}^m = \varphi(W_{ck}^m h_{ck}^{m-1} + b_{ck}^m) \tag{1}$$

where $W_{ck}^m \in \mathbb{R}^{p_{ck} \times p_{c(k-1)}}$ and $b_{ck}^m \in \mathbb{R}^{p_{ck}}$ indicate the weight matrix and bias vector of the k-th layer in the common information extraction sub-network, respectively. φ is a non-linear activation function, commonly include *sigmoid* and *tanh*.

Simultaneously, the output of sample x from the k-th layer of the specific information net-work in the m-th view can be calculated as follows:

$$f_{sk}^m(x) = h_{sk}^m = \varphi(W_{sk}^m h_{sk}^{m-1} + b_{sk}^m) \tag{2}$$

where $W_{sk}^m \in \mathbb{R}^{p_{sk} \times p_{s(k-1)}}$ and $b_{sk}^m \in \mathbb{R}^{p_{sk}}$, represent the weight matrix and bias vector of the k-th layer in the specific information extraction sub-network, respectively.

Therefore, for the i-th sample x_i^m in the m-th view, we can separately obtain corresponding common information and specific information, denoted as $h_{c,i}^m$ and $h_{s,i}^m$.

$$h_{c,i}^m = f_{cn}^m(x_i^m)$$

$$h_{s,i}^m = f_{sn}^m(x_i^m) \tag{3}$$

3.2 Common Information Learning Module

By inputting the multi-view data into common information feature extraction sub-networks, we can obtain common information of the same sample from different views. To maximize the common information extraction from different views, Generative Adversarial Network (GAN) technology is utilized in our framework. Figure 2 shows the structure of the common information learning module. Differ from traditional GAN, we consider the deep common feature extraction network on each view as a generator G, therefore there will be M generators, and input the common feature from the M generators into a for M-categorical discriminator D. The goal of G in this module is to generate feature with similar distributions on different views, so that the discriminator D struggles to determine the feature coming from which view. On the contrary, the goal of D is to determine which view the incoming feature come from through adversarial training. The aim of adversarial learning is to ensure that the extracted common information from different views is as similar as possible. Essentially, this model strives to maximize the extraction of the common information from different views. Through the above analysis, the loss of this module is as follows:

$$L_c = \min_G \max_D \left(\sum_{i=1}^N \sum_{m=1}^M l_i^m log D(G^m(x_i^m)) \right) \tag{4}$$

where, G^m indicates the generator (common information extraction network) on the m-th view, $G^m(x_i^m)$ indicates the feature generated form sample x_i^m generated by generator G^m, and l_i^m indicates the real label for sample x_i^m. The output of $D(G^m(x_i^m))$ is the probability that the generated sample comes from view m:

$$P_i^m = D(G^m(x_i^m)) \tag{5}$$

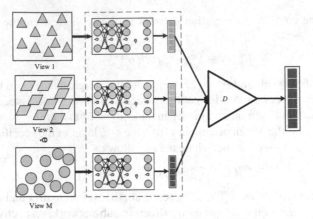

Fig. 2. Common Information Learning Module.

3.3 Specific Information Learning Module

To maximize the extraction of specific information, we minimize the correlation between specific information and common information. In our approach an orthogonal constraint is applied between the specific and common information within each view. The specific information learning network is shown in Fig. 3.

For the i-th sample x_i^m in the m-th view, it is simultaneously input into both the common information extraction sub-network and the specific information extraction sub-network. This results in obtaining the common information feature vector for the sample, denoted as $h_{c,i}^m$, and the specific information feature vector, denoted as $h_{s,i}^m$. Therefore, by incorporating an orthogonal constraint, the loss function is as follows:

$$L_s = \sum_{i=1}^{N} \|(h_{c,i}^m)^T h_{s,i}^m\|^2 \tag{6}$$

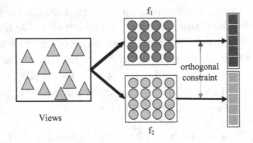

Fig. 3. Specific Information Learning Module.

3.4 Deep Multi-view Clustering Based on Common-Specific Feature Learning

Through the common-specific multi-view feature learning network, the common feature and specific feature can be obtained from each view. For the i-th sample x_i^m in the m-th view, take $h_{c,i}^m$ and $h_{s,i}^m$ as the extracted common and specific feature on view m. Then, by combining the common and specific feature vectors extracted from all views, the common-specific feature h_i for sample i is obtained as follows:

$$h_i = [(h_{c,i})^T, (h_{s,i}^1)^T, (h_{s,i}^2)^T, \ldots, (h_{s,i}^M)^T]^T \tag{7}$$

where $h_{c,i}$ represents the common feature of all views, i.e., multi-view common feature, it can be computed as follows:

$$hh_{c,i} = \frac{1}{M}\sum_{m=1}^{M} h_{c,i}^m \tag{8}$$

Then input the multi-view common-specific feature into the clustering network. Through the iterative training of the common-specific feature learning network and the clustering network, as shown in Fig. 4, to learn a positive clustering structure based on common-specific feature.

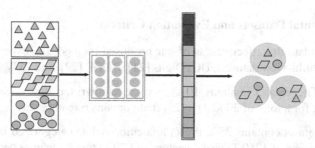

Fig. 4. The Deep Multi-View Clustering based on Common-Specific Feature Learning.

Through the above analysis, the total loss of the imbalanced multi-view clustering algorithm based on common-specific feature learning is designed as follows:

$$L = L_c + \lambda_1 L_s + \lambda_2 L_{clu} \tag{9}$$

where λ_1 and λ_2 are balancing factors used to adjust the weights of each part of the loss in the overall objective function. L_{clu} represents the clustering loss and is computed using the following formula:

$$L_{clu} = \sum_{i=1}^{N}\sum_{j=1}^{K} p_{ij} log\frac{p_{ij}}{q_{ij}} \tag{10}$$

where K is the number of clusters, q_{ij} represents the soft assignment probability of sample i belonging to cluster j, and p_{ij} represents the target probability of sample i belonging to cluster j. q_{ij} and p_{ij} are calculated as follows:

$$q_{ij} = \frac{(1+\|h_i - \mu_j\|^2/\alpha)^{-\frac{\alpha+1}{2}}}{\sum_{j'=1}^{K}(1+\|h_i - \mu_{j'}\|^2/\alpha)^{-\frac{\alpha+1}{2}}} \tag{11}$$

where, u_j indicates the center of cluster j, which can be obtained through clustering algorithms such as k-means applied to the common-specific feature extracted from the samples. α is a parameter variable, in our framework its value is set as 1, following reference [20].

$$p_{ij} = \frac{q_{ij}^2/f_j}{\sum_{j'=1}^{K} q_{ij}^2/f_{j'}} \qquad (12)$$

where f_j is the sum of soft assignment probabilities of all samples belonging to cluster j, specifically:

$$f_j = \sum_{i=1}^{N} q_i^j \qquad (13)$$

4 Experiment

4.1 Experimental Datasets and Evaluation Criteria

In order to validate the effectiveness of our proposed method, the experiments were taken in three multi-view datasets, UCI Digits [21], BDGP [22], and CCV [23] Dataset.

UCI Digits: The dataset contains 10 classes of handwritten digits, each with 200 different digits, for a total of 2000 data [25] that contains 6 feature sets.

BDGP: The dataset contains 2500 drosophila embryo data categorized into 5 classes. Each image consists of 1750-D visual vectors and 79-D textual feature vectors, i.e., the dataset contains 2 modalities.

CCV: The dataset consists of 9317 video segments collected on YouTube and contains 20 different semantic categories.

Evaluation Criteria: In the experiments clustering accuracy (ACC), normalized mutual information (NMI), and clustering purity (Purity) were selected as evaluation criteria.

4.2 Methods of Comparison

In order to fully evaluate algorithm performance, some excellent multi-view clustering methods were collected for comparison. Specifically,

(1) Traditional multi-view clustering methods: BestView [24], ConSC [25], RMSC, MVSC, CSMSC [26], MCNDCL [27].
(2) Multi-view clustering methods based on deep learning: DCCA [28], DMSC [29], DAMC [20].

4.3 Experimental Results

Table 1 shows the results of our method and the compared methods on the UCI Digits. From the results, we can see that the ImMC-CSFL can achieve the best performance on most of the criteria. Different from the BestView and ConSC method which is a single view-based clustering, the ImMC-CSFL effectively utilizes the advantages of multi-view data. The performance of ImMC-CSFL is also significantly improved compared to traditional multi-view methods. The ImMC-CSFL can learn more discriminative and clustering-friendly feature through deep learning techniques.

Table 1. The clustering performance of ImMC-CSFL and compared methods on UCI

Methods	ACC	NMI	Purity
BestView	68.2	66.3	69.9
ConSC	82.8	80.2	83.1
RMSC	86.3	78.0	90.4
MVSC	81.8	85.9	80.2
CSMSC	79.8	76.4	81.2
MCNDCL	90.3	84.6	90.7
DCCA	81.4	78.1	81.4
DMSC	91.6	85.5	91.6
DAMC	96.5	93.2	96.5
ImMC-CSFL	**97.8**	**95.6**	**98.2**

Table 2. The clustering performance of ImMC-CSFL and compared methods on BDGP

Methods	ACC	NMI	Purity
BestView	94.0	89.4	94.2
ConSC	58.4	38.4	58.4
RMSC	60.2	56.3	60.2
MVSC	68.2	56.9	68.4
CSMSC	94.9	84.9	94.8
MCNDCL	88.5	85.7	86.5
DCCA	57.8	40.9	57.8
DMSC	68.1	50.6	73.8
DAMC	98.2	94.6	**98.2**
ImMC-CSFL	**98.5**	**95.8**	97.6

Table 2 shows the evaluation results of all competing methods on the BDGP dataset. It shows that most of the multi-view methods underperform than BestView, our method

ImMC-CSFL is higher than the second best. In addition, the performance of the ImMC-CSFL method is significantly higher than BestView, which indicates that ImMC-CSFL is able to fully integrate the imbalanced multi-view data to improve the clustering performance.

Table 3 shows the experimental results of ImMC-CSFL and other compared methods on the CCV dataset. From the results, it is obvious that all the methods give unsatisfactory results on this dataset. That is because the quality of the samples on each view on this dataset is not high (as can be reflected from the results of BestView). This leads to unsatisfactory performance even after fusing multiple views, and improvement for this aspect will be a direction for subsequent research. Although all the methods underperform on this dataset, the ImMC-CSFL can also alleviate this phenomenon to some extent.

Table 3. The clustering performance of ImMC-CSFL and compared methods on CCV dataset

Methods	ACC	NMI	Purity
BestView	19.5	17.6	22.0
ConSC	10.6	8.6	10.8
RMSC	21.6	18.0	24.1
MVSC	19.3	15.2	21.0
CSMSC	23.9	18.7	27.8
MCNDCL	24.2	19.3	25.6
DCCA	20.7	15.9	21.9
DMSC	17.5	13.5	25.1
DAMC	25.6	22.5	**28.6**
ImMC-CSFL	**28.7**	**25.4**	28.5

5 Summary

In this paper, we propose an imbalanced multi-view clustering algorithm based on common-specific feature learning, called ImMC-CSFL. ImMC-CSFL first extracts the common information feature vector and the specific information feature vector on each view through two deep feature extraction networks, respectively. Then, through GAN technology, maximize common feature extracted from different views. Orthogonal constraints are introduced to minimize the correlation between common feature and specific feature. Finally, input the learned common specific feature into the clustering network for iterative training. Extensive experimental results and validation analyses on three public datasets show that the proposed ImMC-CSFL has better performance than the existing mainstream methods.

Acknowledgments. This work is supported by the National Natural Science Foundation of China (No. 62206092), the Natural Science Foundation of Hunan Province (No. 2023JJ40236, 2023JJ40239 and 2022JJ40129), the Research Foundation of Education Bureau of Hunan Province (No. 21B0582, 21B0565 and 21B0572), the Open Project of Xiangjiang Laboratory (No. 22XJ03012,22XJ03014,22XJ03022).

References

1. Huang, Z., Zhou, J., Peng, X., et al.: Multi-view Spectral Clustering Network. IJCAI, pp. 2563–2569 (2019)
2. Zhu, X., Zhang, S., He, W., et al.: One-step multi-view spectral clustering. IEEE Trans. Knowl. Data Eng.Knowl. Data Eng. **31**(10), 2022–2034 (2019)
3. Yin, H., Hu, W., Li, F., et al.: One-step multi-view spectral clustering by learning common and specific nonnegative embeddings. Int. J. Mach. Learn. Cybern.Cybern. **12**(7), 2121–2134 (2021)
4. Jia, Y., Liu, H., Hou, J., et al.: Multi-view spectral clustering tailored tensor low-rank representation. IEEE Trans. Circuits Syst. Video Technol. **31**(12), 4784–4797 (2021)
5. El Hajjar, S., Dornaika, F., Abdallahde, F., et al.: Multi-view spectral clustering via constrained nonnegative embedding. Inf. Fusion **78**, 209–217 (2022)
6. Gao, H., Nie, mF., Li, X., et al.: Multi-view subspace clustering. In: ICCV 2015, pp. 4238–4246 (2015)
7. Brbic, M., Kopriva, I.: Multi-view low-rank sparse subspace clustering. Pattern Recognit. **73**, 247–258 (2018)
8. Li, R., Zhang, C., Hu, Q., et al.: Flexible multi-view representation learning for subspace clustering. IJCAI 2019, pp. 2916–2922 (2019)
9. Zhang, C., Hu, Q., Fu, H., et al.: Generalized latent multi-view subspace clustering. IEEE Trans. Pattern Anal. Mach. Intell.Intell. **42**(1), 86–99 (2020)
10. Kang, Z., Zhou, W., Zhao, Z., et al.: Large-scale multi-view subspace clustering in linear time. In: AAAI 2020, pp. 4412–4419 (2020)
11. Liu, J., Wang, C., Gao, J., et al.: Multi-view clustering via joint nonnegative matrix factorization. In: SDM 2013, 252–260 (2013)
12. Zhang, Y., Kong, X.W., Wang, Z.F., et al.: Cluster analysis based on multi-view matrix decomposition. J. Autom. **2018**(44).12, 2160–2169 (2018)
13. Mekthanavanh, V., Li, T., Meng, H., et al.: Social web video clustering based on multi-view clustering via nonnegative matrix factorization. Int. J. Mach. Learn. Cybern.Cybern. **10**(10), 2779–2790 (2019)
14. Nie, F., Shi, S., Li, X.: Auto-weighted multi-view co-clustering via fast matrix factorization. Pattern Recogn.Recogn. **102**, 107207 (2020)
15. Liu, S.S., Lin, L.: Integrative clustering of multi-view data by nonnegative matrix factorization. ArXiv, abs/2110.13240 (2021)
16. Liu, X., Dou, Y., Yin, J., et al.: Multiple kernel k-means clustering with matrix-induced regularization. In: AAAI 2016, pp. 1888–1894
17. Zhou, S., Liu, X., Li, M., et al.: Multiple kernel clustering with neighbor-kernel subspace segmentation. IEEE Trans. Neural Networks Learn. Syst. **31**(4), 1351–1362 (2020)
18. Liu, X., Wang, L., Zhu, X., et al.: Absent Multiple Kernel Learning Algorithms. IEEE Trans. Pattern Anal. Mach. Intell.Intell. **42**(6), 1303–1316 (2020)
19. Zhang, X., Ren, Z., Sun, H., et al.: Multiple kernel low-rank representation-based robust multi-view subspace clustering. Inf. Sci. **551**, 324–340 (2021)

20. Li, Z., Wang, Q., Tao, Z., Gao, Q., Yang, Z.: Deep adversarial multi-view clustering network. In: Twenty-Eighth International Joint Conference on Artificial Intelligence, Macao, China, pp. 2952–2958 (2019)

21. Rai, N., Negi, S., Chaudhury, S., Deshmukh, O.: Partial multi-view clustering using graph regularized NMF. In: 23rd International Conference on Pattern Recognition, Cancún, Mexico, pp. 2192–2197 (2016)

22. Cai, X., Wang, H., Huang, H., Ding, C.H.Q.: Joint stage recognition and anatomical annotation of drosophila gene expression patterns. Bioinform. **28**(12), 16–24 (2012)

23. Jiang, Y., Ye, G., Chang, S., Ellis, D.P.W., Loui, A.C.: Consumer video understanding: a benchmark database and an evaluation of human and machine performance. In: 1st International Conference on Multimedia Retrieval, Trento, Italy, pp. 1–8 (2011)

24. Ng, Y., Jordan, M.I., Weiss, Y.: On spectral clustering: analysis and an algorithm. In: Advances in Neural Information Processing Systems, Vancouver, Canada, pp. 849–856 (2001)

25. Kumar, A., Rai, P., III, H.D.: Co-regularized multi-view spectral clustering. In: Annual Conference on Neural Information Processing Systems, Granada, Spain, pp. 1413–1421 (2011)

26. Luo, S., Zhang, C., Zhang, W., Cao, X.: Consistent and specific multi-view subspace clustering. In: Thirty-Second AAAI Conference on Artificial Intelligence, New Orleans, Louisiana, USA, pp. 3730–3737 (2018)

27. Li, X., Zhou, K., Li, C., et al.: Multi-view clustering via neighbor domain correlation learning. Neural Comput. Applic.Applic. **33**, 3403–3415 (2021). https://doi.org/10.1007/s00521-020-05185-y

28. Andrew, G., Arora, R., Bilmes, J.A., Livescu, K.: Deep canonical correlation analysis. In: 30th International Conference on Machine Learning, Atlanta, GA, USA, vol. 28 of JMLR Workshop and Conference Proceedings, pp. 1247–1255 (2013)

29. Abavisani, M., Patel, V.M.: Deep multimodal subspace clustering networks. J. Sel. Topics Signal Processing **12**(6), 1601–1614 (2018)

Multivariate Beta Mixture Model: Probabilistic Clustering with Flexible Cluster Shapes

Yung-Peng Hsu and Hung-Hsuan Chen[✉]

National Central University, Taoyuan City, Taiwan
hhchen1105@acm.org

Abstract. This paper introduces the multivariate beta mixture model (MBMM), a new probabilistic model for soft clustering. MBMM adapts to diverse cluster shapes because of the flexible probability density function of the multivariate beta distribution. We introduce the properties of MBMM, describe the parameter learning procedure, and present the experimental results, showing that MBMM fits diverse cluster shapes on synthetic and real datasets. The code is released anonymously at https://github.com/hhchen1105/mbmm/.

Keywords: Mixture model · EM algorithm · Clustering

1 Introduction

Data clustering groups data points into components so that similar points are within the same component. Data clustering is commonly used for data exploration and is sometimes used as a preprocessing step for later analysis [11].

In this paper, the multivariate beta mixture model (MBMM), a new probabilistic model for soft clustering, is proposed. As the MBMM is a mixture model, it shares many properties with the Gaussian mixture model (GMM), including its soft cluster assignment and parametric modeling. In addition, the MBMM allows the generation of new (synthetic) instances based on a generative process. Because the beta distribution is highly flexible (e.g., unimodal, bimodal, straight line, or exponentially increasing or decreasing), MBMM can fit data with versatile shapes. Figure 1 shows that various cluster shapes can be obtained with a bivariate beta distribution. On the contrary, the shape of a Gaussian distribution is symmetric and unimodal, which limits its fitting capacity.

The multivariate beta distribution is defined in different ways. In some studies, the Dirichlet distribution is considered a multivariate beta distribution (e.g., [9]) because the beta distribution is a special case of the Dirichlet distribution with two parameters. However, we apply the definition provided in [6], which is even more general than the Dirichlet distribution. The relationship between the Dirichlet distribution and our multivariate beta distribution will be discussed in Sect. 2.1 when we introduce the details of the multivariate beta distribution.

© The Author(s), under exclusive license to Springer Nature Singapore Pte Ltd. 2024
D.-N. Yang et al. (Eds.): PAKDD 2024, LNAI 14645, pp. 233–245, 2024.
https://doi.org/10.1007/978-981-97-2242-6_19

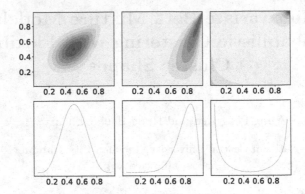

Fig. 1. Examples of the versatile shape of the bivariate beta distribution. The upper row shows three bivariate beta distributions with different parameters. The bottom row shows the marginal distribution of x_1 (i.e., the variable on the horizontal axis in the top row). This distribution can be symmetric unimodal (e.g., the left subfigure), skewed unimodal (e.g., the middle subfigure), or bimodal (e.g., the right subfigure).

This paper presents several contributions. First, we propose a new probabilistic model for soft clustering. Our model is similar to the Gaussian Mixture Model (GMM), but the shape of each cluster is more versatile than those generated by GMM. Second, we compare MBMM with well-known clustering algorithms on synthetic and real datasets to demonstrate its effectiveness. Finally, we release the code for reproducibility. Our implemented class offers `fit()`, `predict()`, and `predict_proba()`, the common methods provided by `scikit-learn`'s clustering algorithms, making it convenient to apply MBMM to new domains.

The rest of the paper is organized as follows. Section 2 introduces the multivariate beta distribution and the proposed MBMM. Section 3 describes experiments on synthetic and real datasets. Section 4 reviews previous work on data clustering. We conclude by discussing the limitations of the MBMM and the ongoing and future work on the MBMM in Sect. 5.

2   Multivariate Beta Mixture Model

2.1   Multivariate Beta Distribution

The probability density function (PDF) of a multivariate beta distribution (MB) has been defined in different ways [2,6]. Here, we apply the definition in [6]: given an instance $x = [x_1, \ldots, x_M]^T$ with M variates (i.e., features) and the shape parameters $a_m > 0, b > 0$ $(m = 1, \ldots, M)$, its PDF is given by

$$MB(x|a_{1:M}, b) = \frac{1}{Z} \times \frac{\prod_{m=1}^{M} \frac{x_m^{a_m-1}}{(1-x_m)^{a_m+1}}}{(1 + \sum_{k=1}^{M} \frac{x_k}{1-x_k})^{a_1+\ldots+a_M+b}}, \tag{1}$$

where $x_m \in (0,1), a_m > 0, b > 0$, and the normalizer Z is defined by

$$Z = \frac{\Gamma(b) \prod_{m=1}^{M} \Gamma(a_m)}{\Gamma(b + \sum_{j=1}^{M} a_j)}, \tag{2}$$

where Γ is the gamma function.

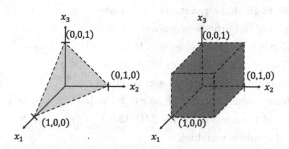

Fig. 2. A comparison of the support of the Dirichlet distribution (left) and our multivariate beta distribution (right) with 3 variates. The Dirichlet distribution is only defined on $x_i \in (0,1)$ such that $x_1 + x_2 + x_3 = 1$ (the standard 2-simplex in R^3). On the contrary, our multivariate beta distribution is defined on $(0,1)^3$ (the unit cube in R^3), which is a superset of the Dirichlet distribution.

In some previous studies, the Dirichlet distribution was treated as a multivariate generalization of the beta distribution (e.g., [9]) since the Dirichlet distribution falls back to the beta distribution when the number of parameters is 2. However, we describe a more general definition of the multivariate beta distribution that regards the Dirichlet distribution as a special case. The relationship between the Dirichlet distribution and the proposed multivariate beta distribution is illustrated in Fig. 2. Specifically, the support of an n-variate Dirichlet distribution is restricted to a standard $(n-1)$-simplex. However, the support of our multivariate beta distribution is a hypercube in an n-dimensional space with a length of 1 on each side. In other words, the Dirichlet distribution is the multivariate beta distribution subject to $\|x\|_1 = 1$.

2.2 MBMM Density Function and Generative Process

In Table 1, we list the notations that will be used in this paper hereafter.

In MBMM, it is assumed that the data points are generated from a mixture of multivariate beta distributions (whose PDF is defined in Eq. 1). Consequently, the probability of the MBMM given C components is

$$p(\boldsymbol{x}_n|\boldsymbol{\theta}) = \sum_{c=1}^{C} \pi_c MB(\boldsymbol{x}_n|\boldsymbol{\theta}_c). \tag{3}$$

Table 1. Notation list

Indices:	
M	Dimensions of an observed instance ($m \in \{1, \dots, M\}$)
C	Number of clusters ($c \in \{1, \dots, C\}$)
N	Number of instances ($n \in \{1, \dots, N\}$)
Parameters:	
$a_{c,m}$	the m-th shape parameter for cluster c; $a_{c,m} > 0$
b_c	the $(M+1)$-th shape parameter for cluster c; $b_c > 0$
π_c	Mixture weight of cluster c, $0 < \pi_c < 1, \sum_{c=1}^{C} \pi_c = 1$
z_n	Cluster that $\boldsymbol{x}_n$ belongs to; $z_n \in \{1, \dots, C\}$
$\gamma_{n,c}$	Probability that $\boldsymbol{x}_n$ belongs to cluster c; $\sum_{c=1}^{C} \gamma_{n,c} = 1$
$\boldsymbol{\theta}_c$	Set of parameters for cluster c; $\boldsymbol{\theta}_c = \{a_{c,1}, \dots, a_{c,M}, b_c\}$
$\boldsymbol{\theta}$	Set of parameters for the MBMM; $\boldsymbol{\theta} = \{a_{1:C,1:M}, b_{1:C}, \pi_{1:C}\}$
Observed random variables:	
$\boldsymbol{x}_n$	Observed instance; $\boldsymbol{x}_n = [x_{n,1}, \dots, x_{n,M}]^T \in R^M$

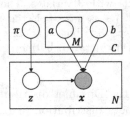

Fig. 3. Graphical representation of the multivariate beta mixture model

The parameter π_c determines the probability that a random instance $\boldsymbol{x}_n$ belongs to cluster c (before knowing the values of the variates in $\boldsymbol{x}_n$), and $MB(\boldsymbol{x}_n | \boldsymbol{\theta}_c)$ gives the PDF if $\boldsymbol{x}_n$ indeed belongs to cluster c.

Figure 3 shows a graphical representation of the multivariate beta mixture model. To generate a sample $\boldsymbol{x}_n$, we first sample a latent variable z_n (the cluster ID of the sample $\boldsymbol{x}_n$) from a multinomial distribution with parameters $\pi_1, \dots, \pi_C$. Suppose that $z_n = c$ after sampling, we further sample an instance $\boldsymbol{x}_n$ from the multivariate beta distribution with parameters $\boldsymbol{\theta}_c$: $MB(\boldsymbol{x}_n | \boldsymbol{\theta}_c) = MB(\boldsymbol{x}_n | a_{c,1}, \dots, a_{c,M}, b_c)$.

2.3 Parameter Learning for the MBMM

In reality, we do not know the values of the parameters $\boldsymbol{\theta} = \{a_{1:C,1:M}, b_{1:C}, \pi_{1:N}\}$ (referring to Fig. 3). We hope to recover these parameters based on the observed $\boldsymbol{x}_n$-s to maximize the likelihood function:

Data: Input data $x_1, \ldots, x_N$, cluster number C
Result: Parameters $\theta = \{\pi_{1:C}, a_{1:C,1:M}, b_{1:C}\}$
Initialize θ randomly;
while not converge **do**
    // E-step
    for $n \leftarrow 1$ **to** N **do**
        for $c \leftarrow 1$ **to** C **do**
            | Update $\gamma_{n,c}$ by Equation 7;
        end
    end
    // M-step
    Update $a_{1:C,1:M}$ and $b_{1:C}$ with the SQP solver [10];
    Update $\pi_{1:C}$ by Equation 8;
    if iteration count reaches a pre-defined value **then**
        | Exit while loop;
    end
end

Algorithm 1: Parameter learning algorithm for the MBMM

$$L(\theta) = p(x_{1:N}, z_{1:N}|\theta) = \prod_{n=1}^{N} \prod_{c=1}^{C} [\pi_c MB(x_n|\theta_c)]^{I(z_n=c)}, \qquad (4)$$

where I is the indicator function.

As the likelihood function (Eq. 4) involves the multiplication of $N \times C$ terms, the result is numerically unstable. Instead, we compute the log-likelihood function to convert multiplications to additions, as shown in Eq. 5. As a result, the computation is more numerically stable.

$$\log L(\theta) = \sum_{n=1}^{N} \sum_{c=1}^{C} I(z_n = c) \left(\log \pi_c + \log MB(x_n|\theta_c)\right). \qquad (5)$$

However, since we cannot observe the latent z_n in practice, direct optimization of Eq. 5 is difficult. As an alternative, we compute the expected value of the log-likelihood function with respect to the latent variables $z_{1:N}$, which involves the expected (but not the true) values of z_n:

$$E_{z_{1:N}}\left[\log L(\theta)\right] = \sum_{n=1}^{N} \sum_{c=1}^{C} \gamma_{n,c} \left(\log \pi_c + \log MB(x_n|\theta_c)\right). \qquad (6)$$

After the above reformulation, the parameters $(\theta_{1:C}, \pi_{1:C})$ that are used to maximize the expected value of the log-likelihood function (Eq. 6) can be learned via the EM algorithm, as given by the pseudocode in Algorithm 1. In the E-step, we compute $\gamma_{n,c}$ (the probability that instance x_n belongs to cluster c) that maximizes Eq. 6 by assuming that the randomly initialized or currently estimated $\pi_{1:C}$ and $\theta_{1:C}$ are correct. The assignment of $\gamma_{n,c}$ has a simple closed-form solution, as shown below

$$\gamma_{n,c} = \frac{\pi_c MB(\boldsymbol{x}_n|\boldsymbol{\theta}_c)}{\sum_{k=1}^{C} \pi_k MB(\boldsymbol{x}_n|\boldsymbol{\theta}_k)}. \tag{7}$$

In the M-step, we search for the parameters $\pi_{1:C}$, $a_{1:C,1:M}$, and $b_{1:C}$ by assuming that the estimated $\gamma_{n,c}$ values in the E-step are correct. However, since the $a_{1:C,1:M}, b_{1:C}$ parameters seem to lack a closed-form solution, we resort to numerical optimization strategies, specifically the sequential quadratic programming (SQP) iterative method, as the minimization strategy [10] because SQP allows linear constraints on the parameters (i.e., $a_{c,m} > 0$ and $b_c > 0 \; \forall c, m$). For parameters $\pi_1, \ldots, \pi_C$, we rely the efficient closed-form solution:

$$\pi_c = \frac{1}{N} \sum_{n=1}^{N} \gamma_{n,c}. \tag{8}$$

We compute the difference between the log-likelihood estimation in successive rounds for the convergence check. Additionally, if the number of iterations reaches a predefined value, we terminate the loop.

2.4   The Similarity Score Between Data Points

Most clustering algorithms define the distance between two samples by converting them into a non-negative real value, i.e., given $\boldsymbol{x}_i, \boldsymbol{x}_j \in R^M$, the distance function is represented by $d_{i,j} : \boldsymbol{x}_i \times \boldsymbol{x}_i \rightarrow \{0, R^+\}$. However, if we define the distance of two samples based on their coordinates and assign samples to the closest cluster centroid, the output shapes of the clusters are inevitably convex.

In MBMM, we define the distance between two data points from a different perspective. Since the PDF of a data point is an affine combination of C multivariate beta distributions (Eq. 3), we consider $MB(\cdot|\boldsymbol{\theta}_1), \ldots, MB(\cdot|\boldsymbol{\theta}_C)$ as the basis to form a function space. Consequently, the coordinate of the data point $\boldsymbol{x}_n$ becomes $\boldsymbol{\gamma}_n = [\gamma_{n,1}, \gamma_{n,2}, \ldots, \gamma_{n,C}]^T$ with respect to the basis functions. The vector $\boldsymbol{\gamma}_n$ is a discrete probability distribution since $\sum_{c=1}^{C} \gamma_{n,c} = 1$. Thus, we can define the distance between the data points $\boldsymbol{x}_i$ and $\boldsymbol{x}_j$ as the distance between the discrete probability distributions $\boldsymbol{\gamma}_i$ and $\boldsymbol{\gamma}_j$. We use the Kullback-Leibler divergence (KL divergence) to determine this distance:

$$d_{i,j}^{KL} := \sum_{c=1}^{C} \gamma_{i,c} \log \left(\frac{\gamma_{i,c}}{\gamma_{j,c}} \right). \tag{9}$$

3   Experiments

3.1   Comparisons on the Synthetic Datasets

Baseline Models. Clustering algorithms can be classified into four types: centroid-based, density-based, distribution-based, and hierarchical clustering algorithms. We select representative models from each of these categories. For centroid-based models, we choose the k-means algorithm, which is likely the most widely used clustering algorithm. We select the DBSCAN algorithm for density-based models, which received the Test of Time award at KDD 2014. We choose the GMM as the distribution-based model and the agglomerative clustering (AC) algorithm as the hierarchical clustering model.

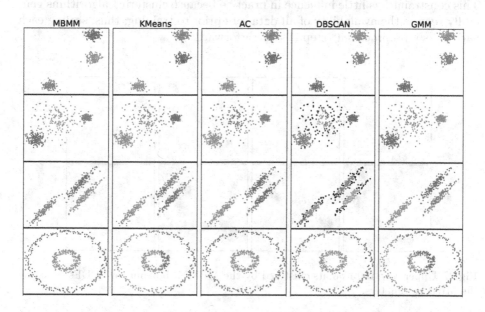

Fig. 4. Clustering algorithms on synthetic datasets

Synthetic Dataset Generation. We generate synthetic datasets for the experiments. First, we create data points from three isotropic 2D Gaussian distributions whose means are distant and whose variances are small in each dimension. Consequently, a data point is close to other data points of the same Gaussian distribution but far from those in other Gaussian distributions (Fig. 4, first row). This dataset represents an ideal case for data clustering.

Second, we generate two distant 2D Gaussian distributions with small variances in each dimension. However, the third distribution has a large variance. Consequently, several data points sampled from the third distribution are mixed with those from other distributions (Fig. 4, second row).

Third, we generate three 2D Gaussian distributions with isolated means as before. However, we introduce a high correlation between two covariates for each Gaussian distribution. Consequently, data points are sometimes closer to data points generated from other distributions (Fig. 4, third row).

Fourth, we generate concentric circles (i.e., one circle within another). If a point from the outer circle is selected, the most distant data point is located on the other side of the same circle. Consequently, such a synthetic dataset is highly challenging for centroid-based and distribution-based methods to group the outer circle as one cluster (Fig. 4, fourth row).

We scale the range of every feature to within 0 and 1 because the support of each variate in the MB distribution is between 0 and 1, as explained in Sect. 2.1. This constraint has little influence in practice because clustering algorithms generally require the availability of all data sets prior to training; thus, scaling each variate as a preprocessing step is straightforward.

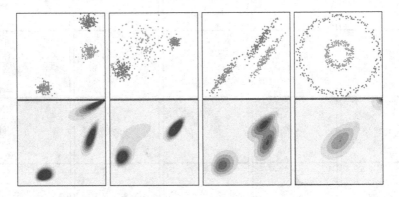

Fig. 5. Upper row: data points clustered by the MBMM; bottom row: PDFs based on the fitted parameters.

Visualizing Clustering Results. Figure 4 shows a visualized comparison of the clustering algorithms for the four synthetic datasets.

All algorithms compared perform well on the first synthetic dataset. However, some data points belonging to the middle cluster are incorrectly grouped as the right cluster for the second dataset when using k-mean and AC. This is because the middle cluster has a wider spread, making the centroid far away from some points in the same cluster. With DBSCAN, many data points from the middle cluster are regarded as outliers because density-based algorithms usually have difficulty when the intracluster distance (the distance between members of a cluster) differs greatly. For similar reasons, unsatisfied performance is obtained using the k means and DBSCAN on the third dataset. Mediocre results are also obtained using the AC algorithm, likely because Ward's linkage function merges the wrong groups. On the concentric circles dataset, poor performance is observed for the k-means and GMM algorithms because they can only group geometrically neighboring nodes in one cluster. Since the AC and DBSCAN algorithms can recursively group adjacent points into the same cluster, it is possible to group two distant points into the same cluster, so reasonable performance is achieved. Excellent results are obtained using our proposed MBMM on all synthetic datasets because the shape of a multivariate beta distribution is versatile. In particular, for the fourth dataset, because a multivariate beta distribution can be bimodal, the MBMM can group the data points in the outer circle as one cluster even though they are geometrically distant.

We visualize the PDFs by fitting the MBMM to the four synthetic datasets (Fig. 5). The upper row shows the data points, and the bottom row shows the PDFs estimated by MBMM, which indeed fits these datasets adequately.

3.2 Comparison on the Real Datasets

Real Datasets. We used two open real datasets. The first dataset is MNIST, which includes grayscale images of handwritten digits. The size of each image is 28×28. Since image pixels should have spatial correlations, directly using the pixel values as input features for clustering algorithms could be problematic. Eventually, we reduce the dimension of each image to 2 dimensions using the following procedure. First, we train a vanilla convolutional neural network (ConvNet) using the Fashion MNIST dataset (not the MNIST dataset). Then, we feed each MNIST image into the Fashion MNIST-trained ConvNet and take the hidden layer before the output (a vector with 512 neurons) as the image representation. Finally, we use a standard autoencoder to reduce the vector into 2 dimensions, which are the inputs of the clustering algorithms. Ultimately, we include only images of number 1 and number 9 in MNIST for experiments.

The second dataset, the breast cancer Wisconsin (diagnostic) dataset, consists of 569 instances. Each instance includes 32 attributes and a binary class label indicating the status of the tumor (benign or malignant). We download the dataset from the UCI Machine Learning Repository.

Results. We compare clustered IDs with ground truth labels to calculate the adjusted Rand index (ARI) [12,16] and adjusted mutual information (AMI) [15], two standard metrics for clustering evaluation. If a clustering result perfectly matches the referenced clusters (labels), both metrics return a score of 1. However, ARI and AMI are biased toward different types of clustering results: ARI

Table 2. Comparison of the clustering algorithms on the MNIST dataset and the breast cancer dataset (mean ± standard deviation)

	MNIST		breast cancer	
	ARI	AMI	ARI	AMI
MBMM	**.937** ± .000	**.884** ± .000	.664 ± .000	.558 ± .000
k-means	.913 ± .000	.850 ± .000	.491 ± .000	.464 ± .000
AC	.933 ± .000	.878 ± .000	**.689** ± .000	**.568** ± .000
DBSCAN	.854 ± .009	.745 ± .011	.554 ± .018	.447 ± .010
GMM	.909 ± .000	.846 ± .000	.664 ± .000	.558 ± .000

prefers balanced partitions (clusters with similar sizes), and AMI prefers unbalanced partitions [13]. For a fair comparison, we report both metrics.

Table 2 shows the results. We repeat each experiment five times and report the mean ± standard deviation. We highlight each metric's first and second highest values in bold and underlined. For MNIST, the top 3 methods are our MBMM, followed by AC, and then k-means. For the cancer dataset, the best performance is achieved using the AC algorithm, followed by our MBMM and GMM. In general, our MBMM and AC perform best among all.

3.3   Distance Between Data Points

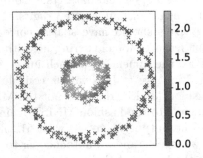

Fig. 6. Distance from red point to other points in concentric circles dataset (Color figure online)

Table 3. Popular clustering algorithms and their properties

	MBMM	k-means	DBSCAN	AC	GMM
Type	Distribution-based	Centroid-based	Density-based	Hierarchical clustering	Distribution-based
Assignment	Soft	Hard	Hard	Hard	Soft
Cluster shape	Versatile	Convex	Versatile	Versatile	Convex
Generative/ Discriminative	Gen	Discr	Discr	Discr	Gen

As explained in Sect. 2.4, we define the distance between x_i and x_j based on the KL divergence between $[\gamma_{i,1},\dots,\gamma_{i,C}]^T$ and $[\gamma_{j,1},\dots,\gamma_{j,C}]^T$. Consequently, even if x_i and x_j are distant based on the Euclidean distance, they could still have a small distance score if $\gamma_{i,c} \approx \gamma_{j,c}$ for most c-s.

We illustrate the red data point's distances to others using concentric circles in Fig. 6. Outer circle points are closer, showing that MBMM's distance function can assign small values even with large Euclidean distances.

4   Related Work

Clustering algorithms can be classified into four types based on how they partition data points: hierarchical, centroid-based, density-based, and distribution-based clustering.

Hierarchical clustering algorithms are top-down or bottom-up, corresponding to the iterative division of each cluster into smaller clusters and the aggregation of smaller clusters into larger clusters. Hierarchical clustering algorithms allow dynamically adjusting the cluster numbers. However, users must define the distance between not only data points but also between clusters, which could sometimes be counterintuitive. Well-known hierarchical clustering algorithms include agglomerative clustering (AC) and BIRCH [17].

Centroid-based algorithms represent each cluster using a centroid, assigning each data point to the closest cluster. However, these algorithms generate clusters with convex shapes, eliminating the possibility of fitting a bimodal cluster. Well-known algorithms include k-means, k-medoids, k-medians, and k-means++.

Density-based algorithms determine clusters by assuming that densely distributed areas are clusters. Typical algorithms include DBSCAN [4] and OPTICS [1]. Although they discover clusters of various shapes, hyperparameter tuning can be time-consuming and heavily influence clustering results [7]. Additionally, density-based algorithms sometimes have difficulty clustering data points when the distances between different data points vary widely.

Distribution-based models assume that each cluster follows a probability distribution. One well-known is GMM, which assumes that each cluster follows a Gaussian distribution. Distribution-based models naturally generate synthetic data points by sampling a cluster ID and then one data point from the data distribution of cluster i. One problem with GMM is that the shape of each cluster must be convex since this is a fundamental property of a Gaussian distribution. As a result, GMM cannot differentiate between inner and outer circles.

Table 3 gives an overview of these clustering algorithms and their properties. Both MBMM and GMM are distribution-based models, which thus allow for soft clustering and synthetic generation of the data points. DBSCAN and AC allow non-convex cluster shapes in which each data point within a cluster is close to only a few data points from the same cluster. MBMM also supports non-convex cluster shapes due to the versatility of the multivariate beta distribution.

The beta mixture model has been studied in the bioinformatics and biochemical domains [5,14]. However, they assumed that each cluster follows a standard beta distribution, which limits the practical usage of these models because each data point must be univariate. This constraint is probably due to the fact that the definition of a multivariate beta distribution is still ambiguous. Our study is more practical because we allow each data point to be multivariate.

5   Discussion

This paper proposes a new probabilistic model, the multivariate beta mixture model, for data clustering. We demonstrate MBMM's effectiveness by thorough experiments on synthetic and real datasets. Furthermore, MBMM is a generative model that allows for the generation of new data points. Compared to another famous generative clustering algorithm, the Gaussian mixture model, MBMM allows for a more flexible cluster shape. To ensure reproducibility, we have released our experimental code and encapsulated the MBMM module as a class with typical class methods supported by the clustering algorithms in `scikit-learn`, facilitating the utilization of the MBMM in various applications.

Although MBMM has these nice properties, we believe that different clustering algorithms should be used in combination to jointly partition data points for the following reasons. First, data clustering is ill-defined due to the lack of ground truth labels during both training and testing, making the choice of training objective and evaluation ad hoc [3]. Additionally, it has been shown that, under reasonably general conditions, no single clustering algorithm can satisfy the three fundamental properties introduced in [8].

The capacity of MBMM is limited by the need for a positive correlation among all variates due to parameter b [6]. We plan to explore multivariate beta distributions allowing both positive and negative correlations based on [2].

References

1. Ankerst, M., Breunig, M.M., Kriegel, H.P., Sander, J.: OPTICS: ordering points to identify the clustering structure. ACM SIGMOD Rec. **28**(2), 49–60 (1999)
2. Arnold, B.C., Ng, H.K.T.: Flexible bivariate beta distributions. J. Multivar. Anal. **102**(8), 1194–1202 (2011)
3. Caruana, R., Elhawary, M., Nguyen, N., Smith, C.: Meta clustering. In: International Conference on Data Mining, pp. 107–118. IEEE (2006)
4. Ester, M., Kriegel, H., Sander, J., Xu, X.: A density-based algorithm for discovering clusters in large spatial databases with noise. In: Simoudis, E., Han, J., Fayyad, U.M. (eds.) International Conference on Knowledge Discovery and Data Mining, pp. 226–231. AAAI Press (1996)
5. Ji, Y., Wu, C., Liu, P., Wang, J., Coombes, K.R.: Applications of beta-mixture models in bioinformatics. Bioinformatics **21**(9), 2118–2122 (2005)
6. Jones, M.: Multivariate t and beta distributions associated with the multivariate f distribution. Metrika **54**(3), 215–231 (2002)
7. Karami, A., Johansson, R.: Choosing dbscan parameters automatically using differential evolution. Int. J. Comput. Appl. **91**(7), 1–11 (2014)
8. Kleinberg, J.: An impossibility theorem for clustering. Adv. Neural Inf. Process. Syst. **15**, 463–470 (2003)
9. Kotz, S., Balakrishnan, N., Johnson, N.L.: Continuous Multivariate Distributions, Volume 1: Models and Applications. John Wiley & Sons, Hoboken (2004)
10. Kraft, D.: A software package for sequential quadratic programming. Wiss. Berichtswesen d. DFVLR (1988)
11. Lien, C.Y., Bai, G.J., Chen, H.H.: Visited websites may reveal users' demographic information and personality. In: International Conference on Web Intelligence, pp. 248–252. IEEE (2019)
12. Rand, W.M.: Objective criteria for the evaluation of clustering methods. J. Am. Stat. Assoc. **66**(336), 846–850 (1971)
13. Romano, S., Vinh, N.X., Bailey, J., Verspoor, K.: Adjusting for chance clustering comparison measures. J. Mach. Learn. Res. **17**(1), 4635–4666 (2016)
14. Schröder, C., Rahmann, S.: A hybrid parameter estimation algorithm for beta mixtures and applications to methylation state classification. Algor. Molec. Biol. **12**(1), 1–12 (2017)

15. Vinh, N.X., Epps, J., Bailey, J.: Information theoretic measures for clusterings comparison: variants, properties, normalization and correction for chance. J. Mach. Learn. Res. **11**, 2837–2854 (2010)
16. Wagner, S., Wagner, D.: Comparing clusterings: an overview. Universität Karlsruhe, Fakultät für Informatik Karlsruhe (2007)
17. Zhang, T., Ramakrishnan, R., Livny, M.: Birch: an efficient data clustering method for very large databases. ACM SIGMOD Rec. **25**(2), 103–114 (1996)

AutoClues: Exploring Clustering Pipelines via AutoML and Diversification

Matteo Francia[ID], Joseph Giovanelli[✉][ID], and Matteo Golfarelli[ID]

University of Bologna, Cesena, Italy
{m.francia,j.giovanelli,matteo.golfarelli}@unibo.it

Abstract. AutoML has witnessed effective applications in the field of supervised learning – mainly in classification tasks – where the goal is to find the best machine-learning pipeline when a ground truth is available. This is not the case for unsupervised tasks that are by nature exploratory and they are performed to unveil hidden insights. Since there is no right result, analyzing different configurations is more important than returning the best-performing one. When it comes to exploratory unsupervised tasks – such as cluster analysis – different facets of the datasets could be interesting for the data scientist; for instance, data items can be effectively grouped together in different subspaces of features. In this paper, AutoClues explores and returns a dashboard of both relevant and diverse clusterings via AutoML and diversification. AutoML ensures that the explored pipelines for cluster analysis (including pre-processing steps) compute good clusterings. Then, diversification selects, out of the explored clusterings, the ones conveying different clues to the data scientists.

Keywords: AutoML · Clustering · Diversification

1 Introduction

Thanks to the abundant presence of data, machine learning (ML) has been employed in a variety of fields (e.g., urban mobility [27]). Depending on the scope of the analysis, the task is defined either supervised (i.e., leveraging a ground truth; e.g., classification) or unsupervised (i.e., without ground truth; e.g., cluster analysis). Data scientists design the workflow as a *ML pipeline*; namely, a series of pre-processing steps (e.g., features selection) are in charge of shaping the data so that the final analysis step can produce the best result. For each step of the pipeline, there are alternative algorithms (e.g., SPEC [31] for feature selection, KMeans [1] for cluster analysis), each with hyperparameters to tune (e.g., the number of features or clusters, respectively).

It is well known that, given the exponential search space, the tuning process is tedious and overwhelming. *Automated Machine Learning* (AutoML) aids in a smart exploration of the hyperparameter search space [14] and lets data scientists focus on analyzing and interpreting the extracted results. AutoML has

© The Author(s), under exclusive license to Springer Nature Singapore Pte Ltd. 2024
D.-N. Yang et al. (Eds.): PAKDD 2024, LNAI 14645, pp. 246–258, 2024.
https://doi.org/10.1007/978-981-97-2242-6_20

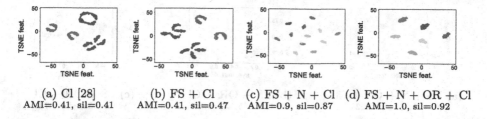

(a) Cl [28]
AMI=0.41, sil=0.41

(b) FS + Cl
AMI=0.41, sil=0.47

(c) FS + N + Cl
AMI=0.9, sil=0.87

(d) FS + N + OR + Cl
AMI=1.0, sil=0.92

Fig. 1. Approach motivation. Feature Selection (FS) - Normalization (N) - Outlier Removal (OR) - Clustering (Cl)

been proven to be effective on supervised tasks where the ground truth eases the evaluation of the hyperparameters optimality [8,25]; yet, when it comes to unsupervised tasks, the road is not paved yet [2]. In this paper, we focus on the unsupervised task of *(crisp) cluster analysis* [1] that returns a partitioning of the original dataset based on the similarity of data items. Cluster analysis is exploratory by nature since, given that no ground truth is available, there is no "correct" result [16] and the aim for the data scientist is to uncover clues hidden in the data. The two main limitations of the AutoML approaches in the literature are (i) auto-tuning is applied to the ML step only, disregarding the pre-processing ones, and (ii) only the most-performing pipeline configuration is returned; although this is reasonable in the supervised context, it provides limited information in unsupervised one due the exploratory nature of the analysis.

To overcome the previous limitations, we devise AutoClues: an *end-to-end* AutoML approach that provides a *dashboard* of *relevant* and *different* clusterings. In particular, we focus on the following contributions:

- *generalizing* AutoML formulation to deal with unsupervised ML pipelines;
- *tuning* a thorough ML pipeline to discover clusterings that would have been unrevealed otherwise;
- *diversifying* the generated clusterings to ensure that the dashboard is both high-quality and leads to different insights (i.e., disclose something new);
- *providing* a customizable generator of synthetic datasets for benchmarking in (crisp) cluster analysis.

To let the reader appreciate the novelty of AutoClues, we rely on a synthetic 10-dimensional (10D) dataset that includes 6 natural clusters. Figure 1a shows the t-SNE [19] visualization[1] of the clustering obtained applying AutoML4Clust [28], an approach of the literature, that solely tunes the clustering step Cl. Figures 1b to 1d show the clusterings obtained by tuning an ML pipeline that incrementally includes feature selection (FS), normalization (N), and outlier removal (OR). In (b), FS identifies the most relevant features; in (c), N standardizes such features thus avoiding bias due to different domain ranges; in (d), OR drops any data items that are not representative. It is apparent how the tuning improves

---

[1] This dimensionality reduction visualizes high-dimensional clusterings in 2D, preserving distance proportions. We apply it with the default Scikit-learn hyperparameters.

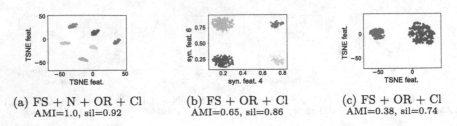

(a) FS + N + OR + Cl
AMI=1.0, sil=0.92

(b) FS + OR + Cl
AMI=0.65, sil=0.86

(c) FS + OR + Cl
AMI=0.38, sil=0.74

Fig. 2. Relevant and diverse clusterings returned by AutoClues. Feature Selection (FS) - Normalization (N) - Outlier Removal (OR) - Clustering (Cl)

throughout the different steps, making it possible for Cl to properly detect the 6 natural clusters. To quantitatively understand the improvements, we rely on the silhouette index $sil \in [-1, 1]$ (the higher the better), an estimation of the "goodness" of a clustering considering solely the data itself; namely, the *separability* between the clusters and their *cohesion* [32]. While in real-case unsupervised problems the ground truth is not available, for our synthetic example we can also measure how the returned clusters match the synthetic ones through the adjusted mutual information [30] $AMI \in [0, 1]$ (the higher the better).

As to the limitation of returning only the most-performing pipeline configuration, Fig. 2 depicts an example of an AutoClues dashboard with three different facets (clustering). Along with the expected clusters (a), the representation in (b) unveils 4 macro clusters in 2 original features, while the representation in (c) highlights a large gap of 2 main clusters. In real-world problems, those facets may help the data scientist to understand the data.

2   Related Works

Although the active development of data pre-processing techniques in cluster analysis (e.g., outlier removal, feature selection), and the evidence of their benefits, there is no trace of automatic solutions that tunes a thorough ML pipeline.

Former approaches employed model-free techniques (i.e., without a model to drive the optimization) to tune the combination of number of clusters and number of features. Evolutionary algorithms are population-based heuristics inspired by biological evolution mechanisms (e.g., reproduction, mutation) or physical phenomena (e.g., particle swarm, black holes). The population is intended as the search space, individuals are configurations, and a mutation mechanism allows the modification of the current candidate, hence the exploration. Recent works that follow this modus operandi (i.e., MOGA [5], MODE-cf [12], TPE-AutoClust [6]) compose the candidate with a string that encodes information about both the feature selection and the clustering. Authors in [23] leverage simulated annealing, an algorithm that comes from a technique involving heating and controlled cooling of material in metallurgy. In [22], it is leveraged a gravitational search algorithm, inspired by the theory of Newtonian gravity.

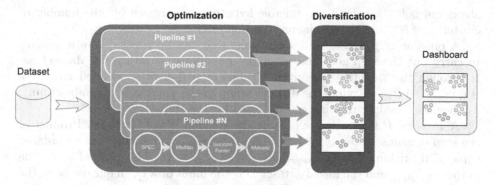

Fig. 3. Overview of AutoClues.

Model-based techniques leverage past evaluations to fit a model and visit the most prominent configurations. Such techniques have been proven to achieve extremely good results, specifically in the (supervised) AutoML field where a pipeline has to be instantiated with different algorithms and hyperparameters. Authors in [18,21,28] apply such techniques on cluster analysis, but do not consider any pre-processing phase. In MALSS [15], the authors apply AutoML with simple (no-clustering-oriented) pre-processing, and still for the aim of suggesting the number of clusters. In [7], authors introduce a pipeline for unsupervised clustering, yet specifically tailored for multivariate time series originating from HPC job monitoring. Finally, such approaches retrieve just one solution.

3 AutoClues

AutoClues aims at (i) tuning the whole ML pipeline and (ii) suggesting multiple clusterings to provide the data scientist different facets of the datasets. In literature, these two problems are addressed through *optimization* and *diversification* techniques respectively. Figure 3 shows an overview of the architecture.

3.1 Formalization

Optimization. We seek for the most "correct" pipeline. A pipeline is a sequence of steps that process a dataset in order to return a clustering and, at each step, an algorithm can be picked among several alternatives.

More formally, a *dataset D* is a collection of $|D|$ data items that is characterized by a set of features $\mathcal{F}$ (i.e., columns). An *Algorithm A* is a function that transforms a dataset D' into a new dataset D'' and has a set of (possibly empty) *hyperparameters* that regulates its behavior. Each hyperparameter has a domain, and the possible algorithm configurations are represented as the Cartesian product of all the hyperparameter domains, denoted with Λ_A. We call *algorithm instance* $\lambda_A \in \Lambda_A$ an (ordered) tuple in which each hyperparameter has been assigned with a value from its domain. For instance, a clustering

250   M. Francia et al.

algorithm is KMeans and two tunable hyperparameters can be: the number of clusters $k \in \mathbb{N}^+$ and the maximum iterations $iter \in \mathbb{N}^+$.

A pipeline *step* S is a set of algorithms that can be selected as alternatives to carry out the step goal, including the possibility to return the dataset as-is through the "Identity" algorithm I. The step domain is defined as $\Lambda_S = \Lambda_{A_1} \uplus \ldots \uplus \Lambda_{A_{|S|}}$, therefore the disjoint union of the domain of each algorithm[2]. λ_S denotes the algorithm instance selected for the step.

A *Pipeline* P is a sequence of steps, its domain is the Cartesian Product of the step domains $\Lambda_P = \Lambda_{S_1} \times \ldots \times \Lambda_{S_{|P|}}$, and a *pipeline instance* is an ordered tuple of algorithm instances $\lambda_P = (\lambda_{S_1}, \ldots, \lambda_{S_{|P|}}) \in \Lambda_P$. The input of λ_{A_i} is the output of $\lambda_{A_{i-1}}$ and the initial dataset D is the input of λ_{A_1}. In our use case, the last step is mandatory since it fulfills cluster analysis by returning a clustering.

A *(crisp) Clustering* is a partition of the dataset into a set of non-overlapping clusters $C = \{c_1, \ldots, c_{|C|}\}$ (i.e., groups of data items) by minimizing the distance of data items in the same cluster and maximizing the distance of different clusters. Given a pipeline domain Λ_P and a dataset D, the optimal instance is

$$\lambda_P^* = argmax_{\lambda_P \in \Lambda_P} rel(C) \tag{1}$$

where rel is a goodness metric for C obtained by applying λ_P on D.

Diversification. The process of returning a set of relevant and diverse solutions – clusterings in our case – is known as diversification, a multi-objective optimization problem that can be formulated as follows. Let $\mathcal{C}$ be the set of clusterings that have been explored in the pipeline optimization process (Eq. (1)), then our goal is selecting a set of α clusterings $\mathcal{C}^* \subseteq \mathcal{C}$ maximizing a *score* that represents a tradeoff between finding *relevance* and *diversity*.

$$\mathcal{C}^* = argmax_{\hat{\mathcal{C}} \subseteq \mathcal{C}, \alpha = |\hat{\mathcal{C}}|} score(\beta, \hat{\mathcal{C}}) \tag{2}$$

$$score(\beta, \hat{\mathcal{C}}) = (1 - \beta) \, rel(\hat{\mathcal{C}}) + \beta \, div(\hat{\mathcal{C}}) \tag{3}$$

where $\beta \in [0..1]$ is the tradeoff parameter and

$$rel(\hat{\mathcal{C}}) = (|\hat{\mathcal{C}}| - 1) \sum_{C \in \hat{\mathcal{C}}} rel(C) \tag{4}$$

$$div(\hat{\mathcal{C}}) = \sum_{C_i \in \hat{\mathcal{C}}} \sum_{C_j \in (\hat{\mathcal{C}} \setminus C_i)} div(C_i, C_j) \tag{5}$$

$div(\hat{\mathcal{C}})$ is the sum of pairwise clustering diversity comparisons and $rel(\hat{\mathcal{C}})$ is the sum of clustering relevance; $rel(\hat{\mathcal{C}})$ entails a multiplication factor $|\hat{\mathcal{C}}| - 1$ to make $rel(\hat{\mathcal{C}})$ and $div(\hat{\mathcal{C}})$ comparable.

3.2 Implementation

Table 1 reports the pipeline with steps, algorithms, and hyperparameters; AutoClues is available on GitHub[3]. The first step is *Feature selection*. Since clus-

---

[2] If an algorithm has no hyperparameters ($\Lambda_A = \varnothing$), we set a placeholder $\Lambda_A = \{1\}$.
[3] https://github.com/big-unibo/autoclues.

Table 1. Steps and algorithms optimized by AutoClues.

Step	Algorithm	#Hyper	$|\Lambda_A|$
Feature selection	SPEC [31]	1	$|\mathcal{F}| - 1$
	WKMeans [13]	2	$3 \cdot (|\mathcal{F}| - 1)$
	Pearson Filtering	1	10
Normalization	Standardization	0	1
	Robust Scaling	3	12
	MinMax	0	1
Outlier removal	Local Outlier Factor [3]	1	3
	Isolation Forest [17]	1	3
Clustering	KMeans [1]	1	$\sqrt{|D|}$
	Agglomerative clustering [20]	1	$\sqrt{|D|}$

ter analysis is particularly sensitive to non-informative and correlated features, we leverage algorithms from spectral family and based on the Pearson correlation, respectively. Follows, *Normalization* to adjust values on different scales and ensure that all the features contribute equally to the cluster formation. Here, the literature has a considerable consensus on the well-known techniques such as Standardization, Robust scaling, and Min-max. The last pre-processing step is *Outlier Removal* for discarding any data points that are not representative of the cluster, we leverage Local Outlier Factor [3] and Isolation Forest [17]. Finally, the *Clustering* step. Since we focus on *crisp spherical* clustering algorithms, we consider KMeans [1] and Agglomerative Clustering [20] with complete linkage. As to the order of steps, in [11], the authors reduce the combinations of the former steps; we further constrained the order of the steps by computing several experiments and observing the impact of the alternatives.

Optimization. To explore promising pipeline instances, we leverage Bayesian Optimization (BO) [14]. BO is iterative: as it explores hyperparameter configurations, it progressively builds an accurate model of the domain to decide the next configuration to explore. The exploration continues until a budget in terms of either iterations or time is reached.

Selecting the optimal pipeline instance requires the relevance metric (i.e., $rel(C)$) to evaluate the goodness of the retrieved clustering. In AutoClues, such a metric is customizable. Well-known metrics leveraged for spherical clustering are: the silhouette index (SIL) [32], contrasting the average distance to elements in the same cluster with the average distance to elements in other clusters, and the Davies-Bouldin Index (DBI) [4], computing the average similarity between clusters by considering their own size. However, clustering metrics show a bias toward lower dimensionalities i.e., yielding higher scores when fewer features are chosen [12,16]. To overcome this, we employ t-SNE [19] (with default Scikit-learn hyperparameters) to project the clusterings to a latent 2D space. Distances in this latent space are preserved, and we can compute the chosen metric atop, enabling a fair evaluation of clusterings across diverse feature spaces.

Diversification. Implementing diversification in cluster analysis involves assessing the extent to which two clusterings differ from each other, hence how the returned dashboard looks. It is crucial to rely on a metric that considers not only shared cluster membership but also structural interrelationships.

Information theory introduces the concept of Mutual Information, quantifying the degree of dependence between two variables and, more specifically, Adjusted Mutual Information (AMI) allows for chance agreement[4]—providing more robust and meaningful measures. When applied to cluster analysis, AMI $\in [0,1]$ considers the labels assigned to data points within clusters, assuming higher values when the clusters in one partition align with those in another. Since we need a diversity metric, we adapt the formula as in:

$$div(C_i, C_j) = 1 - AMI(C_i, C_j)$$

where C_i, C_j clusterings coming from different pipeline instances.

Finally, considering that the diversification problem is NP-hard, we compute it by exploiting the MMR heuristic solution[5] [29] that selects the best-performing clustering and iteratively adds the clusterings that most diversify the outcome.

4   Benchmark Generation and Empirical Evaluation

Evaluation in cluster analysis consists of assessing the approach performance in finding well-separated clusters and – if available – their alignment with a hypothetical ground truth. It is crucial to test on datasets that conform with the leveraged clustering algorithms, e.g., in our case, containing crisp spherical clusters. Yet, there is a lack of benchmarks (i.e., suites of datasets for fair comparisons) and the few available [9,10,26] are tailored to their specific scenarios. This translates into approaches relying upon datasets from supervised tasks, with no guarantees on the underlying clusters' shape.

In Sect. 4.1, we introduce a benchmarking generator and a suite of synthetic datasets. In Sect. 4.2, we leverage such a suite to assess the effectiveness and efficiency of AutoClues. Finally, in Sect. 4.3, we rely on real datasets to provide a comparison against other approaches in the literature.

4.1   Benchmark Generation

The synthetic benchmarking generator is available at https://github.com/big-unibo/clustering_benchmarking. We create datasets of $|D|$ instances, defining $|C|$ natural hyper-spherical clusters in a space of $|\mathcal{F}|$ features. Then, we blur such clusters by posing common challenges faced by clustering algorithms. This

---

[4] In statistics, it serves as a baseline for assessing the significance in random variations.
[5] We use the default hyperparameter $\beta = 0.5$, and set α according to the test at hand.

Table 2. Dataset characteristics and performance achieved by AutoClues.

Dataset	Characteristics							AutoClues Performance			
	$\|D\|$	$\|\mathcal{F}\|$	$\|\mathcal{C}\|$	$\sigma(D)$	$\sigma(\mathcal{F})$	SIL_N	SIL_B	SIL	AMI	Score	Div. time (s)
syn1	2905	2	3	0.15	0.50	0.72	0.48	0.79	1.0	4.12	$1.64 \cdot 10^3$
syn2	264	5	3	0.25	0.20	0.64	0.4	0.7	0.83	4.76	$1.74 \cdot 10^2$
syn3	900	7	12	0.11	0.14	0.88	0.6	0.92	1.0	4.39	$1.51 \cdot 10^3$
syn4	1446	2	13	0.22	1.00	0.59	0.31	0.85	0.95	3.92	$2.99 \cdot 10^3$
syn5	1673	4	8	0.17	0.25	0.71	0.54	0.81	0.99	4.22	$2.18 \cdot 10^3$
syn6	2905	2	3	0.15	0.50	0.72	0.61	0.84	0.87	4.63	$1.07 \cdot 10^3$
syn7	264	5	3	0.25	0.20	0.64	0.41	0.72	0.86	3.78	$1.67 \cdot 10^2$
syn8	1639	8	21	0.13	0.00	0.87	0.69	0.91	1.0	4.45	$3.13 \cdot 10^3$
syn9	525	3	2	0.21	0.00	0.66	0.42	0.69	0.87	3.96	$3.67 \cdot 10^2$
syn10	1446	2	13	0.22	1.00	0.59	0.32	0.86	0.97	4.03	$4.09 \cdot 10^3$
syn11	4813	10	3	0.27	0.40	0.46	0.17	0.09	0.42	2.92	$1.08 \cdot 10^2$
syn12	2905	2	3	0.15	0.50	0.72	0.57	0.79	1.0	4.48	$1.62 \cdot 10^3$
syn13	264	5	3	0.25	0.20	0.64	0.4	0.74	0.89	4.37	$2.75 \cdot 10^2$
syn14	525	3	2	0.21	0.00	0.66	0.22	0.71	0.9	4.38	$5.81 \cdot 10^2$
syn15	2905	2	3	0.15	0.50	0.72	0.61	0.81	1.0	4.06	$1.13 \cdot 10^3$
syn16	264	5	3	0.25	0.20	0.64	0.49	0.7	0.88	3.45	$1.84 \cdot 10^2$
syn17	900	7	12	0.11	0.14	0.88	0.41	0.92	1.0	4.62	$1.45 \cdot 10^3$
syn18	525	3	2	0.21	0.00	0.66	0.3	0.69	0.84	4.34	$2.85 \cdot 10^2$
syn19	2905	2	3	0.15	0.50	0.72	0.53	0.86	0.87	4.79	$1.28 \cdot 10^3$
syn20	264	5	3	0.25	0.20	0.64	0.44	0.7	0.88	3.74	$2.13 \cdot 10^2$

includes noise on instances $\sigma(D)$, such as the presence of outliers, and noise on features $\sigma(F)$, such as irrelevant, correlated, or distorted features.

To obtain datasets with different characteristics, we set boundaries for each of these dimensions and sample within them according to the Sobol sequence [24], a quasi-random low-discrepancy search converging to an equi-distributed coverage. In particular, the defined boundaries are: $|D|$ between 100 and 5000, $|F|$ between 2 and 10, $|C|$ between 2 and $\sqrt{|D|}$, $\sigma(D)$ between 0.1 and 0.3, and $\sigma(F)$ between 0 and 1. Table 2 provides a suite of 20 synthetic datasets.

Dataset complexity can be examined via the silhouette $SIL \in [0, 1]$, the higher the simpler. SIL_N measures the cohesion and separability of the natural clusters C in their original feature space, while SIL_B measures the silhouette of blurred clusters (i.e. after introducing noise (σ)). The former SIL_N indicates the presence of well-separated clusters. With the first quartile $Q1 = 0.64$, median $Q2 = 0.66$, and third quartile $Q3 = 0.72$, we observe that 25% of datasets are complex already at this stage. The latter SIL_B registers significantly lower values: syn11 emerges as an especially complex dataset with $SIL_B = 0.17$ but, overall, we confirm the presence of a good distribution between more and less challenging datasets: $Q1 = 0.36$, $Q2 = 0.43$, and $Q3 = 0.55$.

4.2 Effectiveness and Efficiency

We test AutoClues on the suite of generated synthetic datasets. For the optimization, we adopted the silhouette SIL index as a relevance objective metric and a budget of 7200 s (2 h); for the diversification we set $\alpha = 3$ and $\beta = 0.5$, resulting in a dashboard of 3 clusterings where relevance and diversity are weighted equally. Tests are run on a single core of an Intel Core i7 machine at 3.20 GHz and 64 GB of main memory. Given an AutoClues dashboard, Table 2 provides the maximum $SIL \in [0, 1]$ as the cohesion of the found clusters and the maximum $AMI \in [0, 1]$ as the alignment with the natural ones[6]. Besides, we report the dashboard score of Equation (2), summarizing the overall relevance and diversity in the dashboard, and the diversification computation time.

Effectiveness. The achieved silhouette not only shows AutoClues' ability to find well-separated clusters in 19 cases out of 20, but it also demonstrates to overcome the silhouette of natural clusters SIL_N. This is achieved through preprocessing such as projecting natural clusters into more compact feature subsets and mitigating potential noise. The only critical exception is syn11, already highlighted in the previous section. AMI also confirms strong agreement between retrieved and natural clusters ($Q1 = 0.87$, $Q2 = 0.9$, $Q3 = 1$).

As to the dashboard, considering the experimental setting ($\alpha = 3$, $\beta = 0.5$, $rel, div \in [0, 1]$), the score is bounded in $[0, 6]$. Notably, given the inherent trade-off relationship between rel and div in the context of a diversification problem, it is noteworthy that scores at the boundaries are less likely to manifest, with values around 4 already acknowledged as high-quality [29]. Analogously, quartile values of the score ($Q1 = 4$, $Q2 = 4.34$, $Q3 = 4.47$) confirm Autoclues' ability to find relevant and diverse clusterings.

Efficiency. Table 2 reports the computation time for diversification, when the optimization budget is set to 2 h. We can observe 25% of datasets compute the dashboard in less than $Q1 = 2.44 \cdot 10^2$ s (4 min), 50% in less than $Q2 = 1.07 \cdot 10^3$ s (18 min), and 75% in less than $Q3 = 1.56 \cdot 10^3$ s (26 min). Besides, reducing the optimization budget leads to a decrease in diversification time, maintaining high performance. Figure 4 shows how AutoClues converges to the values reported in Table 2. Dashboards are generated at different snapshots during the 2-h optimization process, and the same metrics are computed: SIL, AMI, dashboard score, and diversification time. We summarize the information by plotting the mean and standard deviation among the whole suite, the convergence is quantified as a progress ratio relative to the final achieved performance. Notably, the optimization time is illustrated in a logarithmic scale, highlighting already fast convergence. After 60 s of optimization, we have clusterings with SIL and AMI as good as 80% and 95% of the optimal and a dashboard almost 60% of the final within a negligible diversification time—roughly 2% of the total, 30 s on average.

Within 300 s of optimization, an average of 90% of SIL, 97% of AMI, and 75% of the dashboard score are registered with a diversification cost of 10%—on average, 1 min and half. The trends of SIL and AMI saturated 100% right

---

[6] Metrics are computed on the original dataset (i.e., no t-SNE distortion).

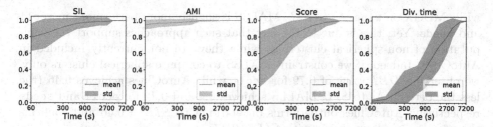

Fig. 4. AutoClues convergence through time (in log scale).

Table 3. Comparison with other approaches in the literature.

Dataset	Characteristics			Performance					
	$\|D\|$	$\|\mathcal{F}\|$	$\|C\|$	DBI ↓				SIL ↑	
				MOGA	MODE-cf	MALSS	AutoClues	MALSS	AutoClues
blood	748	4	2	–	–	0.3	**0**	0.73	**1**
breast	106	9	6	–	0.7	1.6	**0.54**	0.16	**0.60**
ecoli	327	7	5	–	0.92	**0.35**	0.46	**0.72**	0.46
iris	150	4	3	0.39	0.67	0.6	**0.38**	0.57	**0.71**
seeds	210	7	3	–	–	0.8	**0.4**	**0.45**	0.37
thyroid	215	5	3	–	–	0.64	**0.2**	0.6	**0.92**
vehicle	846	18	4	–	–	0.6	**0.15**	0.61	**0.72**
wine	178	13	3	**0.77**	1.22	1.4	1.01	0.28	**0.38**

afterward, while both dashboard score and cost increase linearly until 900 s of optimization, in which the dashboard achieves 90% of its score within a cost of 30%—5 min on average. After such a threshold, improvements in the score are not considered worth it for the computation cost. This is due to the increasing number of solutions to be evaluated during diversification, while relevant and diverse clusterings are already present in the dashboard.

4.3   Comparison

We compare AutoClues with state-of-the-art approaches in the literature against real datasets, considered as standard benchmarks. We selected the ones that provided either the performance on classification datasets (MOGA [5], MODE-cf [12]) or code for reproducibility (MALSS [15]). The former two approaches are evolutionary algorithms, and the latter is a general-purpose AutoML tool. These approaches do not provide a dashboard but only the best-performing clustering. Thus, for a fair comparison, we set $\alpha = 1$ to return the best clustering, and we adopt as relevance the same metric used in the competing approaches. MOGA and MODE-cf measure performance solely through the Davies-Bouldin Index (DBI [4], the less the better), MALLS also provides SIL (the higher the better).

Table 3 shows that AutoClues outperforms the reported approaches in 6 out of 8 datasets. According to DBI, MALSS achieves better performance on **seeds**, while MOGA on `wine`. As to SIL, MALSS is slightly more performant in `ecoli` and `seeds`. Yet, this is due to the fact that such approaches support the computation of non-spherical clusterings while they are not currently included in AutoClues. Indeed, if we constraint MALSS to compute spherical clusters only, we observe a DBI value of 0.79 for `ecoli` while AutoClues achieves 0.46 (the less the better). As to SIL, MALLS achieves 0.4 and 0.45 for `ecoli` and `seeds` respectively; AutoClues outperforms on `ecoli` with $SIL = 0.46$ and confirms the previous result on `seeds` with $SIL = 0.37$.

5 Conclusion and Future Work

We introduced AutoClues, an end-to-end cluster analysis approach that leverages AutoML techniques to provide a diverse and relevant dashboard of clusterings. Our findings demonstrate that optimizing pre-processing significantly enhances performance, allowing AutoClues to overcome current state-of-the-art approaches. For future research, we plan to explore (i) meta-learning approaches to identify more effective pre-processing steps, (ii) integrating human feedback in the loop, and (iii) providing automatic explanations for the retrieved dashboard.

References

1. Arthur, D., Vassilvitskii, S.: k-means++: The advantages of careful seeding. Technical report, Stanford (2006)
2. Barlow, H.B.: Unsupervised learning. Neural Comput. **1**(3), 295–311 (1989)
3. Breunig, M.M., Kriegel, H.P., Ng, R.T., Sander, J.: LoF: identifying density-based local outliers. In: Proceedings of the 2000 ACM SIGMOD International Conference on Management of Data. , pp. 93–104 (2000)
4. Davies, D.L., Bouldin, D.W.: A cluster separation measure. IEEE Trans. Pattern Anal. Mach. Intell. **PAMI-1**(2), 224–227 (1979)
5. Dutta, D., Dutta, P., Sil, J.: Simultaneous continuous feature selection and k clustering by multi objective genetic algorithm. In: 2013 3rd IEEE International Advance Computing Conference (IACC), pp. 937–942 (2013)
6. ElShawi, R., Sakr, S.: TPE-autoclust: a tree-based pipline ensemble framework for automated clustering. In: 2022 IEEE International Conference on Data Mining Workshops (ICDMW), pp. 1144–1153 (2022)
7. Enes, J., Expósito, R.R., Fuentes, J., Cacheiro, J.L., Touriño, J.: A pipeline architecture for feature-based unsupervised clustering using multivariate time series from HPC jobs. Inf. Fusion **93**, 1–20 (2023)
8. Francia, M., Giovanelli, J., Pisano, G.: Hamlet: a framework for human-centered automl via structured argumentation. Futur. Gener. Comput. Syst. **142**, 182–194 (2023)
9. Fränti, P., Sieranoja, S.: K-means properties on six clustering benchmark datasets (2018)
10. Gagolewski, M.: A framework for benchmarking clustering algorithms. SoftwareX **20**, 101270 (2022)
11. Giovanelli, J., Bilalli, B., Abelló, A.: Data pre-processing pipeline generation for autoETL. Inf. Syst. **108**, 101957 (2022)

12. Hancer, E.: A new multi-objective differential evolution approach for simultaneous clustering and feature selection. Eng. Appl. Artif. Intell. **87**, 103307 (2020)
13. Huang, J., Ng, M., Rong, H., Li, Z.: Automated variable weighting in k-means type clustering. IEEE Trans. Pattern Anal. Mach. Intell. **27**(5), 657–668 (2005)
14. Hutter, F., Hoos, H.H., Leyton-Brown, K.: Sequential model-based optimization for general algorithm configuration. In: Coello, C.A.C. (ed.) LION 2011. LNCS, vol. 6683, pp. 507–523. Springer, Heidelberg (2011). https://doi.org/10.1007/978-3-642-25566-3_40
15. Kamoshida, R., Ishikawa, F.: Automated clustering and knowledge acquisition support for beginners. Procedia Comput. Sci. **176**, 1596–1605 (2020)
16. Lensen, A., Xue, B., Zhang, M.: Using particle swarm optimisation and the silhouette metric to estimate the number of clusters, select features, and perform clustering. In: Squillero, G., Sim, K. (eds.) EvoApplications 2017. LNCS, vol. 10199, pp. 538–554. Springer, Cham (2017). https://doi.org/10.1007/978-3-319-55849-3_35
17. Liu, F.T., Ting, K.M., Zhou, Z.H.: Isolation-based anomaly detection. ACM Trans. Knowl. Discov. Data (TKDD) **6**(1), 1–39 (2012)
18. Liu, Y., Li, S., Tian, W.: AutoCluster: meta-learning based ensemble method for automated unsupervised clustering. In: Karlapalem, K., et al. (eds.) PAKDD 2021. LNCS (LNAI), vol. 12714, pp. 246–258. Springer, Cham (2021). https://doi.org/10.1007/978-3-030-75768-7_20
19. Van der Maaten, L., Hinton, G.: Visualizing data using t-SNE. J. Mach. Learni. Res. **9**(11) (2008)
20. Murtagh, F., Contreras, P.: Algorithms for hierarchical clustering: an overview. Wiley Interdisc. Rev. Data Min. Knowl. Discov. **7**(6) (2017)
21. Poulakis, Y., Doulkeridis, C., Kyriazis, D.: Autoclust: a framework for automated clustering based on cluster validity indices. In: ICDM, pp. 1220–1225. IEEE (2020)
22. Prakash, J., Singh, P.K.: Gravitational search algorithm and k-means for simultaneous feature selection and data clustering: a multi-objective approach. Soft. Comput. **23**(6), 2083–2100 (2019)
23. Saha, S., Spandana, R., Ekbal, A., Bandyopadhyay, S.: Simultaneous feature selection and symmetry based clustering using multiobjective framework. Appl. Soft Comput. **29**(C), 479–486 (2015)
24. Sobol, I.: The distribution of points in a cube and the accurate evaluation of integrals (in Russian) zh. Vychisl. Mat. i Mater. Phys **7**, 784–802 (1967)
25. Thornton, C., Hutter, F., Hoos, H.H., Leyton-Brown, K.: Auto-Weka: combined selection and hyperparameter optimization of classification algorithms. In: Proceedings of the 19th ACM SIGKDD, pp. 847–855 (2013)
26. Thrun, M.C., Ultsch, A.: Clustering benchmark datasets exploiting the fundamental clustering problems. Data Brief **30**, 105501 (2020)
27. Toch, E., Lerner, B., Ben-Zion, E., Ben-Gal, I.: Analyzing large-scale human mobility data: a survey of machine learning methods and applications. Knowl. Inf. Syst. **58**(3), 501–523 (2019)
28. Tschechlov, D., Fritz, M., Schwarz, H.: Automl4clust: efficient autoML for clustering analyses, pp. 343–348 (2021)
29. Vieira, M.R., et al.: On query result diversification. In: 27th IEEE International Conference on Data Engineering (ICDE), pp. 1163–1174. IEEE (2011)
30. Vinh, N.X., Epps, J., Bailey, J.: Information theoretic measures for clusterings comparison: is a correction for chance necessary? In: Proceedings of the 26th Annual International Conference on Machine Learning, pp. 1073–1080 (2009)

31. Zhao, Z., Liu, H.: Spectral feature selection for supervised and unsupervised learning. In: Proceedings of the 24th International Conference on Machine Learning (2007)
32. Zhu, L., Ma, B., Zhao, X.: Clustering validity analysis based on silhouette coefficient. J. Comput. Appl. **30**(2), 139–141 (2010)

Local Subsequence-Based Distribution
for Time Series Clustering

Lei Gong[1,2]($\boxtimes$), Hang Zhang[1,2], Zongyou Liu[1,2], Kai Ming Ting[1,2], Yang Cao[3],
and Ye Zhu[3]

[1] National Key Laboratory for Novel Software Technology, Nanjing University,
Nanjing, China
[2] School of Artificial Intelligence, Nanjing University, Nanjing, China
{gongl,zhanghang,liuzy,tingkm}@lamda.nju.edu.cn
[3] School of Information Technology, Deakin University, Burwood, VIC, Australia
{charles.cao,ye.zhu}@ieee.org

Abstract. Analyzing the properties of subsequences within time series
can reveal hidden patterns and improve the quality of time series clustering. However, most existing methods for subsequence analysis require
point-to-point alignment, which is sensitive to shifts and noise. In this
paper, we propose a clustering method named CTDS that treats time
series as a set of independent and identically distributed (iid) points
in $\mathbb{R}^d$ extracted by a sliding window in local regions. CTDS utilises
a distributional measure called Isolation Distributional Kernel (IDK)
that can capture the subtle differences between probability distributions
of subsequences without alignment. It has the ability to cluster large
non-stationary and complex datasets. We evaluate CTDS on UCR time
series benchmark datasets and demonstrate its superior performance
than other state-of-the-art clustering methods.

Keywords: Time series · Unsupervised clustering · Distributional
treatment · Kernel methods

1 Introduction

Time series data, which is prevalent in various domains such as finance, engineering and meteorology [26], can reveal valuable patterns and trends through
clustering. Clustering is an unsupervised machine learning technique that does
not rely on any labels or annotations [9]. However, clustering time series is challenging due to the presence of noise, amplitude change, offset translation and
phase shift in data. Therefore, many studies have proposed new similarity measures or representation methods that can capture the essential features of time
series and preserve their similarity [7], thereby improving clustering.

The presence of noise and shifts in real-world time series data make it difficult to spot similarities within the same class and significant differences between
different classes, especially when looking at the global characteristics of the time
series to compute pairwise similarities. This is why methods that directly measure entire long series perform poorly in clustering [20]. However, a more effective approach is to focus on the local information of time series, characterized

© The Author(s), under exclusive license to Springer Nature Singapore Pte Ltd. 2024
D.-N. Yang et al. (Eds.): PAKDD 2024, LNAI 14645, pp. 259–270, 2024.
https://doi.org/10.1007/978-981-97-2242-6_21

by subsequences. These subsequences can be viewed as patterns that potentially hold discriminative information crucial for clustering. Many state-of-the-art algorithms, e.g., U-shapelets [25], SUSh [20] and SPF [9], employ this strategy. They transform pre-defined length subsequences into symbolic patterns and compare them with randomly chosen patterns, showcasing promising clustering on time series data. However, these methods suffer from the drawbacks of using the Symbolic Aggregate Approximation (SAX) technique to discretize subsequences, which lose information during the transformation process [20].

A recent distributional approach [16] in time series based on Isolation Distributional Kernel (IDK) [16] views a subsequence as a set of iid one-dimensional points generated from an unknown distribution in the $\mathbb{R}$ domain. The authors demonstrate the power of their approach for detecting anomalies in stationary time series. However, this approach has two shortcomings. First, the scheme of the $\mathbb{R}$ domain breaks the temporal order in a subsequence, which is detrimental to capturing temporal differences that are significant for time series mining. Secondly, the method relies on the stationary properties of time series, making it challenging to discriminate complex and non-stationary time series effectively.

To overcome the first issue, we propose a new $\mathbb{R}^d$ domain treatment. The intuition is that when considering subsequences of length d as points in a d-dimensional space, the distributions of these points extracted from homogeneous time series tend to be inherently similar. The proposed treatment can preserve the sequential order of values in each subsequence, which leads to superior performance in time series clustering. Furthermore, we propose a novel time series clustering algorithm called CTDS, i.e., **C**lustering **T**ime series by comparing the **D**istribution of **S**ubsequences. CTDS is effective and runs with a linear time complexity. In addition, it can deal with long and non-stationary time series datasets by dividing them into several relatively stationary parts and efficiently measuring the similarity within each part individually. It enables the calculation of similarity between time series to focus on paired local regions, which is advantageous for time series clustering.

We evaluate the performance of CTDS[1] on the UCR archive [4] and compare it against existing SOTA time series clustering methods. The results show that our proposed distribution-based method achieves superior clustering performance. The contributions of this work can be summarized as follows:

1. Providing valuable insight into analysing a time series as a distribution of subsequences in $\mathbb{R}^d$ domain. This perspective enables applying more effective and efficient measurement based on distribution without point-to-point alignment for similarity measuring between different time series. We show that measuring the distribution of subsequences between different time series is a more effective way to distinguish them than using multiple unsupervised shapelets or discretized subsequence partitions for time series clustering.
2. Proposing a novel and efficient method for clustering time series based on their distributional characteristics. Our method discriminates the distribu-

---

[1] The source code is available at https://github.com/LeisureGong/CTDS.

tions of subsequences extracted from time series, and leverages the advantages of treating them in the $\mathbb{R}^d$ domain. Furthermore, our method preserves the pairwise similarity between time series in local regions that can handle non-stationary time series.
3. Conducting empirical evaluation on 85 benchmark time series datasets. The results demonstrate the superiority of our CTDS in terms of both accuracy and efficiency, compared to state-of-the-art time series clustering methods.

The rest of this paper is organized as follows. In Sect. 2, we present the related work. In Sect. 3, we provide a detailed explanation of our motivation, while Sect. 4 describes our approach. We present the experimental results and analyses, including a runtime comparison, a parameter study and a visualized explanation in Sect. 5. Section 6 provides conclusions.

2   Related Work

The existing time series clustering methods are broadly divided into three main categories: distance-based methods, representation-based methods and distribution-based methods [1,7].

Most distance-based methods measure the point dissimilarity between time series for clustering. Kshape adapts K-means [12] by using Shape-based Distance (SBD) as the distance measure for clustering time series [14]. Another commonly used distance measure for time series clustering is Dynamic Time Warping (DTW), which applies dynamic programming to find the optimal temporal alignment between two time series [2]. However, distance-based methods often encounter challenges such as noise, amplitude scaling, longitudinal scaling and temporal drift, which are inherent characteristics of time series data.

Regarding the representation-based approach, raw time series are transformed into lower-dimensional feature vectors. Then, a conventional clustering algorithm is applied to the extracted representations [10]. Previous studies on shapelets [22] have shown that if typical subsequences can be identified within a time series, utilizing the Euclidean distance to measure the similarity of these subsequences can effectively distinguish between different classes of time series. SPIRAL [8] proposes a scalable representation learning framework that preserves pairwise similarities of time series and combines with KMeans-DTW for clustering. DTC [13] integrates dimensionality reduction and temporal clustering into a single end-to-end learning framework to learn non-linear features.

Distribution-based methods offer an alternative way to analyze time series by measuring the similarity between probability distributions of points in subsequences. These methods provide a natural solution for effectively handling time series data with skewness or noise [16]. WTK uses Wasserstein distance [21] to measure the similarity between subsequence sets across different time series. In TNC [19], the neighbourhood distribution of a subseries is defined as the set of all window subsequences with centroids sampled from a normal distribution, facilitating the clustering of multivariate time series data by incorporating local and global information.

3 $\mathbb{R}^d$ Domain Treatment: A Series Is Treated as a Set of Sliding-Window Points

Let $T = \{t_1, t_2, ..., t_l\}$ represent a time series with length l, where each $t_i \in \mathbb{R}$. The proposed approach involves representing the time series as a set of sliding window points $\mathbf{x} \in \mathbb{R}^d$, with d denoting the window size. This treatment allows the time series T to be treated as a distribution in $\mathbb{R}^d$, comprising at most $(l - d + 1)$ independent and identically distributed (i.i.d.) points. Each point corresponds to a local pattern within the original series.

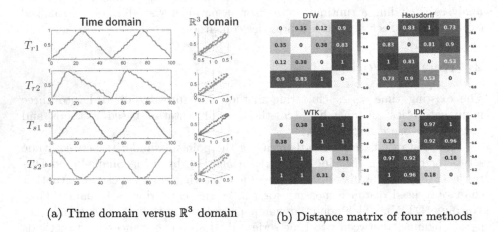

(a) Time domain versus $\mathbb{R}^3$ domain   (b) Distance matrix of four methods

Fig. 1. (a) A motivation example; (b) The (normalized) distance matrix of DTW (Time domain) versus Hausdorff, WTK and IDK ($\mathbb{R}^3$ domain).

This $\mathbb{R}^d$ domain treatment has two main advantages over the existing time domain treatment which employs measures such as Hausdorff and DTW. One significant advantage is that the $\mathbb{R}^d$ domain treatment eliminates the need for point-to-point alignment, which is a recurring issue in the time domain. Existing measures have attempted to address this problem but with limited success. Yet, many still have trouble with time-shift, where a time-shift version of time series T is measured to have non-zero (sometimes significantly large) distance from T. Additionally, the measures in time domain often require a point-to-point distance as their core computation after point-to-point alignment. This is the root cause of their high time complexities.

By using the proposed $\mathbb{R}^d$ domain treatment and a recent Distributional Kernel called IDK [17], we show that it requires no point-to-point alignment and point-to-point distance measure. As a result, IDK produces a more effective distance measurement across time series with low time complexity.

Figure 1a hows an example of two pairs of time series. The first pair of time series T_{r1} & T_{r2} have triangular shapes that differ by the location of their peaks only. The second pair of time series T_{s1} & T_{s2} are sine waves, where one is a 180-degree time shift of the other. Hausdorff and DTW distances are typical measures used in the time domain treatment. Their distance confusion matrices for the two pairs of time series are given in Fig. 1b (DTW) and Fig. 1b (Hausdorff).

Note that Hausdorff distance measures T_{s1} & T_{s2} to have a significantly large non-zero distance, though they have the shortest distance, compared with any of T_{r1} & T_{r2}. There are other aberrations, e.g., (a) T_{r1} has a shorter distance wrt T_{s2} than T_{r2}; and (b) T_{r2} has a shorter distance wrt T_{s1} than T_{r1}. Different kinds of aberrations occur with DTW: (i) T_{s1} has the largest distance wrt T_{s2}, where the former is just a time-shift version of the latter; and (ii) T_{r1} has a shorter distance wrt T_{s1} than T_{r2}.

In contrast, using the $\mathbb{R}^d$ domain treatment with IDK shown in Fig. 1b (IDK), the expected distances are obtained, i.e., the pair T_{r1} & T_{r2}, so as the pair T_{s1} & T_{s2}, are judged to be most similar to each other. It is interesting to note that, although sliding window has been used widely in dealing with time series, the resulting set of points has not been treated as a distribution. This is because the time domain treatment is unquestionably adhered to. The sliding window is used as a convenience means for point-to-point alignment when comparing two subsequences extracted from two time series [23, 25].

Though from a different motivation, Wasserstein Time series Kernel (WTK) [3] can be interpreted to have used the $\mathbb{R}^d$ domain treatment and it uses Wasserstein distance to compute the distance between two time series, after the $\mathbb{R}^d$ domain conversion via sliding window. The use of Wassertsein distance (WD) has two key shortcomings: (a) high time complexity and (b) WD does not have the identity property, i.e., WD cannot ensure that dissimilar time series will always exhibit greater distances than similar ones. The second shortcoming elucidates why, in Fig. 1b, the time series T_{s1} and T_{s2} exhibit the same distance to T_{r1} as measured by WTK, even though T_{s1} is perceived to be more similar to T_{r1} than T_{s2}.

4   Method

4.1   Isolation Distributional Kernel

Isolation Kernel [18] is a data-dependent similarity measure that adapts to the local distribution of the data, i.e. two points in a sparse region are more similar than two points of equal inter-point distance in a dense region. The key idea of Isolation Kernel is to use a space partitioning strategy $\mathbb{H}_\psi(D)$ to split the whole data space into ψ non-overlapping partitions based on a random sample of ψ points from the given dataset D.

Isolation Kernel. For any two points x, y $\in \mathbb{R}^d$, the similarity between x and y wrt D is defined as the expectation taken over the probability on all partitioning $H \in \mathcal{H}_\psi(D)$ that both x and y fall into the same isolating partition $\theta \in H$:

$$\mathcal{K}_I(x, y | D) = \mathbb{E}_{\mathcal{H}_\psi(D)}[\mathbb{I}(x, y \in \theta | \theta \in H)] \tag{1}$$

where $\mathbb{I}(B)$ is the indicator function which outputs 1 if B is true, otherwise, $\mathbb{I}(B) = 0$.

Feature Map of Isolation Kernel. [17] For point $x \in \mathbb{R}^d$, the feature mapping $\Phi : x \rightarrow \{0,1\}^{t \times \psi}$ of $\mathcal{K}_I$ is a vector that represents the partitions in all the partitioning $H_i \in \mathbb{H}_{\Phi}(D), i = 1, ..., t$; where x falls into one of the ψ partitions in each partitioning H_i.

Let $T^{(i,d)} = \{t_i, t_i + 1, t_i + 2, ..., t_i + d - 1\}$ be a subsequence of a time series T, where i and d are the starting positing and length of the subsequence, respectively. Given a time series T, let $S_T^d = \{T^{(i,d)} | i = 0, 1, ..., l - d + 1\}$ be all possible subsequences of length d.

To measure the similarity between two time series T_p and T_q, we can calculate the average pairwise similarity between their subsequences as follows.

Definition 1. *IDKS Isolational Distributional Kernel of two time series T_p and T_q based on subsequences distribution is given as:*

$$\mathcal{K}_I(T_p, T_q \mid D) = \frac{1}{t|S_{T_p}^m||S_{T_q}^m|} \sum_{T_p^{(i,d)} \in S_{T_p}^m} \sum_{T_q^{(i,d)} \in S_{T_q}^m} \left\langle \Phi(T_p^{(i,d)}|\mathbb{S}), \Phi(T_q^{(i,d)}|\mathbb{S}) \right\rangle \tag{2}$$

$$= \frac{1}{t} \left\langle \widehat{\Phi}(\mathcal{P}_{S_{T_p}^m}|\mathbb{S}), \widehat{\Phi}(\mathcal{P}_{S_{T_q}^m}|\mathbb{S}) \right\rangle$$

where $\widehat{\Phi}(\mathcal{P}_S) = \frac{1}{|S|} \sum_{s_i \in S} \Phi(s_i)$ is Kernel Mean Map (KME), and $\mathbb{S} = S_{T_p}^d \cup S_{T_q}^d$.

Isolation distribution kernel requires a partition mechanism, we employ the Voronoi Diagram [15] to implement the partitioning H based on the nearest neighbour search method, i.e., the similarity between two subsequences is the probability of how likely they share the same nearest neighbour from a random subset of ψ subsequences. If both share the same nearest neighbour, then they are located in the same Voronoi cell.

Definition 2. *Nearest Neighbour The nearest neighbor of a subsequence $T_p^{(i,d)}$ w.r.t. a subsequence set S is :*

$$nn(s; S) = \arg\min_{T_q^{(j,d)} \in S} d(T_p^{(i,d)}, T_q^{(j,d)}) \tag{3}$$

where $d(T^{(i,d)}, T^{(j,d)}) = \sqrt{\sum_{k=0}^{d-1} (t_{i+k} - t_{j+k})^2}$ is the distance between two equal-length subsequences $T^{(i,d)}$ and $T^{(j,d)}$.

4.2   Time Series Clustering Based on Distributional Kernel

Figure 2 provides an overview of the workflow of our proposed method. The detailed algorithm, Clustering Time series based on Distribution of Subsequences (CTDS), is outlined in Algorithm 1 and Algorithm 2.

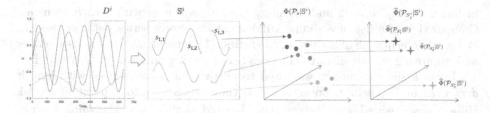

Fig. 2. An illustration of the proposed CTDS algorithm running on a local relatively stationary segment of a time series dataset with three instances. Each time series in D^i produces three subsequences, which are mapped into Isolation Kernel feature space and then represented as mean embeddings. The whole feature map for each time series data is the concatenation of the Kernel Mean Embedding (KME) values from all segments.

Algorithm 1. IDKS

Input: Time series subsequences set $\mathbb{S} = \{S_i, i = 1, ..., n\}$; sample sizes ψ;

Output: Kernel mean embedding $\widehat{\Phi}(\mathcal{P}_{S_i}|\mathbb{S})$ of T_i.

1: Get the feature map $\Phi(\cdot|\mathbb{S})$ of $\mathbb{S} \in \mathbb{R}^d$ using isolation kernel;   ▷ Treat each subsequence as a d-dimensional point in $\mathbb{R}^d$

2: **for** i=1,...,n **do**

3:    $\widehat{\Phi}(\mathcal{P}_{S_i}|\mathbb{S}) = \frac{1}{|S_i|}\sum_{s\in S_i}\Phi(s|\mathbb{S})$;   ▷ Calculate the KME for each subsequence

4: **end for**

5: Return $\widehat{\Phi}(\mathcal{P}_{S_i}|\mathbb{S})$;

Algorithm 2. CTDS

Input: A dataset D of n time series with length l ; Segment length m;

Output: C clusters.

1: Split D into w equal-length segments $\{D^i, i = 1, \ldots, w\}$, where $w = \lceil\frac{l}{m}\rceil$. Each D^i includes n series with m points.

2: **for** $i = 1, ..., w$ **do**

3:    **for** $j = 1, ..., n$ **do**

4:        Extract all subsequences of length $d = m/2$ from each series in D^i, denoted as S_j^i

5:    **end for**

6:    $\mathbb{S}^i = \bigcup_{i=1}^{n} S_j^i$

7:    $\{\widehat{\Phi}(\mathcal{P}_{S_j^i}|\mathbb{S}^i), j = 1, ..., n\}$=IDKS($\mathbb{S}^i$);   ▷ Get the KME of each series in D^i

8:    $\forall_{j\in[1,...,n]}, E_j = E_j \parallel \widehat{\Phi}(\mathcal{P}_{S_j^i}|\mathbb{S}^i)$;   ▷ Concatenate the KMEs for each time series to form E_j

9: **end for**

10: Use K-means on $\mathbf{E}$ and get the cluster result C.

Real-world time series are often long and non-stationary, meaning that the distributions of subsequences change significantly over time, reflecting the evolving patterns and features of the time series. To effectively capture the differences among the dataset, CTDS splits the whole time series into w smaller sub-regions in line 1 of Algorithm 2 and examines the subsequences within each segment. Compared to the use of a sliding window on the entire series data, as used in WTK, this technique offers two key benefits: (1) generating fewer subsequences and improving the algorithm's efficiency; (2) focusing on local pattern match.

Then, a sliding window was used to extract all subsequences with length $d = m/2$ in D^i, before converting each time series T_i into a point in level-1 Hilbert space using kernel feature map, denoted as $\widehat{\Phi}(\mathcal{P}_{S_i}|\mathbb{S})$. The whole feature map for each time series data is the concatenation of the KME values from all segments, denoted as $\mathbf{E} = \{E_1, E_2, ..., E_n\}$.

In the final stage, similar to existing time series representation learning methods [7,10], our algorithm can employ a conventional clustering algorithm (e.g. K-means) to perform the clustering task.

5   Experiments

We compared CTDS with other 8 clustering methods through the empirical evaluation on 85 datasets from the UCR time series archive [4].[2] The parameters of each algorithm were searched in a reasonable range following their original paper. We fix $\psi = 128$, and $m = 120$ for CTDS to evaluate its general performance on all datasets.[3] Since some clustering algorithms have uncertainty, we reported the average result of each algorithm over five trials on each dataset, with the best parameter setting w.r.t mutual information (NMI).[4] The comparison algorithms include: (a) distance-based method Kshape [14]; (b) representation-based methods SUSh [20], SPF [9], SPIRAL [8], DTC [13] and TS2Vec [24]; (c) distribution-based method WTK [3].

For methods that rely on subsequences, we applied the z-score normalization following the existing work [6]. Details of the experimental settings are given in the supplementary material[5].

5.1   Experimental Results

The overall comparison results are shown in Fig. 3a. we can observe that the clustering performance of CTDS outperforms all other methods in terms of both the average and median value of NMI scores over the 85 datasets.

We also conducted the Friedman test with the post-hoc Nemenyi test [5] on the NMI results. It shows that CTDS is ranked first and is significantly better than all others except TS2Vec.

---

[2] The training and testing subsets of each dataset are merged for clustering evaluation.
[3] For short time series data with length less than 150, we directly generate all subsequences without split the data into segments.
[4] We also evaluated their performance using RI and found a similar result. All experimental details can be found in the supplementary file.
[5] For methods that solely produce representations or kernel matrices, we use K-means or Kernel K-means as the clustering algorithm.

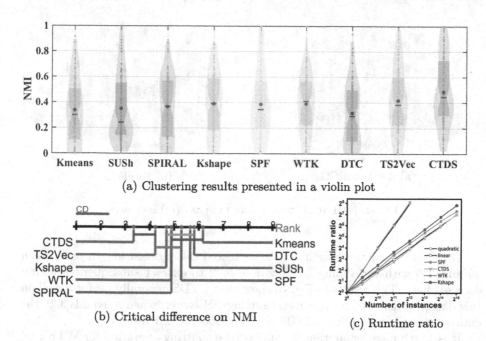

(a) Clustering results presented in a violin plot

(b) Critical difference on NMI

(c) Runtime ratio

Fig. 3. (a) A violin plot of the clustering results measured by NMI on 85 UCR datasets. The black bar and black dot denote the median and mean values, respectively; (b) Friedman-Nemenyi tests (at significance level 0.1) of the comparison on NMI; (c) CPU runtime ratio comparison on the CBF dataset.

In addition, we compare the efficiency of 4 competitive algorithms including CTDS, SPF, WTK and Kshape. Figure 3c shows the comparison result in terms of runtime ratio, on the CBF dataset for different numbers of instances, ranging from 512 to 32,768. CTDS is the only algorithm with a linear complexity. In contrast, WTK has a quadratic trend as the number of time series increases. Both SPF and Kshape show a superlinear increase.

5.2   Parameter Sensitivity Analysis

CTDS has three key parameters, including ψ as the number of partitions, m as the number of segments and d as the size of the sliding window. We find CTDS is not sensitive to ψ in most cases. Here we use two datasets $TwoLeadECG$ and $FaceFour$ to explore the effects of those parameters.

$TwoLeadECG$ is a short time series with a length of 84. In this case, the dataset was not split, and a sliding window of varying length d was used against different ψ. The result is shown in Fig. 4a. We can see from the results that a larger ψ normally produces a better clustering performance while the d should be around 40, i.e., half of the data length. Other papers also suggest that candidate subsequences should typically be set to a length greater than 20 [20] to capture more information on complex shapes.

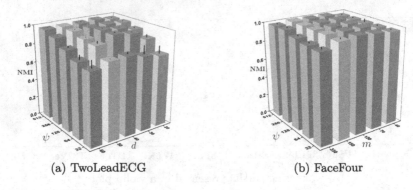

(a) TwoLeadECG   (b) FaceFour

Fig. 4. NMI at different values of ψ as $d(m)$ increases.

On *FaceFour* has a length of 350, we tested the segment length m between 40 and 350 with the sliding window $d = m/2$ to extract subsequences within each segment. The results in Fig. 4b show that CTDS generally perform well on this dataset with different parameter settings. However, when m equals 350, the entire dataset is not split, and CTDS yields poorer NMI results.

It is worth mentioning that we applied the splitting operation for WTK and improved its clustering results as well, the results are provided in the supplementary material. This confirms our assumption that clustering time series should focus on pattern matches on local regions rather than treat the series as a whole.

5.3   Visualization

Figure 5 presents t-SNE [11] visualizations of the original *StarLightCurves* dataset, the representation of TS2Vec, and the embedding obtained from CTDS. It is one of the largest-scale time series datasets which has 9236 instances with a length of 1024 in the UCR archive. When using Euclidean distance on the original dataset for t-SNE, two different classes of time series are significantly mixed together on the two-dimensional plot, making it challenging to distinguish

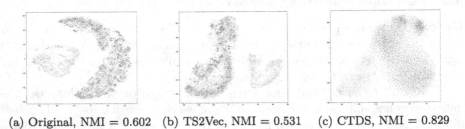

(a) Original, NMI = 0.602   (b) TS2Vec, NMI = 0.531   (c) CTDS, NMI = 0.829

Fig. 5. The visualization with t-SNE on the original *StarLightCurves* dataset and the representations produced by TS2Vec and CTDS. The NMI results are obtained through K-means clustering. The colors of the points indicate the actual labels.

them shown in Fig. 5a. However, CTDS based on IDK makes different clusters easier to extract as their boundaries become clearer and better than using the SOTA representation learning method TS2Vec, as shown in Fig. 5c. This implies that CTDS can capture more local information and subtle differences between complex time series.

6 Conclusion

The paper proposes a novel clustering method that treats time series as a set of points in the $\mathbb{R}^d$ domain. It use a data-dependent distributional measure IDK to quantify the distribution of subsequences, which enables effective similarity measurement without requiring point-to-point alignment. CTDS has linear time complexity and focuses on local regions to discriminate non-stationary time series data for clustering. We demonstrate that CTDS can produce significantly better clustering results than state-of-the-art on 85 datasets. The visualization of the embedding obtained by CTDS also validates its capability to capture subtle differences in complex datasets.

Acknowledgements. This project is supported by National Natural Science Foundation of China (Grant No. 62076120).

References

1. Aghabozorgi, S., Shirkhorshidi, A.S., Wah, T.Y.: Time-series clustering - a decade review. Inf. Syst. **53**, 16–38 (2015)
2. Begum, N., Ulanova, L., Wang, J., Keogh, E.: Accelerating dynamic time warping clustering with a novel admissible pruning strategy. In: Proceedings of the 21th ACM SIGKDD International Conference on Knowledge Discovery and Data Mining, pp. 49–58 (2015)
3. Bock, C., Togninalli, M., Ghisu, E., Gumbsch, T., Rieck, B., Borgwardt, K.: A Wasserstein subsequence kernel for time series. In: 2019 IEEE International Conference on Data Mining (ICDM), pp. 964–969. IEEE (2019)
4. Dau, H.A., et al.: The UCR time series archive. IEEE/CAA J. Automatica Sinica **6**(6), 1293–1305 (2019)
5. Demšar, J.: Statistical comparisons of classifiers over multiple data sets. J. Mach. Learn. Res. **7**, 1–30 (2006)
6. He, Y., Chu, X., Wang, Y.: Neighbor profile: bagging nearest neighbors for unsupervised time series mining. In: 2020 IEEE 36th International Conference on Data Engineering (ICDE), pp. 373–384. IEEE (2020)
7. Lafabregue, B., Weber, J., Gançarski, P., Forestier, G.: End-to-end deep representation learning for time series clustering: a comparative study. Data Min. Knowl. Disc. **36**(1), 29–81 (2022)
8. Lei, Q., Yi, J., Vaculin, R., Wu, L., Dhillon, I.S.: Similarity preserving representation learning for time series clustering. In: Proceedings of the Twenty-Eighth International Joint Conference on Artificial Intelligence, IJCAI (2017)
9. Li, X., Lin, J., Zhao, L.: Time series clustering in linear time complexity. Data Min. Knowl. Disc. **35**, 2369–2388 (2021)

10. Ma, Q., Zheng, J., Li, S., Cottrell, G.W.: Learning representations for time series clustering. In: Advances in Neural Information Processing Systems, vol. 32 (2019)
11. Van der Maaten, L., Hinton, G.: Visualizing data using t-SNE. J. Mach. Learn. Res. **9**(11) (2008)
12. MacQueen, J., et al.: Some methods for classification and analysis of multivariate observations. In: Proceedings of the Fifth Berkeley Symposium on Mathematical Statistics and Probability, Oakland, CA, USA, vol. 1, pp. 281–297 (1967)
13. Madiraju, N.S.: Deep temporal clustering: fully unsupervised learning of time-domain features. Ph.D. thesis, Arizona State University (2018)
14. Paparrizos, J., Gravano, L.: k-shape: Efficient and accurate clustering of time series. In: Proceedings of the 2015 ACM SIGMOD International Conference on Management of Data, pp. 1855–1870 (2015)
15. Qin, X., Ting, K.M., Zhu, Y., Lee, V.: Nearest-neighbour-induced isolation similarity and its impact on density-based clustering. In: Proceedings of the 33rd AAAI Conference on AI (AAAI 2019). AAAI Press (2019)
16. Ting, K.M., Liu, Z., Zhang, H., Zhu, Y.: A new distributional treatment for time series and an anomaly detection investigation. Proc. VLDB Endow. **15**(11), 2321–2333 (2022)
17. Ting, K.M., Xu, B.C., Washio, T., Zhou, Z.H.: Isolation distributional kernel: a new tool for kernel based anomaly detection. In: Proceedings of the 26th ACM SIGKDD International Conference on Knowledge Discovery and Data Mining, pp. 198–206 (2020)
18. Ting, K.M., Zhu, Y., Zhou, Z.H.: Isolation kernel and its effect on SVM. In: Proceedings of the 24th ACM SIGKDD International Conference on Knowledge Discovery and Data Mining, pp. 2329–2337 (2018)
19. Tonekaboni, S., Eytan, D., Goldenberg, A.: Unsupervised representation learning for time series with temporal neighborhood coding. In: International Conference on Learning Representations (2021)
20. Ulanova, L., Begum, N., Keogh, E.: Scalable clustering of time series with u-shapelets. In: Proceedings of the 2015 SIAM International Conference on Data Mining, pp. 900–908. SIAM (2015)
21. Vallender, S.: Calculation of the Wasserstein distance between probability distributions on the line. Theory Probab. Appl. **18**(4), 784–786 (1974)
22. Ye, L., Keogh, E.: Time series shapelets: a new primitive for data mining. In: Proceedings of the 15th ACM SIGKDD International Conference on Knowledge Discovery and Data Mining, pp. 947–956 (2009)
23. Yeh, C.C.M., et al.: Matrix profile i: all pairs similarity joins for time series: a unifying view that includes motifs, discords and shapelets. In: 2016 IEEE 16th International Conference on Data Mining (ICDM), pp. 1317–1322. IEEE (2016)
24. Yue, Z., et al.: Ts2vec: towards universal representation of time series. In: Proceedings of the AAAI Conference on Artificial Intelligence, vol. 36, pp. 8980–8987 (2022)
25. Zakaria, J., Mueen, A., Keogh, E.: Clustering time series using unsupervised-shapelets. In: 2012 IEEE 12th International Conference on Data Mining, pp. 785–794. IEEE (2012)
26. Zhao, Y., Ye, L., Li, Z., Song, X., Lang, Y., Su, J.: A novel bidirectional mechanism based on time series model for wind power forecasting. Appl. Energy **177**, 793–803 (2016)

Distributed MCMC Inference for Bayesian Non-parametric Latent Block Model

Reda Khoufache[1]([⊠]), Anisse Belhadj[1], Hanene Azzag[2], and Mustapha Lebbah[1]

[1] Paris-Saclay University, UVSQ, David Lab, 78035 Versailles, France
{reda.khoufache,mutapha.lebbah}@uvsq.fr,
med-anisse.belhadj@outlook.com
[2] LIPN (UMR CNRS 7030), Sorbonne Paris Nord University, Villetaneuse, France
azzag@univ-paris13.fr

Abstract. In this paper, we introduce a novel Distributed Markov Chain Monte Carlo (MCMC) inference method for the Bayesian Non-Parametric Latent Block Model (DisNPLBM), employing the Master/Worker architecture. Our non-parametric co-clustering algorithm divides observations and features into partitions using latent multivariate Gaussian block distributions. The workload on rows is evenly distributed among workers, who exclusively communicate with the master and not among themselves. DisNPLBM demonstrates its impact on cluster labeling accuracy and execution times through experimental results. Moreover, we present a real-use case applying our approach to co-cluster gene expression data. The code source is publicly available at https://github.com/redakhoufache/Distributed-NPLBM

Keywords: Co-clustering · Bayesian non-parametric · Distributed computing

1 Introduction

Given a data matrix, where rows represent observations and columns represent variables or features, co-clustering aims to infer a row partition and a column partition simultaneously. The resulting partition is composed of homogeneous blocks. When a dataset exhibits a dual structure between observations and variables, co-clustering outperforms conventional clustering algorithms which only infers a row partition without considering the relationships between observations and variables. Co-clustering is a powerful data mining tool for two-dimensional data and is widely applied in various fields such as bioinformatics [12].

To tackle the co-clustering problem, the Latent Block Model (LBM) was introduced by [10]. This probabilistic model assumes the existence of hidden block components, such that elements that belong to the same block independently follow identical distribution. A Bayesian Non-Parametric extension of the LBM (NPLBM) was introduced in [14]. This model makes two separate priors on the proportions and a prior on the block component distribution, which allows to automatically estimate the number of co-clusters during the inference process. In

© The Author(s), under exclusive license to Springer Nature Singapore Pte Ltd. 2024
D.-N. Yang et al. (Eds.): PAKDD 2024, LNAI 14645, pp. 271–283, 2024.
https://doi.org/10.1007/978-981-97-2242-6_22

[9], the authors present a BNP Functional LBM which extends the recent LBM [2] to a BNP framework to address the multivariate time series co-clustering.

To infer parameters of NPLBM, the Collapsed Gibbs sampler introduced in [16], is a Markov Chain Monte Carlo (MCMC) algorithm that iteratively updates the column partition, given the row partition, and vice versa. It samples row and column memberships sequentially based on their respective marginal posterior probabilities. The collapsed Gibbs sampler is an efficient MCMC algorithm because the co-cluster parameters are analytically integrated away during the sampling process. MCMC methods have the good property of producing asymptotically exact samples from the target density. However, these techniques are known to suffer from slow convergence when dealing with large datasets.

Distributed computing consists of distributing data across multiple computing nodes (workers), which allows parallel computations to be performed independently. Distributed computing offers the advantage of accelerating computations and overcoming memory limitations. Existing programming paradigms for distributed computing, such as Map-Reduce consist of a map and reduce functions. The map function applies the needed transformations on data and produces intermediate key/value pairs. The reduce function merges the map's function results to form the output value. The Map-Reduce job is executed on master/workers architecture, where the master coordinates the job execution, and workers execute the map and reduce tasks in parallel.

This paper proposes a new distributed inference for NPLBM when the number of observations is too large. We summarize our contributions: **(1)** We have developed a new distributed MCMC inference of the NPLBM using the Master/Worker architecture. The rows are evenly distributed among the workers which only communicate with the master. **(2)** Each worker infers a local row partition given the global column partition. Then, sufficient statistics associated with each local row cluster are sent to the master. **(3)** At the master, the global row partition is estimated. Then, given the global row partition, the column partition is estimated. This allows the estimation of the global co-clustering structure. **(4)** Theoretical background and computational details are provided.

2   Related Work

Numerous scalable co-clustering algorithms have been proposed in the literature. The first distributed co-clustering (DisCo) using Hadoop is introduced in [18]. In [7], authors devised a parallelized co-clustering approach, specifically designed to tackle the high-order co-clustering problem with heterogeneous data. Their methodology extends the approach initially proposed in [11], enabling the computation of co-clustering solutions in a parallel fashion, leveraging a Map-Reduce infrastructure. In [6], a parallel simultaneous co-clustering and learning (SCOAL) approach is introduced, also harnessing the power of Map-Reduce. This work focuses on predictive modeling for bi-modal data. In [4], introduces a distributed framework for data co-clustering with sequential updates (Co-ClusterD). The authors propose two distinct approaches to parallelize sequential updates for alternate minimization co-clustering algorithms. However, it's worth noting that these approaches are parametric and assume knowing a priori the true numbers of row and column clusters, respectively, which are unknowable in real-life applications. One of the main challenges in distributing Bayesian Non-Parametric co-clustering lies in efficiently handling and discovering new block components.

3   Bayesian Non-parametric Latent Block Model

3.1   Model Definition

Let n, p, and d be positive integers, and let $X = (x_{i,j})_{n,p} \in \mathbb{R}^{n \times p \times d}$ be the observed dataset. Here, n represents the number of rows, p is the number of columns, and d denotes the dimension of the observation space. Let $\mathbf{z} = (z_i)_n$ be the row membership vector (row partition), where each z_i is a latent variable such that $z_i = k$ signifies that the i-th row $x_{i,\cdot}$ belongs to the row cluster k. Similarly, let $\mathbf{w} = (w_j)_p$ be the column membership vector (column partition), where $w_j = l$ indicates that the j-th column $x_{\cdot,j}$ belongs to the column cluster l. The NPLBM is defined as follows:

$$x_{i,j} \mid \{z_i, w_j, \theta_{z_i,w_j}\} \sim F\left(\theta_{z_i,w_j}\right),$$
$$\theta_{z_i,w_j} \sim G_0, \ z_i \mid \pi \sim \mathrm{Mult}(\pi), \ w_j \mid \rho \sim \mathrm{Mult}(\rho),$$
$$\pi \sim \mathrm{SB}(\alpha), \ \rho \sim \mathrm{SB}(\beta).$$

According to this definition, the observation $x_{i,j}$ is sampled by first generating the row proportions $\pi \sim \mathrm{SB}(\alpha)$ and column proportions $\rho \sim \mathrm{SB}(\beta)$ according to the Stick-Breaking (SB) process [19] parameterized by concentration parameters $\alpha > 0$ and $\beta > 0$ respectively. Secondly, sampling the row and column memberships $\mathbf{z}$ and $\mathbf{w}$ from the Multinoulli distribution (Mult) parameterized by π and ρ, respectively. Then, sampling the block component parameter θ_{z_i,w_j} from the base distribution G_0. Finally, drawing the cell value $x_{i,j}$ that belongs to the block (z_i, w_j) from the component distribution $F(\theta_{z_i,w_j})$. We assume that F is the multivariate Gaussian distribution (i.e., $\theta_{k,l} = (\mu_{k,l}, \Sigma_{k,l})$, with $\mu \in \mathbb{R}^d$ and $\Sigma_{k,l} \in \mathbb{R}^{d \times d}$ a positive semi-definite matrix), and G_0 is the Normal Inverse Wishart [15] (NIW) conjugate prior with hyper-parameters $(\mu_0, \kappa_0, \Psi_0, \nu_0)$.

3.2   Inference

The goal is to estimate the row and column partitions $\mathbf{z}$ and $\mathbf{w}$ given the dataset X, the prior G_0, and the concentration parameters α and β, by sampling from the joint posterior distribution $\mathrm{p}(\mathbf{z}, \mathbf{w} | X, G_0, \alpha, \beta)$. However, direct sampling from this distribution is intractable but can be achieved using the collapsed Gibbs Sampler introduced in [16]. Given initial row and column partitions. The inference process consists of alternating between updating the row partition given the column partition and then updating the column partition given the row partition. At each iteration, to update the row partition $\mathbf{z}$, each z_i is updated sequentially by sampling from $\mathrm{p}(z_i | \mathbf{z}_{-i}, \mathbf{w}, X, G_0, \alpha)$, where $\mathbf{z}_{-i} = \{z_r | r \neq i\}$.

The column partition update is similar to the row partition update. The complete algorithm and computation details of the inference process are given in [8].

4   Proposed Inference

The main objective of our method is to make the inference scalable when the number of observations becomes too large. The rows are distributed evenly over the workers. At each iteration, we alternate between two levels:

4.1   Worker Level

Let E be the number of workers, n^e be the number of rows in worker e, $X^e = (x_{i,j}^e)_{n^e \times p} \in \mathbb{R}^{n^e \times p \times d}$ the local dataset in worker e, each cell $x_{i,j}^e$ is a d-dimensional vector. Let $\mathbf{z}^e = (z_i^e)_{n^e}$ be the local row partition (i.e. $z_i^e = k$ means that the i-th row of e-th worker belongs to the k-th local row cluster). At this level, each local row membership z_i^e is updated given other local row memberships $\mathbf{z}_{-i}^e = \{z_r^e | r \neq i\}$ by sampling from $\mathrm{p}(z_i^e | \mathbf{z}_{-i}^e, \mathbf{w}, X^e, G_0, \alpha) \propto$:

$$\begin{cases} n_k^e \mathrm{p}\left(x_{i,\cdot}^e \mid \mathbf{w}, \mathbf{x}_{k,\cdot}^e, G_0\right) & \text{existing row cluster k,} & \text{(1a)} \\ \alpha \mathrm{p}\left(x_{i,\cdot}^e \mid \mathbf{w}, G_0\right), & \text{new row cluster,} & \text{(1b)} \end{cases}$$

where n_k^e is the size of local row cluster k in worker e, $x_{i,\cdot}^e$ is the i-th row of worker e, and $\mathbf{x}_{k,\cdot}^e = \{x_{i,\cdot}^e | z_i^e = k\}$ the content of local row cluster k in worker e. Since G_0 is a prior conjugate to F, the joint prior and posterior predictive distributions needed in 1a and 1b are computed analytically [9]. After having updated the local row partition, for a given row cluster k in worker e, for each column j, we compute the following sufficient statistics:

$$T_{k,j}^e = \frac{1}{n_k^e} \sum_{i=1, z_i^e=k}^{n^e} x_{i,j}^e \in \mathbb{R}^d, \tag{2}$$

$$S_{k,j}^e = \sum_{i=1, z_i^e=k}^{n^e} (x_{i,j}^e - T_{k,j}^e)(x_{i,j}^e - T_{k,j}^e)^T \in \mathbb{R}^{d \times d}, \tag{3}$$

where $(\cdot)^T$ denotes the transpose operator. We let $\mathcal{S}^e = \Big\{ (T_{k,j}^e, S_{k,j}^e) \mid (k,j) \in \{1, \cdots, K^e\} \times \{1, \cdots, p\} \Big\}$, the set of sufficient statistics, where K^e is the number of row clusters inferred in worker e. Finally, the sufficient statistics and sizes of each cluster are sent to the master. The DisNPLBM inference process at the worker level is described in Algorithm 1, which represents the Map function.

4.2   Master Level

At this level, the objective is to estimate the global row and column partition given sufficient statistics, local cluster sizes, and the prior. In the following, we detail these two steps:

Algorithm 1. DisNPLBM inference at worker level

1: **Input:** $X^e_{n^e \times p \times d}$, α, G_0, $\mathbf{z}^e$, and $\mathbf{w}$.
2: **For** $i \leftarrow 1$ **to** n^e **do:**
3:     Remove $x^e_{i,\cdot}$ from the its local row cluster.
4:     Sample z^e_i according to Eq. 1a and Eq. 1b.
5:     Add $x^e_{i,\cdot}$ to its new local row cluster.
6: **For** $k \leftarrow 1$ **to** K^e **do:**
7:     **For** $j \leftarrow 1$ **to** p **do:**
8:         Compute $T^e_{k,j}$ and $S^e_{k,j}$ as defined in Eq. 2 and Eq. 3, respectively.
9: **Output:** Updated row partition $\mathbf{z}^e$, sufficient statistics S^e, sizes of each cluster.

Global Row Partition Estimation. The global row membership $\mathbf{z}$ is estimated by clustering the local row clusters. Instead of assigning the rows sequentially and individually to their row cluster, we assign the batch of rows that already share the same local row cluster to a global row cluster. Hence, the rows assigned to the same global row cluster will share the same global row membership. Since the workers operate asynchronously, the results are joined in a streaming way using the Reduce function without waiting for all workers to finish their tasks.

Let $S^{e_1}, S^{e_2}, K^{e_1}$ and K^{e_2} be the sets of sufficient statistics and the number of local row clusters returned by two workers e_1 and e_2 respectively. The goal is to cluster the local row clusters $\{\mathbf{x}^{e_1}_{1,\cdot}, \cdots, \mathbf{x}^{e_1}_{K^{e_1},\cdot}\}$ and $\{\mathbf{x}^{e_2}_{1,\cdot}, \cdots, \mathbf{x}^{e_2}_{K^{e_2},\cdot}\}$. To perform such clustering, we proceed as follows: we first set the initial cluster partition equal to the local partition inferred in cluster e_1. Then, for each $h \in \{1, \cdots, K^{e_2}\}$, we sample $z^{e_2}_h$, the membership of $\mathbf{x}^{e_2}_{h,\cdot}$ from $p(z^{e_2}_h \mid \mathbf{z}^{e_2}_{-h}, X, G_0, \alpha) \propto$

$$\begin{cases} n_k p(\mathbf{x}^{e_2}_{h,\cdot} \mid z^{e_2}_h = k, \mathbf{X}_{k,\cdot}, G_0), & \text{existing row cluster } k, & (4a) \\ \alpha p(\mathbf{x}^{e_2}_{h,\cdot} \mid G_0) & \text{new row cluster,} & (4b) \end{cases}$$

where n_k is the size of global row cluster k, $\mathbf{X}_{k,\cdot}$ the content of global row cluster k, and $\mathbf{z}^{e_2}_{-h} = \{z^{e_2}_{h'} | h' \neq h\}$. The joint posterior and the joint prior predictive distributions (Eq. 4a, and Eq. 4b respectively) are computed analytically by only using sufficient statistics, i.e., without having access to the content of local and global clusters:

$$p\left(\mathbf{x}^e_{h,\cdot} \mid G_0\right) = \pi^{-n^e_h \frac{d}{2}} \cdot \frac{\kappa_0^{d/2}}{(\kappa^e_h)^{d/2}} \cdot \frac{\Gamma_d\left(\nu^e_h/2\right)}{\Gamma_d\left(\nu_0/2\right)} \cdot \frac{|\Psi_0|^{\nu_0/2}}{|\Psi^e_h|^{\nu^e_h/2}}$$

where $|\cdot|$ is the determinant, Γ denotes the gamma function, and the hyperparameter $(\mu^e_h, \kappa^e_h, \Psi^e_h, \nu^e_h)$ are updated using the sufficient statistics:

$$\mu^e_h = \frac{\kappa_0 \mu_0 + n^e_h T^e_h}{\kappa^e_h}, \quad \kappa^e_h = \kappa_0 + n^e_h, \quad \nu^e_h = \nu_0 + n^e_h,$$

$$\Psi^e_h = \Psi_0 + S^e_h + \frac{\kappa_0 n^e_h}{\kappa^e_h}\left(\mu_0 - T^e_h\right)\left(\mu_0 - T^e_h\right)^T,$$

where $T_h^e = \frac{1}{p}\sum_{j=1}^{p} T_{h,j}$ and $S_h^e = \frac{1}{p}\sum_{j=1}^{p} S_{h,j}^e$. Moreover, we have

$$p(\mathbf{x}_{h,\cdot}^e \mid z_h^e = k, \mathbf{X}_{k,\cdot}, G_0) = \pi^{\frac{-dn_h^e}{2}} \cdot \frac{\kappa_k^{d/2}}{(\kappa_h^e)^{d/2}} \cdot \frac{\Gamma_d(\nu_h^e/2)}{\Gamma_d(\nu_k/2)} \cdot \frac{|\Psi_k|^{\nu_k/2}}{|\Psi_h^e|^{\nu_h^e/2}}$$

where the posterior distribution parameters $(\mu_k, \kappa_k, \Psi_k, \nu_k)$ associated to the global cluster k are updated from the prior as follows:

$$\mu_k = \frac{\kappa_0\mu_0 + n_k T_k}{\kappa_k}, \quad \kappa_k = \kappa_0 + n_k, \quad \nu_k = \nu_0 + n_k,$$

$$\Psi_k = \Psi_0 + S_k + \frac{\kappa_0 n_k}{\kappa_k}(\mu_0 - T_k)(\mu_0 - T_k)^T,$$

with T_k and S_k the aggregated sufficient statistics when local clusters are assigned to the same global cluster. They are given by:

$$T_k = \frac{1}{n_k}\sum_{e,h|\,\mathbf{z_h^e}=\mathbf{k}} n_h^e \cdot T_h^e, \tag{5}$$

$$S_k = \sum_{e,h|\,\mathbf{z_h^e}=\mathbf{k}} S_h^e + \sum_{e,h|\,\mathbf{z_h^e}=\mathbf{k}} \left(n_e^h \cdot T_h^e \cdot T_h^{eT}\right) - n_k \cdot T_k \cdot T_k^T. \tag{6}$$

This step consists of joining workers' local row clusters in a streaming way. The recursive joining process stops when the global row partition is estimated. If $K^{(e_1,e_2)}$ is the number of inferred global row clusters, then the process stops when $\sum_{k=1}^{K^{(e_1,e_2)}} n_k = n$. The procedure is detailed in the Algorithm 2.

Algorithm 2. Join workers results (Reduce Function)

1: **Input:** $\mathcal{S}^{e_1}$, $\mathcal{S}^{e_2}$, α and prior G_0.
2: Initialize global membership $\mathbf{z}$ according to $\mathbf{z}^{e_1}$.
3: **For** each $h \in K^{e_2}$ **do:**
4:     Sample $z_h^{e_2}$ according to Eq. 4a and Eq. 4a.
5:     Add $\mathbf{x}_h^{e_2}$ to its new global row cluster.
6:     Update the membership vector $\mathbf{z}$.
7: **For** $k \leftarrow 1$ **to** $K^{(e_1,e_2)}$**do:**
8:     Compute S_k and T_k according to Eq. 5 and Eq 6.
9: **Output:** Updated row partition $\mathbf{z}$, aggregated sufficient statistics and clusters sizes.

Column Memberships Estimation. Given the sufficient statistics $\mathcal{S}^1, \mathcal{S}^2, \cdots, \mathcal{S}^E$, the global row partition $\mathbf{z}$, the prior G_0, and the concentration parameter β, the objective is to update the column partition $\mathbf{w} = (w_j)_p$, each w_j is drawn according to $p(w_i|\mathbf{w}_{-j}, \mathbf{z}, X, G_0, \beta) \propto$

$$\begin{cases} p_k p(x_{\cdot,j} \mid \mathbf{z}, \mathbf{w}_{-j}, X_{-j}, G_0, \beta), & \text{existing column cluster } l, \quad (7a) \\ \beta p(x_{\cdot,j} \mid \mathbf{z}, G_0) & \text{new column cluster,} \quad (7b) \end{cases}$$

with $x_{.,j}$ the j-th column and X_{-j} the dataset without column j. Similarly, the joint posterior predictive and the joint prior predictive distributions (Eq. 7a, and Eq. 7b respectively) are computed analytically without having access to the columns, but only by using sufficient statistics. In fact, we have:

$$ \mathrm{p}\left(x_{.,j} \mid \mathbf{z}, G_0\right) = \prod_{k=1}^{K} \mathrm{p}\left(\mathbf{x}_{k,j} | G_0\right) $$

with K the global number of inferred row clusters, and $\mathbf{x}_{k,j}$ the element of column j that belong to the row cluster k. We have:

$$ \mathrm{p}\left(\mathbf{x}_{k,j} | G_0\right) = \pi^{-p_{k,j} \times \frac{d}{2}} \cdot \frac{\kappa_0^{d/2}}{\kappa_{k,j}^{d/2}} \cdot \frac{\Gamma_d(\nu_{k,j}/2)}{\Gamma_d(\nu_0/2)} \cdot \frac{|\Psi_0|^{\nu_0/2}}{|\Psi_{k,j}|^{\nu_{k,j}/2}} $$

with $p_{k,j}$ the cardinal of $\mathbf{x}_{k,j}$. The updated hyper-parameters are obtained with:

$$ \mu_{k,j} = \frac{\kappa_0\mu_0 + p_{k,j}T_{k,j}}{\kappa_{k,j}}, \quad \kappa_{k,j} = \kappa_0 + p_{k,j}, \quad \nu_{k,j} = \nu_0 + p_{k,j}, $$

$$ \Psi_{k,j} = \Psi_0 + S_{k,j} + \frac{\kappa_0 p_{k,j}}{\kappa_{k,j}}\left(\mu_0 - T_{k,j}\right)\left(\mu_0 - T_{k,j}\right)^T. $$

Moreover, the posterior predictive distribution is computed as follows:

$$ \mathrm{p}(x_{.,j} \mid \mathbf{z}, \mathbf{w}_{-j}, X_{-j}, G_0, \beta) = \prod_{k=1}^{K} \mathrm{p}\left(\mathbf{x}_{k,j} | G_{k,l}\right) $$

with $G_{k,l}$ the posterior distribution associated with block (k, l) (i.e., row cluster k and column cluster l). We have:

$$ \mathrm{p}\left(\mathbf{x}_{k,j} | G_{k,l}\right) = \pi^{-p_{k,j} \times \frac{d}{2}} \cdot \frac{\kappa_{k,l}^{d/2}}{\kappa_{k,j}^{d/2}} \cdot \frac{\Gamma_d(\nu_{k,j}/2)}{\Gamma_d(\nu_{k,l}/2)} \cdot \frac{|\Psi_{k,l}|^{\nu_{k,l}/2}}{|\Psi_{k,j}|^{\nu_{k,j}/2}} $$

with $(\mu_{k,l}, \kappa_{k,l}, \Psi_{k,l}, \nu_{k,l})$ the block posterior distribution parameters given by:

$$ \mu_{k,l} = \frac{\kappa_0\mu_0 + p_{k,l}T_{k,l}}{\kappa_{k,l}}, \quad \kappa_{k,l} = \kappa_0 + p_{k,l}, \quad \nu_{k,l} = \nu_0 + p_{k,l}, $$

$$ \Psi_{k,l} = \Psi_0 + S_{k,l} + \frac{\kappa_0 p_{k,l}}{\kappa_{k,l}}\left(\mu_0 - T_{k,l}\right)\left(\mu_0 - T_{k,l}\right)^T. $$

With $T_{k,l}$ and $S_{k,l}$, the aggregated sufficient statistics obtained when local clusters are assigned to the same global block (k, l), and they are computed as follows:

$$ T_{k,l} = \frac{1}{p_{k,l}} \sum_{e,h|\, \mathbf{z}_h^e = \mathbf{k}, \mathbf{w} = 1} p_{h,l}^e \cdot T_{h,l}^e $$

$$ S_{k,l} = \sum_{e,h|\, \mathbf{z}_h^e = \mathbf{k}, \mathbf{w} = 1} S_{h,l}^e + \sum_{e,h|\, \mathbf{z}_h^e = \mathbf{k}, \mathbf{w} = 1} \left(p_{h,l}^e \cdot T_{h,l}^e \cdot T_{h,l}^{e\ T}\right) - p_{k,l} \cdot T_{k,l} \cdot T_{k,l}^T $$

where $p_{k,l}$ is the number of cells in the global cluster (k, l). The column partition update is detailed in algorithm 3.

Algorithm 3. Column clustering

1: **Input**: Sufficient statistics, row partition, β and prior G_0.
2: **For** $j \leftarrow 1$ **to** p **do**:
3:     Remove $x_{.,j}$ from its column cluster.
4:     Sample w_j according to Eq. 7a, and Eq. 7b.
5:     Add $x_{.,j}$ to its new column cluster.
6: **Output**: Column-partition **w**.

---

5   Experiments

To evaluate our approach, we conducted several experiments. Firstly, we compare our distributed algorithm with other state-of-the-art co-clustering and clustering algorithms in terms of row clustering performance on synthetic and real-world datasets. Secondly, we compare the execution time and clustering performance of our distributed algorithm DisNPLBM and the centralized NPLBM [9] on synthetic datasets with different row sizes. Lastly, we investigate the scalability of DisNPLBM by increasing the number of nodes while keeping the number of rows fixed. The clustering performance is evaluated using the clustering metrics Adjusted Rand Index (ARI) [13] and Normalized Mutual Information (NMI) [20].

5.1   Experiment Settings

In the following experiments, we use an uninformative prior NIW as in [9]. Therefore, we set the NIW hyper-parameters as follows: μ_0, and the matrix precision Ψ_0 are respectively set to be empirical mean vector and covariance matrix of all data. κ_0 and ν_0 are set to their lowest values, which are 1 and $d+1$, respectively, where d is the dimension of the observation space. The initial partition consists of a single cluster, and the algorithms run for 100 iterations.

The distributed algorithm is executed on the Neowise machine (1 CPU AMD EPYC 7642, 48 cores/CPU) and Gros machines (1 CPU Intel Xeon Gold 5220, 18 cores/CPU), both hosted by Grid5000[1]. For enhanced portability and deployment flexibility, DisNPLBM is containerized using the Docker image bitnami/spark 3.3.0. We employ Kubernetes for orchestrating Docker images and deploy the Kubernetes cluster on Grid5000 using Terraform[2].

5.2   Clustering Performance

We first evaluate the row clustering performance of our algorithm on both synthetic and real-world datasets; we compare its results with two co-clustering algorithms, NPLBM [9] and LBM [10], and two clustering algorithms, K-means and Gaussian mixture model (GMM). We applied the algorithms to 4 datasets:

---

[1] https://www.grid5000.fr/.
[2] https://www.terraform.io/.

Table 1. The mean and the standard deviation of ARI and NMI over 10 runs on different datasets. The best result within each row is marked as bold.

Dataset		DisNPLBM	NPLBM	LBM	GMM	K-means
Synthetic	ARI	**1.00 ± 0.00**	**1.00 ± 0.00**	0.42 ± 0.03	0.38 ± 0.05	0.39 ± 0.01
	NMI	**1.00 ± 0.00**	**1.00 ± 0.00**	0.78 ± 0.02	0.70 ± 0.02	0.71 ± 0.01
Wine	ARI	0.52 ± 0.03	**0.56 ± 0.04**	0.56 ± 0.07	0.51 ± 0.07	0.50 ± 0.04
	NMI	0.59 ± 0.02	**0.65 ± 0.03**	0.64 ± 0.03	0.64 ± 0.03	0.64 ± 0.02
Chowdary	ARI	**0.78 ± 0.01**	0.07 ± 0.01	0.65 ± 0.00	0.74 ± 0.01	0.75 ± 0.01
	NMI	**0.68 ± 0.02**	0.11 ± 0.01	0.58 ± 0.01	0.63 ± 0.01	0.64 ± 0.01
Nutt	ARI	**0.58 ± 0.01**	0.56 ± 0.02	0.54 ± 0.04	0.08 ± 0.00	0.11 ± 0.04
	NMI	**0.74 ± 0.01**	0.74 ± 0.01	0.68 ± 0.02	0.28 ± 0.02	0.30 ± 0.00

Synthetic dataset of size 150×150, generated from 10×3 Gaussian components. *Wine* dataset [1] represents a chemical analysis of three types of wines grown in the same region. The dataset consists of 178 observations, 12 features, and 3 clusters. We also apply the algorithms to two bioinformatics datasets *Chowdary* (104 samples, 182 genes, and 2 clusters) [5] and *Nutt* (22 samples, 1152 genes, and 2 clusters) [17]. Each sample's gene expression level is measured using the Affymetrix technology leading to strictly positive data ranging from 0 to 16000. We apply the Box-Cox transformation [3], to make the data Gaussian-like. Since the number of Genes is much greater than the number of samples we distribute the columns across the workers to achieve scalability, this is legitimate since the row and column clustering are symmetric in our case.

Table 1 presents the mean and standard deviation of ARI and NMI across 10 launches for each method on each dataset. Our method outperformed other approaches in the Bioinformatics datasets. Additionally, it has estimated the true clustering structure in the synthetic dataset. While NPLBM and LBM slightly outperform our method on the *Wine* dataset, our approach still yields satisfying results, surpassing traditional methods like GMM and K-means. It's crucial to note that this experiment focuses on comparing clustering performance, without considering execution times due to different inference algorithms. Figure 1 illustrates the Heatmaps of Chowdary data before DisNPLBM and reordered data after DisNPLBM. In the recorded data, there is a visible checkerboard pattern distinguishing co-clusters. Co-clustering simultaneously clusters samples and genes, revealing groups of highly correlated genes with distinct correlation structures among different sets of individuals, such as between disease and healthy individuals or different types of disease. This may allow to identify which genes are responsible for some diseases.

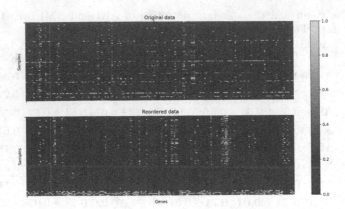

Fig. 1. Heatmaps of Chowdary data. The first row represents the original data. The second row represents the reordered data after DisNPLBM.

5.3   Comparison of the Distributed and Centralized Approaches

We compare the execution times and clustering performance of the distributed and the centralized NPLBM. We execute both algorithms on synthetic datasets of sizes $n \times p \times d$, where $n \in \{20K, \cdots, 100K\}$, $p = 90$ and $d = 1$ generated from $K \times L$ Gaussian components, with $K = 10$ and $L = 3$ (i.e., $K = 10$ row clusters and $L = 3$ column clusters). We stop at $n = 100K$ because the centralized version is too slow; running over 100K observations would take too much time. The distributed algorithm is executed on the Neowise machine in local mode using 24 cores. The centralized algorithm is executed on the same machine using one core.

Table 2. ARI, NMI, number of inferred block clusters $(\hat{K} \times \hat{L})$, and the running time in seconds achieved by the distributed (Dis.) and centralized (Cen.) algorithms.

n	ARI		NMI		$\hat{K} \times \hat{L}$		Running time (s)	
	Dis.	Cen.	Dis.	Cen.	Dis.	Cen.	Dis.	Cen.
20K	1.0	1.0	1.0	1.0	30	30	400.21	2265.69
40K	1.0	1.0	1.0	1.0	30	30	693.02	6452.78
60K	1.0	1.0	1.0	1.0	30	30	1122.80	10511.01
80K	1.0	1.0	1.0	1.0	30	30	1373.04	19965.01
100K	1.0	1.0	1.0	1.0	30	30	1572.90	41897.12

Table 2 reports the clustering metrics ARI, NMI, number of inferred block clusters $(\hat{K} \times \hat{L})$, and the running times obtained by the centralized and distributed inference algorithms on datasets with different row sizes. The results show that our approach considerably reduces the execution time. For example, it is reduced by a factor of 26 for a dataset with 100K rows. On the other hand, we remark that both the distributed and centralized methods performed very well in terms of clustering with values of 1 indicating perfect clustering. Moreover, both methods inferred the true number of clusters. Overall, the distributed approach runs much faster than the centralized method without compromising the clustering performance which makes it more efficient in terms of computational time.

5.4 Distributed Algorithm Scalability

We now investigate the scalability of our approach by increasing the number of cores up to 64 in a distributed computing environment. We employ a dataset with $n = 500K$ rows, $p = 20$ columns, and $d = 1$ (representing the observation space dimension). The dataset is generated from $K \times L$ Gaussian components, where $K = 10$ is the number of row clusters and $L = 3$ is the number of column clusters. To conduct this evaluation, we deploy a Kubernetes cluster using up to 6 Gros Machines.

Table 3. ARI, NMI, number of inferred clusters, and the running time in seconds achieved by the distributed approach when distributing on different numbers of cores.

Cores	ARI	NMI	$\hat{K} \times \hat{L}$	Running time (s)
2	1.0	1.0	30	88943.45
4	1.0	1.0	30	27964.73
8	0.99	0.99	30	16202.15
32	0.98	0.99	33	2715.80
64	0.98	0.99	33	1861.85

Table 3 presents clustering metrics ARI and NMI, the number of inferred block clusters, and running time as the number of cores increases. The running time significantly decreases with an increasing number of cores, with the execution time reduced by a factor of 48 when using 64 cores compared to two cores. This demonstrates the efficient scalability of our algorithm with the number of workers. It's worth noting a slight overestimation of the number of clusters with more cores. Additionally, there is a slight decrease in ARI and NMI scores. Nevertheless, our approach still achieves very high clustering metrics and accurately estimates the number of clusters.

6 Conclusion

This article presents a novel distributed MCMC inference for NPLBM. NPLBM has the advantage of estimating the number of row and column clusters. However, the inference process becomes too slow when dealing with large datasets. Our proposed method achieves high scalability without compromising the clustering performance. Our future research will explore the potential extension of this method to the multiple Coclustering model.

Acknowledgements. This work has been supported by the Paris Île-de-France Région in the framework of DIM AI4IDF. I thank Grid5000 for providing the essential computational resources and the start-up HephIA for the invaluable exchange on scalable algorithms.

References

1. Aeberhard, S., Forina, M.: Wine. UCI Machine Learning Repository (1991). https://doi.org/10.24432/C5PC7J
2. Ben Slimen, Y., Allio, S., Jacques, J.: Model-based co-clustering for functional data. Neurocomputing **291**, 97–108 (2018)
3. Box, G.E., Cox, D.R.: An analysis of transformations. J. R. Stat. Soc. Ser. B Stat Methodol. **26**(2), 211–243 (1964)
4. Cheng, X., Su, S., Gao, L., Yin, J.: Co-clusterd: a distributed framework for data co-clustering with sequential updates. IEEE Trans. Knowl. Data Eng. **27**(12), 3231–3244 (2015)
5. Chowdary, D., Lathrop, J., Skelton, J., Curtin, K., Briggs, T., Zhang, Y., Yu, J., Wang, Y., Mazumder, A.: Prognostic gene expression signatures can be measured in tissues collected in rnalater preservative. J Mol Diagn **8**(1), 31–39 (2006)
6. Deodhar, M., Jones, C., Ghosh, J.: Parallel simultaneous co-clustering and learning with map-reduce. In: 2010 IEEE International Conference on Granular Computing, pp. 149–154 (2010)
7. Folino, F., Greco, G., Guzzo, A., Pontieri, L.: Scalable parallel co-clustering over multiple heterogeneous data types, pp. 529 – 535, August 2010
8. Goffinet, E.: Multi-Block Clustering and Analytical Visualization of Massive Time Series from Autonomous Vehicle Simulation. Theses, Université Paris 13 Sorbonne Paris Nord, December 2021
9. Goffinet, E., Lebbah, M., Azzag, G., Loic, G., Coutant, A.: Non-parametric multivariate time series co-clustering model applied to driving-assistance systems validation. In: International Workshop on Advanced Analysis & Learning on Temporal Data (2021)
10. Govaert, G., Nadif, M.: Clustering with block mixture models. Pattern Recogn. **36**, 463–473 (2003)
11. Greco, G., Guzzo, A., Pontieri, L.: Coclustering multiple heterogeneous domains: Linear combinations and agreements. IEEE Trans. Knowl. Data Eng. **22**(12), 1649–1663 (2010)
12. Hanisch, D., Zien, A., Zimmer, R.: Co-clustering of biological networks and gene expression data. Bioinformatics **18**, 05 (2002)
13. Hubert, L., Arabie, P.: Comparing partitions. J. Classification **2**(1), 193–218 (1985)
14. Meeds, E., Roweis, S., Meeds, E., Roweis, S.: Nonparametric bayesian biclustering (2007)
15. Murphy, K.P.: Conjugate bayesian analysis of the gaussian distribution. def $1(2\sigma2)$, 16 (2007)

16. Neal, R.M.: Markov chain sampling methods for Dirichlet process mixture models. J. Comput. Graph. Stat. **9**(2), 249–265 (2000)
17. Nutt, C.L., et al.: Gene expression-based classification of malignant gliomas correlates better with survival than histological classification. Cancer Res. **63**(7), 1602–1607 (2003)
18. Papadimitriou, S., Sun, J.: Disco: distributed co-clustering with map-reduce: a case study towards petabyte-scale end-to-end mining. In: 2008 Eighth IEEE International Conference on Data Mining, pp. 512–521 (2008)
19. Sethuraman, J.: A constructive definition of Dirichlet priors. Stat. Sin. **4**(2), 639–650 (1994)
20. Strehl, A., Ghosh, J.: Cluster ensembles - a knowledge reuse framework for combining multiple partitions. J. Mach. Learn. Res. **3**, 583–617 (2002)

Towards Cohesion-Fairness Harmony: Contrastive Regularization in Individual Fair Graph Clustering

Siamak Ghodsi[1]([✉]) [iD], Seyed Amjad Seyedi[2] [iD], and Eirini Ntoutsi[3] [iD]

[1] L3S Research Centre, Leibniz University Hannover, Hannover, Germany
ghodsi@l3s.de
[2] University of Kurdistan, Sanandaj, Iran
amjadseyedi@uok.ac.ir
[3] RI CODE, University of the Bundeswehr Munich, Munich, Germany
eirini.ntoutsi@unibw.de

Abstract. Conventional fair graph clustering methods face two primary challenges: i) They prioritize balanced clusters at the expense of cluster cohesion by imposing rigid constraints, ii) Existing methods of both individual and group-level fairness in graph partitioning mostly rely on eigen decompositions and thus, generally lack interpretability. To address these issues, we propose *iFairNMTF*, an individual Fairness Nonnegative Matrix Tri-Factorization model with contrastive fairness regularization that achieves balanced and cohesive clusters. By introducing fairness regularization, our model allows for customizable accuracy-fairness trade-offs, thereby enhancing user autonomy without compromising the interpretability provided by nonnegative matrix tri-factorization. Experimental evaluations on real and synthetic datasets demonstrate the superior flexibility of iFairNMTF in achieving fairness and clustering performance.

Keywords: Fair Graph Clustering · Fair-Nonnegative Matrix Factorization · Fair Unsupervised Learning · Individual Fairness

1 Introduction

Graph-structured data is ubiquitous in various real-world applications including recommender systems, e-commerce, social networks, and neural networks. Graph clustering is essential for identifying meaningful patterns within graphs. Despite the advancements in algorithmic fairness for supervised learning scenarios which are mostly tailored for independent and identically distributed (i.i.d.) data [20], the topic of fairness is less explored in the unsupervised learning domain and especially for graphs. A motivating example comes from the educational domain [22]: how to divide students in a classroom into smaller groups for collaborative assignments. It is demanded to diversify group members from different genders or races while respecting existing friendship networks and maintaining connections. Graphs comprise non-i.i.d. data; thus, the broad literature

© The Author(s), under exclusive license to Springer Nature Singapore Pte Ltd. 2024
D.-N. Yang et al. (Eds.): PAKDD 2024, LNAI 14645, pp. 284–296, 2024.
https://doi.org/10.1007/978-981-97-2242-6_23

on fairness for i.i.d. data is generally not applicable to graphs [6]. However, some approaches mitigate bias by converting graph data into tabular form and leveraging existing methods. Additionally, there exist bias mitigation approaches that transform tabular data into hypergraphs based on dataset similarities, e.g. [8].

In the realm of fairness in i.i.d. clustering, the pioneering work of [3] introduced balance score, a fairness measure rooted in statistical parity [7], aiming at clusters of balanced demographic subgroups given a sensitive feature. Inspired by this work, [13] proposed a spectral graph clustering (SC) framework promoting group fairness that was later extended in [26] to scaled networks. However, there is not much literature on clustering with individual fairness, which prioritizes treating similar individuals (nodes in our context) similarly. A spectral model based on PageRank was proposed in [12] introducing a notion of individual fairness but for supervised node-classification tasks, whereas [27] introduces an individual-fair model for multi-view graph clustering. Only in [10], an (unsupervised) graph partitioning method employed an individual fairness approach which constrains a spectral clustering with a representation graph constructed solely based on sensitive information of individuals. SC methods are based on minimizing either the Ratio-cut or the Normalized-cut heuristic that generally tend to minimize the number of links pointing outside each cluster [17]. These cut-based heuristics do not guarantee to discover the optimal graph partitioning. Thus, incorporating (hard) fairness constraints into these rigid frameworks, which is usually also not a trivial and straightforward process, makes achieving the optimal solution challenging such that usually a relaxed form of the problem is being solved as in [10,13]. In addition, since the solution to these hard-constrained spectral approaches is based on the eigen-decomposition of the graph, it lacks interpretability.

To address the identified issues, we introduce a versatile fairness-aware model for graph clustering, the so-called individually-Fair Symmetric Nonnegative Matrix Tri-Factorization (*iFairNMTF*) model with contrastive regularization. Building on the symmetric NMF [14,16], a model tailored for graph clustering, the NMTF [21] extends its capabilities inheriting its intrinsic interpretability through non-negativity and direct clustering, while other models require steps like graph and/or node embedding [4,5], representation learning [25], or eigen-decomposition [12,13,26] before performing the final clustering. Additionally, NMTF provides better clustering and also introduces an explicit interpretability factor for inter-cluster interactions. We integrate these capabilities with a novel soft individual fairness regularization in *iFairNMTF* with an adjustable parameter λ for balancing both fairness and clustering objectives. Our key contributions include: i) A flexible joint learning framework with adjustable fairness regularization, accommodating customization of fairness enforcement in relation to clustering quality. The framework supports the linear integration of fairness and other problem-specific constraints via a customizable cost function. ii) Introduction of a contrastive fairness regularization, promoting the distribution of similar individuals across clusters based on sensitive attribute membership while ensuring distinct representation of dissimilar individuals within each cluster. iii) Retention of SNMF advantages, providing an interpretable data representation due to non-negativity and direct clustering. iv) Integration of an explicit interpretability factor, exposing inter-cluster relationships. v) Extensive experiments demonstrating the efficacy of our model with soft-fairness constraints and emphasizing the significance of the adjustable trade-off optimization.

To the best of our knowledge, our proposed joint learning contrastive framework is the first attempt to integrate an NMF model into a fairness-aware learning framework. The rest of this paper is organized as follows: In Sect. 2, we review related work. Our method is introduced in Sect. 3. The experimental evaluation is presented in Sect. 4. Conclusions and outlook are discussed in Sect. 5.

2   Background and Related Works

Problem Formulation. Let us assume an undirected graph $\mathcal{G} = (V, E)$ where $V = \{v_1, v_2, \ldots, v_n\}$ is the set of n nodes and $E \subseteq V \times V$ is the set of edges. The adjacency matrix $A \in \mathbb{R}^{n \times n}$ encodes the edge information; the existence or non-existence of an edge between two nodes v_i, v_j is modeled as $a_{ij} = 1$ and $a_{ij} = 0$, respectively. Also, we assume no self-loops (edge connecting a node to itself), so $a_{ii} = 0$ for all $i \in [n]$. Let us further assume that the set of vertices constitutes m disjoint groups identified based on a sensitive attribute e.g., gender or race, such that $V = \dot{\cup}_{s \in [m]} V_s$. The goal is to find a non-overlapping clustering of V into $k \geq 2$ clusters $V = \{C_1 \dot{\cup} \ldots \dot{\cup} C_k\}$ which is subject to individual fairness.

Individual Fairness. Individual fairness primarily formalized in [7] identifies a model f to be fair if, for any pair of inputs v_i, v_j which are sufficiently close (as per an appropriate metric), the model outputs $f(v_i), f(v_j)$ should also be close (as per another appropriate metric). In other terms, pairwise node distances in the input space and output space should satisfy the Lipschitz continuity Condition. Specifically, it requires the distance of any node pairs in the output space to be smaller or equal to their corresponding distance in the input space (usually re-scaled by a scalar). Given a pair of nodes v_i and v_j, the Lipschitz condition is:

$$D(f(v_i), f(v_j)) \leq L \cdot d(v_i, v_j) \tag{1}$$

where $f(\cdot)$ is the predictive model producing the node-level outputs (e.g., embeddings). $D(\cdot, \cdot)$ and $d(\cdot, \cdot)$ are the distance metrics of output and input space and L is the Lipschitz constant that re-scales the input distance between nodes v_i, v_j. In order to measure individual fairness based on L, [29] proposed consistency on non-graph data with the intuition to measure the average distance of the output between each individual and its k-nearest neighbors such that:

$$1 - \frac{1}{n \cdot k} \sum_{i=1}^{n} \left| f(x_i) - \sum_{j \in kNN(x_i)} f(x_j) \right| \tag{2}$$

where $f(x_i)$ is the probabilistic classification output for node features x_i of node v_i and $kNN()$ is the neighborhood of node v_i. In general, a larger average distance indicates a lower level of individual fairness.

Individual Fairness for Graph Clustering. The notion of individual fairness in graph mining [6] can be divided into three categories by application: i) node pair distance-based fairness, ii) node ranking-based fairness, and iii) individual fairness in graph clustering. The core idea in the first category is the investigation of achieving individual fairness in node representation and node embedding problems based on pairwise node distances. For example, in [15] a notion of consistency is proposed based on a similarity matrix S that characterizes node similarity in input space and can be derived from node attributes, graph topology, or domain experts. Moreover, in [12] a measure is proposed that calculates the similarity-weighted output discrepancy between nodes to measure unfairness. This metric calculates the weighted sum of pairwise node distance in the output space, where the weighting score is the pairwise node similarity. Hence for any graph mining algorithm, a smaller value of the similarity-weighted discrepancy typically implies a higher level of individual fairness.

The second category aims to achieve individual fairness by establishing node rankings. This involves creating two ranking lists in the input and output space, R_1 and R_2 based on a pairwise similarity matrix S in the input space. The satisfaction of individual fairness is determined by the alignment of these ranking lists, ensuring that R_1 and R_2 are identical for each individual [5].

The third category which remains relatively less explored and is the focus of our work, surveys individual-level fairness for graph clustering. In essence, if all neighbors of each node in a graph, are proportionally distributed to each cluster, individual fairness is then fulfilled [9]. One of the pioneering recent works in this direction is the work of Gupta, et.al., [10] according to which a clustering algorithm satisfies individual fairness for node v_i if:

$$\frac{|\{v_j : A_{i,j} = 1 \wedge v_j \in C_k\}|}{|C_k|} = \frac{|\{v_j : A_{i,j} = 1\}|}{|V|} \tag{3}$$

for all clusters C_k. The key intuition is that for each node, the ratio occupied by its one-hop neighbors in its cluster should be the same as the ratio occupied by its one-hop neighbors in the entire population (i.e. the main graph).

3   The iFairNMTF Model

Inspired by the individual fairness of [10], we propose a novel individual fairness regularization for graph clustering. It constitutes a contrastive graph regularization that incorporates positive and negative elements, signifying the attraction and repulsion of individuals towards their similar and dissimilar neighbors, based on a sensitive (node) attribute. By integrating this regularization into a flexible clustering framework, we introduce a unique individually **Fair N**on-negative **Matrix Tri-F**actorization joint learning model (iFairNMTF).

3.1   The iFairNMTF Model Formulation

Symmetric NMF (SNMF) [14] is an extension of the traditional NMF that transforms it into a versatile graph clustering model. This model factorizes an adjacency matrix $A \in \mathbb{R}_+^{n \times n}$ and is based on the assumption that similar samples

($A_{ij} > 0$) should have similar representations ($h_i h_j^\top > 0$) and dissimilar samples ($A_{ij} = 0$) should have opposite representations ($h_i h_j^\top = 0$), where H can be interpreted as the *node-to-cluster membership matrix*. More formally:

$$\min_{H \geq 0} \|A - HH^\top\|_F^2, \tag{4}$$

An extended form of the SNMF is the SNM-Tri-Factorization [21] (we omit the "S" and refer NMTF hereafter) which has been tailored to address graph clustering tasks [1,11]. It takes into account the *cluster-cluster interactions matrix* using an additional factor W such that, $A_{ij} \approx h^{(i)} W h^{(j)\top}$. More formally:

$$\min_{H,W \geq 0} \|A - HWH^\top\|_F^2, \tag{5}$$

where W can be interpreted as the cluster interactions. We build upon this model and extend it into an individual fairness joint learning framework using a contrastive regularization. More formally:

$$\min_{H,W \geq 0} \|A - HWH^\top\|_F^2 + \lambda \mathcal{R}_C(H), \tag{6}$$

Schematically, the model block diagram is illustrated in Fig. 1. The left term comes from Eq. (5) and $\mathcal{R}_C(H)$ is a contrastive regularization constraining the cluster indicator H relatively adjusted by the magnitude of a flexible λ parameter ensuring its alignment with group demographics. The contrastive term $C = P - N$ consists of a positive and a negative component:

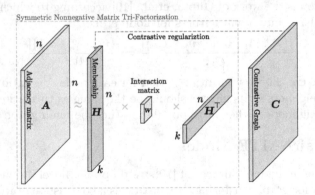

Fig. 1. Schematic representation of the iFairNMTF model with contrastive regularization.

$$\mathcal{N}_{i,j} = \begin{cases} 1, & \text{if } g_i = g_j \\ 0, & \text{otherwise.} \end{cases} \qquad \mathcal{P}_{i,j} = \begin{cases} 1, & \text{if } g_i \neq g_j \\ 0, & \text{otherwise.} \end{cases} \tag{7}$$

$$N_{ij} = \mathcal{N}_{ij}/\sum_{r=1}^n \mathcal{N}_{ir}, \qquad P_{ij} = \mathcal{P}_{ij}/\sum_{r=1}^n \mathcal{P}_{ir},$$

which can be enforced to apply the attraction of different demographic groups into the same cluster, and repulsion of same-group members to ensure diversity of their distribution into different clusters according to Eq. (8):

$$\min_{H} \mathcal{R}_C = \sum_{i=1}^{n} \sum_{j=1}^{n} \|h^{(i)} - h^{(j)}\|^2 C_{ij} = \mathrm{Tr}(H^\top L H). \tag{8}$$

where $L = D - C$ is the graph Laplacian and $D_{ii} = \sum_{j=1}^{n} C_{ij}$. By adding the contrastive regularization $\mathcal{R}_C$ to the NMTF (5), we derive the final objective function (loss function) of iFairNMTF, $\mathcal{L} = \mathcal{L}_{\mathcal{F}} + \lambda \mathcal{R}_C$ as follow:

$$\min_{H,W \geq 0} \|A - HWH^\top\|_F^2 + \lambda \mathrm{Tr}(H^\top L H), \tag{9}$$

The objective function in Eq. (9) is a combination, trading-off between the clustering loss and the constrastive regularization term to ensure individual fairness. The hyper-parameter $\lambda \in [0, +\infty)$ controls the compromise between clustering performance and fairness. Smaller λ implies a higher importance of the clustering performance and prompts the model to prioritize generating strong and cohesive clusters. Conversely, a higher λ prioritizes fairness, prompting the model to create diversified clusters that fairly represent groups of V_s.

3.2   The iFairNMTF Model Optimization

In this section, we focus on solving the iFairNMTF model. The objective function in Eq. (9) is a fourth-order non-convex function with respect to the entries of H and has multiple local minima. For these types of problems, it is difficult to find a global minimum; thus a good convergence property we can expect is that every limit point is a stationary point. Therefore, we adopt multiplicative updating rules to update the membership matrix H and introduce two Lagrangian multiplier matrices of Θ, and Φ to enforce the nonnegative constraints on H, and W respectively, resulting in the following equivalent objective function:

$$\min_{H,W} \mathcal{L} = \|A - HWH^\top\|_F^2 + \lambda \mathrm{Tr}(H^\top L H) - \mathrm{Tr}(\Theta^\top H) - \mathrm{Tr}(\Phi^\top W),$$

which can be further rewritten as follows:

$$\min_{H,W} \mathcal{L} = \mathrm{Tr}(A^\top A - 2A^\top HWH^\top + HW^\top H^\top HWH^\top)$$
$$+ \lambda \mathrm{Tr}(H^\top L H) - \mathrm{Tr}(\Theta^\top H) - \mathrm{Tr}(\Phi^\top W). \tag{10}$$

The partial derivative of $\mathcal{L}$ with respect to H is

$$\frac{\partial \mathcal{L}}{\partial H} = -2A^\top HW - 2AHW^\top + 2HW^\top H^\top HW \tag{11}$$
$$+ 2HWH^\top HW^\top + 2\lambda L H - \Theta.$$

Algorithm 1. Individual Fair Nonnegative Matrix Tri-Factorization (iFairNMTF)

Input: adjacency matrix A, group set g, latent factor k, trade-off parameter λ;
Output: cluster assignment M;

1: Construct the contrastive graph C according to (7);
2: **while** convergence not reached **do**
3:    Update cluster-membership matrix H according to (13);
4:    Update cluster-interaction matrix W according to (16);
5: **end while**
6: Calculate cluster assignment $M_i \leftarrow \arg\max(h^{(i)}), \forall i \in \{1,\ldots,n\}$
7: **return** cluster-membership matrix H and cluster-interaction matrix W;

By setting the partial derivative $\frac{\partial \mathcal{L}}{\partial H}$ to 0, we have:

$$\Theta = -2A^\top HW - 2AHW^\top + 2HW^\top H^\top HW + 2HWH^\top HW^\top + 2\lambda LH. \tag{12}$$

From the Karush-Kuhn-Tucker complementary slackness conditions (KKT), we obtain $H \odot \Theta = 0$ where $\odot$ denotes the element-wise product. This is the fixed point equation that the solution must satisfy at convergence. By solving this equation, we derive the following updating rule for H:

$$H \leftarrow H \odot \left(\frac{A^\top HW + AHW^\top + \lambda L^- H}{HW^\top H^\top HW + HWH^\top HW^\top + \lambda L^+ H} \right)^{\frac{1}{4}}. \tag{13}$$

To guarantee the nonnegativity, we separate the positive and negative elements as $L = L^+ - L^-$. Similarly, we differentiate $\mathcal{L}$ with respect to W such that:

$$\frac{\partial \mathcal{L}}{\partial W} = -2H^\top AH + 2H^\top HWH^\top H - \Phi \tag{14}$$

By setting the partial derivative $\frac{\partial \mathcal{L}}{\partial W}$ to 0, we obtain Φ as:

$$\Phi = -2H^\top AH + 2H^\top HWH^\top H. \tag{15}$$

From the complementary slackness KKT conditions we obtain $W \odot \Phi = 0$. This is another fixed point equation that the solution must satisfy at convergence. Finally, by solving this equation, we derive the following updating rule for W:

$$W \leftarrow W \odot \frac{H^\top AH}{H^\top HWH^\top H}. \tag{16}$$

4 Experimental Evaluation

4.1 Experimental Setup

Datasets. In the paper, six real-world and three synthetic networks are used for benchmarking the performance of the proposed method against competitors. Our synthetic networks are generated according to a generalized Stochastic Block Model (SBM) [13] with equal-sized clusters $|C_l| = n/k$ and groups

$|V_s| = n/g$ randomly distributed among the clusters. We generate three SBM networks of 2K, 5K, and 10K nodes with k = 5 clusters and g = 5 groups. Real datasets include three high school friendship networks [18]: *Facebook, Friendship*, and *Contact-Diaries* which represent connections among a group of French high school students. *DrugNet* [28] is a network encoding acquaintanceship between drug users in Hartford, CT. *LastFMNet* [24] contains mutual follower relations among users of Last.fm, a recommendation-based online radio and music community in Asia. Lastly, *NBA* is a network containing relationships between around 400 NBA basketball players [4]. A detailed description of both real and synthetic datasets, as well as instructions on generating the SBM networks are provided in the supplementary material[1]. Likewise for dataset statistics including size and number of sensitive groups and also details on cleaning the real datasets.

Competitors. We compare *iFairNMTF* with four state-of-the-art graph clustering methods, namely, with two group-fair models: i) *Fair-SC* [13], and its scalable version (ii) *sFair-SC*) [26], iii) an individual-fairness model (*iFair-SC*) [10] and iv) a deep graph neural network (*DMoN*) [25]. The three former models are fairness-aware and have been already discussed. The latter model is one of the very few DNNs developed for pure graph-partitioning problems, but does not consider fairness. This model extends the general graph neural network (GNN) architecture into a deep modularity optimization GNN. It operates on attributed graphs, thus we pass the sensitive attribute as node-attribute to it. The number of layers and learning rate are set according to the official source code provided by the authors (layers = 64 or 512 for small and large networks, $\alpha = 0.001$). The number of epochs for DMoN and our method is 500. To produce reliable results, all experiments are averaged over 10 independent runs.

Evaluation Measures. We use accuracy for measuring clustering assignment quality on synthetic networks. For real-world networks, since the ground truth cluster structures are unknown, we use Newman's modularity (Q) measure [2,19] which analyzes the homogeneity of clusters by calculating the proportion of internal links in each cluster for a given partitioning compared to the expected proportion of edges in a null graph with the same degree distribution. Modularity is preferable over cut-based measures due to its robustness against imbalanced cluster sizes. We measure the fairness of clustering in terms of the popular average balance (B) measure [13,26]: $B = \frac{1}{k} \sum_{l=1}^{k} Balance(C_l)$, where $Balance(C_l)$ calculates the minimum group proportion of C_l according to Eq. (17):

$$Balance(C_l) = \min_{s \neq s' \in [m]} \frac{|V_s \cap C_l|}{|V_{s'} \cap C_l|}, \tag{17}$$

where $l \in [1, k]$ iterates over all the k clusters and V_s identifies each sensitive group of the sensitive attribute. The minimum balance of each cluster can range between $[0, 1]$, thus their average also ranges between $[0, 1]$.

Parameters. Our model has an adjustable hyper-parameter λ to trade-off between the degree of fairness and clustering efficiency (Eq. (9)). The range of λ includes 50 values from $[0, 100]$ with a median of 3 for small and from

---

[1] Link to supplemental file and source codes: Github.com/SiamakGhodsi/iFairNMTF.

[0, 3500] for large datasets (must be set separately for each dataset.). The effect of λ is discussed in Sect. 4.3. The trade-off parameter λ can be set based on user preferences between fairness and clustering quality. A practical way is to select the best value according to the intersection point of B and Q, see Fig. 3.

4.2   Clustering Quality vs Fairness

In real datasets, the ground truth partitioning of the networks is unknown, therefore we report the performance for various number of clusters. Figure 2 illustrates the comparison of our method's results in terms of Q (clustering quality/ modularity) and B (fairness/balance) with those of other models on two datasets, for various numbers of clusters. Dataset balance, highlighted by the yellow dashed line, identifies the proportion of the smallest to the largest group of the sensitive attribute, calculated according to Eq. (17). For iFairNMTF, the best λ values for each k are used. They are selected based on the intersection of Q and B charts as in Fig. 3: $\lambda = 2$ for DrugNet, and $\lambda = 100$ for LastFM.

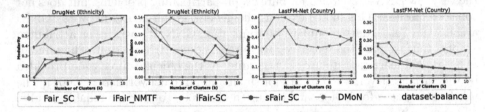

Fig. 2. Performance comparison w.r.t. clustering quality/modularity Q and cluster fairness B (higher values are better for both measures) on DrugNet, and LastFM for different number of clusters $k \in [2, 10]$. $k = 10$ is the convergence point of all models.

As we can see from this figure, our model outperforms the SC-based models in terms of both measures on both datasets. It reports a lower clustering quality Q on LastFM than DMoN which is a neural model primarily focusing on identifying the most modular partitioning of the graph through modularity optimization. DMoN's Q outcomes reveal varied patterns on LastFM and DrugNet, attributed to differences in network size and density. Neural models typically excel in data-intensive learning cycles, yielding better performance on larger datasets. For instance, LastFM, a substantially larger graph with 5k nodes and 20k edges, showcases this advantage compared to the 200-node DrugNet. However, in terms of fairness, DMoN fails to generate diverse clusters w.r.t. the sensitive attribute, as evidenced by low balance (B). In contrast, our model consistently achieves well-distributed clusters, boasting the highest balance scores among all competitors.

Next, we compare all the models, on all the datasets with a fixed number of clusters $k = 5$, the median of our selected number of clusters. The results are presented in Table 1. We distinguish between real and SBM networks based on the measurable accuracy of partitioning quality in SBM networks, where ground-truth clusters are known. Additionally, we present the average modularity (Q) and balance (B) across all clusters for real datasets.

Table 1. Results illustrating modularity (Q) and average balance (B) of real networks, and accuracy (Acc) and average balance (B) results on SBM networks for $k = 5$ clusters.(**Bold-underline**) and underline indicate best and second best B results. Best Q, Acc are highlighted with boldfaced gray .

Network	FairSC		sFairSC		iFairSC		DMoN		iFairNMTF	
	B	Q	B	Q	B	Q	B	Q	B	Q
Diaries	0.708	0.612	**0.809**	**0.684**	0.699	0.647	0.263	0.145	0.648	0.640
Facebook	0.327	0.449	**0.602**	0.500	0.330	0.448	0.268	0.048	0.514	0.509
Friendship	0.391	0.483	0.485	0.627	0.374	0.392	0.183	0.140	**0.631**	0.669
DrugNet	0.052	0.263	0.052	0.270	0.061	0.263	0.000	0.326	**0.124**	0.588
NBA	0.083	0.000	**0.323**	0.113	0.072	0.000	0.036	0.057	0.286	0.150
LastFM	0.065	0.003	0.056	0.035	0.066	0.002	0.000	0.526	**0.069**	0.600
	B	Acc	B	Acc	B	Acc	B	Acc	B	Acc
SBM-2K	0.575	0.588	–	–	0	0.799	–	–	**0.953**	0.958
SBM-5K	**0.995**	0.998	–	–	0	0.799	–	–	0.941	0.962
SBM-10K	0.999	0.999	–	–	0	0.600	–	–	1	1

The results on real networks indicate the superiority of our proposed iFairN-MTF model while reporting the best Q values on $5/6$ (meaning 5 out of 6) datasets and $3/6$ w.r.t. B. Similarly, on SBM networks iFairNMTF stands the best with $2/3$ best accuracy and balance scores. It is worth noting that, in the SBM experiment, DMoN and sFairSC failed to deliver the required number of clusters resulting in empty clusters implying inconsistency in accuracy calculation since the cluster assignments need to be masked to be comparable to true labels.

4.3 Parameter Analysis

This section studies the effect of the λ hyper-parameter on the iFairNMTF model's performance in terms of Q and B for $k = 5$ clusters in comparison with the performance of other models. In this experiment, we also provide the results of the vanilla SC and vanilla NMTF (the same as iFairNMTF with $\lambda = 0$) models. The results are illustrated in Fig. 3. Based on the results, a comparably good value for the λ parameter can be selected in the intersection of the two measures. These twin charts provide a nice opportunity to visualize the distribution of results and make it easy to select. For instance, values in the range $[0.1, 4]$ for Drugnet and $[55, 200]$ for LastFM are suggested. It gives the end-user a desirable autonomy and depends on the user's demands on how to select values for this parameter. Consider that, since LastFM is a much larger network than Drugnet, we increase the range of λ with 100 values from 0 to $\lambda = 5000$. Complementary results can be found in the supplementary material (see footnote 4).

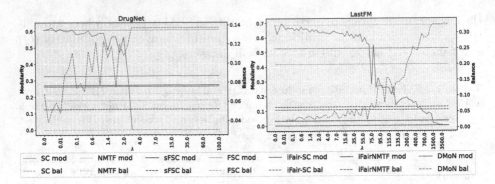

Fig. 3. Parameter λ analysis of the iFairNMTF on Drugnet and LastFM-Net datasets with k = 5 in terms of Q and B for $\lambda \in [0, 100]$. Solid lines depict modularity and dashed lines represent balance. Only the behavior of FairSNMF depends on λ.

4.4 Interpretability Analysis

In this section, in addition to the inherent model interpretability through the direct clustering given by the $\boldsymbol{H}$ factor, we investigate the explicitly interpretable intermediary factor $\boldsymbol{W} \in \mathbb{R}_+^{k \times k}$ of the iFairNMTF model introduced in Eq. (5). This factor is a symmetric square matrix consisting of non-negative scores representing the strength of cluster-cluster interactions. Diagonal elements reflect intra-cluster connectivity such that the score for dense clusters is expected to be higher. An illustrative example of a graph with 40 nodes distributed between 4 clusters and an imbalanced group distribution of 35% (square shape) to 65% (triangle shape) is shown in Fig. 4. We apply Algorithm 1 to this graph with $\lambda = 1$, and the model identifies the true clusters. Entries corresponding to clusters like I−II, which have no interactions (no links) together, are assigned a value of 0. Furthermore, the score for clusters IV−I is notably lower compared to IV−II, reflecting the difference in the number of connecting links between these clusters.

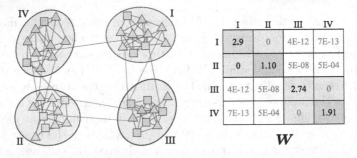

	I	II	III	IV
I	2.9	0	4E-12	7E-13
II	0	1.10	5E-08	5E-04
III	4E-12	5E-08	2.74	0
IV	7E-13	5E-04	0	1.91

$$W$$

Fig. 4. Interpretability of $\boldsymbol{W}$ factor for a 40-node graph divided to 4 clusters. Shapes indicate groups.

5   Conclusion and Outlook

In this paper, we introduce the iFairNMTF model, an individually fair flexible approach for graph clustering that takes sensitive (node) attributes into account. iFairNMTF modifies the NMTF model's objective function by incorporating a contrastive penalty term, ensuring that clustering outcomes align with sensitive demographic information and thereby promoting individually fair cluster representations through the attraction and repulsion advantage of the proposed contrastive regularization term. The trade-off regularization parameter λ empowers users to customize the balance between clustering performance and fairness based on their specific needs. Our experiments on both real and synthetic datasets demonstrate that adjusting the trade-off parameter allows for achieving a desired equilibrium between maximizing clustering cohesion and promoting fairness. Promising directions for future research include exploring multi-objective techniques to effectively balance fairness and cohesion objectives, particularly in complex, multi-dimensional discrimination scenarios [23]. Additionally, developing NMF tailored for group fairness, with an emphasis on integrating both individual and group notions into algorithmic design. Finally, evaluating fair clustering methods, esp. for individual fairness remains an ongoing challenge.

Acknowledgements. This work has received funding from the European Union's Horizon 2020 research and innovation programme under Marie Sklodowska-Curie Actions (grant agreement number 860630) for the project "NoBIAS - Artificial Intelligence without Bias". This work reflects only the authors' views and the European Research Executive Agency (REA) is not responsible for any use that may be made of the information it contains. The research was also supported by the EU Horizon Europe project MAMMOth (GrantAgreement 101070285).

References

1. Abdollahi, R., Amjad Seyedi, S., Reza Noorimehr, M.: Asymmetric semi-nonnegative matrix factorization for directed graph clustering. In: ICCKE, pp. 323–328 (2020)
2. Chakraborty, T., Dalmia, A., Mukherjee, A., Ganguly, N.: Metrics for community analysis: A survey. ACM Comput. Surv. **50**(4), 1–37 (2017)
3. Chierichetti, F., Kumar, R., Lattanzi, S., Vassilvitskii, S.: Fair clustering through fairlets. In: Advances in NeurIPS, pp. 5029–5037 (2017)
4. Dai, E., Wang, S.: Say no to the discrimination: learning fair graph neural networks with limited sensitive attribute information. In: WSDM, pp. 680–688 (2021)
5. Dong, Y., Kang, J., Tong, H., Li, J.: Individual fairness for graph neural networks: a ranking based approach. In: KDD, pp. 300–310. ACM (2021)
6. Dong, Y., Ma, J., Wang, S., Chen, C., Li, J.: Fairness in graph mining: a survey. IEEE Transactions on Knowledge and Data Engineering, pp. 1–22 (2023)
7. Dwork, C., Hardt, M., Pitassi, T., Reingold, O., Zemel, R.S.: Fairness through awareness. In: Proceedings of the 3rd ITCS Conference, pp. 214–226 (2012)
8. Ghodsi, S., Ntoutsi, E.: Affinity clustering framework for data debiasing using pairwise distribution discrepancy. In: EWAF. CEUR Proceedings, vol. 3442 (2023)

9. Gupta, S., Dukkipati, A.: Protecting individual interests across clusters: Spectral clustering with guarantees. CoRR abs/2105.03714 (2021)

10. Gupta, S., Dukkipati, A.: Consistency of constrained spectral clustering under graph induced fair planted partitions. In: Advances in NeurIPS, pp. 13527–13540 (2022)

11. Hajiveiseh, A., Seyedi, S.A., Tab, F.A.: Deep asymmetric nonnegative matrix factorization for graph clustering. Pattern Recognit. **148**, 110179 (2024)

12. Kang, J., He, J., Maciejewski, R., Tong, H.: Inform: individual fairness on graph mining. In: KDD, pp. 379–389. ACM (2020)

13. Kleindessner, M., Samadi, S., Awasthi, P., Morgenstern, J.: Guarantees for spectral clustering with fairness constraints. In: ICML, vol. 97, pp. 3458–3467 (2019)

14. Kuang, D., Park, H., Ding, C.H.Q.: Symmetric nonnegative matrix factorization for graph clustering. In: SDM, pp. 106–117 (2012)

15. Lahoti, P., Gummadi, K.P., Weikum, G.: Operationalizing individual fairness with pairwise fair representations. Proc. VLDB Endow. **13**(4), 506–518 (2019)

16. Li, T., Ding, C.c.: Nonnegative matrix factorizations for clustering: a survey. In: Data Clustering, pp. 149–176. Chapman and Hall/CRC (2018)

17. von Luxburg, U.: A tutorial on spectral clustering. Stat. Comput. **17**(4), 395–416 (2007)

18. Mastrandrea, R., Fournet, J., Barrat, A.: Contact patterns in a high school: a comparison between data collected using wearable sensors, contact diaries and friendship surveys. PLoS ONE **10**(9), e0136497 (2015)

19. Newman, M.E.J.: Modularity and community structure in networks. Proc. Natl. Acad. Sci. **103**(23), 8577–8582 (2006)

20. Ntoutsi, E., et al.: Bias in data-driven artificial intelligence systems-an introductory survey. Wiley Interdisciplinary Reviews: Data Mining and Knowledge Discovery **10**(3), e1356 (2020)

21. Pei, Y., Chakraborty, N., Sycara, K.P.: Nonnegative matrix tri-factorization with graph regularization for community detection in social networks. In: IJCAI, pp. 2083–2089. AAAI Press (2015)

22. Quy, T.L., Friege, G., Ntoutsi, E.: Multi-fair capacitated students-topics grouping problem. In: PAKDD (1). LNCS, vol. 13935, pp. 507–519. Springer (2023)

23. Roy, A., Horstmann, J., Ntoutsi, E.: Multi-dimensional discrimination in law and machine learning - A comparative overview. In: FAccT. pp. 89–100. ACM (2023)

24. Rozemberczki, B., Sarkar, R.: Characteristic functions on graphs: Birds of a feather, from statistical descriptors to parametric models. In: CIKM, pp. 1325–1334 (2020)

25. Tsitsulin, A., Palowitch, J., Perozzi, B., Müller, E.: Graph clustering with graph neural networks. J. Mach. Learn. Res. **24**, 127:1–127:21 (2023)

26. Wang, J., Lu, D., Davidson, I., Bai, Z.: Scalable spectral clustering with group fairness constraints. In: AISTATS, pp. 6613–6629 (2023)

27. Wang, Y., Kang, J., Xia, Y., Luo, J., Tong, H.: ifig: Individually fair multi-view graph clustering. In: IEEE Big Data, pp. 329–338. IEEE (2022)

28. Weeks, M.R., Clair, S., Borgatti, S.P., Radda, K., Schensul, J.J.: Social networks of drug users in high-risk sites: finding the connections. AIDS Behav. **6**, 193–206 (2002)

29. Zemel, R.S., Wu, Y., Swersky, K., Pitassi, T., Dwork, C.: Learning fair representations. In: ICML (3). JMLR Workshop and Conference Proceedings, vol. 28, pp. 325–333. JMLR.org (2013)

Data Mining Processes and Pipelines

NETEFFECT: Discovery and Exploitation of Generalized Network Effects

Meng-Chieh Lee[1(✉)], Shubhranshu Shekhar[2], Jaemin Yoo[3], and Christos Faloutsos[1]

[1] Carnegie Mellon University, Pittsburgh, USA
{mengchil,christos}@cs.cmu.edu
[2] Brandeis University, Waltham, USA
sshekhar@brandeis.edu
[3] KAIST, Seoul, South Korea
jaemin@kaist.ac.kr

Abstract. Given a large graph with few node labels, how can we (a) identify whether there is *generalized network-effects* (GNE) or not, (b) estimate GNE to explain the interrelations among node classes, and (c) exploit GNE efficiently to improve the performance on downstream tasks? The knowledge of GNE is valuable for various tasks like node classification and targeted advertising. However, identifying GNE such as homophily, heterophily or their combination is challenging in real-world graphs due to limited availability of node labels and noisy edges. We propose NETEFFECT, a graph mining approach to address the above issues, enjoying the following properties: (i) Principled: a statistical test to determine the presence of GNE in a graph with few node labels; (ii) General and Explainable: a closed-form solution to estimate the specific type of GNE observed; and (iii) Accurate and Scalable: the integration of GNE for accurate and fast node classification. Applied on real-world graphs, NETEFFECT discovers the unexpected absence of GNE in numerous graphs, which were recognized to exhibit heterophily. Further, we show that incorporating GNE is effective on node classification. On a million-scale real-world graph, NETEFFECT achieves **over 7×** speedup (14 *minutes* vs. 2 hours) compared to most competitors.

Keywords: Network Effects · Heterophily Graphs · Node Classification

1 Introduction

Given a large graph with few node labels and no node features, how to check whether the graph structure is useful for classifying nodes or not? Node classification is often employed to infer labels on large real-world graphs. Since manual labeling is expensive and time-consuming, it is common that only few node labels are available. For example, in a million-scale social network, identifying even a fraction (say 5%) of users' groups is prohibitive, limiting the application of methods that assume many labels are given. Recently, with prevalence of graphs in industry and academia alike, there is a growing need among users to know whether these graph structures provide meaningful information for inference tasks. Therefore, before investing a huge amount of time

© The Author(s), under exclusive license to Springer Nature Singapore Pte Ltd. 2024
D.-N. Yang et al. (Eds.): PAKDD 2024, LNAI 14645, pp. 299–312, 2024.
https://doi.org/10.1007/978-981-97-2242-6_24

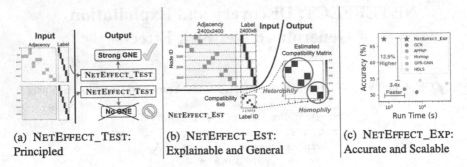

Fig. 1. NETEFFECT works well, thanks to its three novel contributions: (a) NETEFFECT_TEST statistically the existence of GNE. (b) NETEFFECT_EST explains the graph with the X-ophily compatibility matrix. (c) NETEFFECT_EXP wins and is fast.

and resources into potentially unsuccessful experiments, a preliminary test is earnestly needed.

That is to say, we want to know whether the given graph has *generalized network-effects* (GNE) or not. A graph with GNE provides meaningful information through the structure that can be used to identify the labels of nodes. For example, "talkative person tends to make friends with talkative ones" denotes homophily, while "teenagers incline to interact with the ones that have opposite gender on social media" denotes heterophily. It is thus important to distinguish which GNE the graph has, i.e., homophily, heterophily, or both (which we call "x-ophily"), if there is any. Given c classes, an intuitive way to describe GNE is via a $c \times c$ compatibility matrix, which shows the relative influence between each class pair. It can be used to explain the graph property, as well as be exploited to better assign the labels in the graph.

However, identifying GNE is commonly neglected in literature: inference-based methods assume that the relationship is given by domain experts [4,6]; most graph neural networks (GNNs) assume homophily [9,10,24]. Although some previous works [14,16,25] use homophily statistics to analyze the given graph, our work has a very different direction because of three reasons. First, they are designed to identify the absence of homophily, and thus can not clearly distinguish GNE, which includes different non-homophily cases, i.e., heterophily, both, or no GNE. Second, to compute accurate statistics, they use all the node labels in the graph, which is impractical during node classification. Finally, their analysis rely heavily on the results of GNNs, which means in addition to the graph structure, the node features also significantly influence the conclusions of GNE. In contrast, our work aims to answer 3 research questions:

RQ1. **Hypothesis Testing**: How to identify whether the given graph has GNE or not, with only few labels?

RQ2. **Estimation**: How to estimate GNE in a principled way, and explain the graph with the estimation?

RQ3. **Exploitation**: How to efficiently exploit GNE on node classification with only few labels?

We propose NETEFFECT, with 3 contributions as the corresponding solutions:

1. **Principled:** NETEFFECT_TEST uses statistical tests to decide whether GNE exists at all. Figure 1a shows how it works, and Fig. 2 shows its discovery, where many large real-world datasets known as heterophily graphs have little GNE.
2. **General and Explainable:** NETEFFECT_EST explains whether the graph is homophily, heterophily, or X-ophily by precisely estimating the compatibility matrix with the derived closed-form formula. In Fig. 1b, it explains the interrelations of classes by the estimated compatibility matrix, which implies X-ophily.
3. **Accurate and Scalable:** NETEFFECT_EXP efficiently exploits GNE to perform better in node classification. It wins in both accuracy and time on a million-scale heterophily graph "Pokec-Gender", only requiring 14 *minutes* (Fig. 1c).

Reproducibility: The code[1] and the extended version with appendix[2] are made public.

2 Background and Related Work

2.1 Background

Notation. Let G be an undirected and unweighted graph with n nodes and m edges and an adjacency matrix A. Each node i has a unique label $l(i) \in \{1, 2, \ldots, c\}$, where c is the number of classes. Let $E \in \mathbb{R}^{n \times c}$ be the initial belief matrix with the prior information, i.e., the labeled nodes. $E_{ik} = 1$ if $l(i) = k$, and $E_{ik} = 0$ if $l(i) \neq k$. For the nodes without labels, their entries are set to $1/c$. $H \in \mathbb{R}^{c \times c}$ is a row-normalized compatibility matrix, where H_{ku} is the relative influence of class l on class u. The residual of a matrix around k is $\hat{Y} = Y - k \times \mathbf{1}$, where $\mathbf{1}$ is matrix of ones.

Belief Propagation (BP). FABP [11] and LINBP [6] accelerate BP by approximating the final belief assignment. In particular, LINBP approximates the final belief as:

$$\hat{B} = \hat{E} + A\hat{B}\hat{H}, \tag{1}$$

where $\hat{B}$ is a residual final belief matrix, initialized with zeros. The compatibility matrix H and initial beliefs E are centered around $1/c$ to ensure convergence. HOLS [4] is a BP-based method, which propagates the labels by weighing with higher-order cliques.

2.2 Related Work

Table 1 presents qualitative comparison of state-of-the-art approaches against our proposed NETEFFECT. Notice that only NETEFFECT fulfills all the specs.

---

[1] https://github.com/mengchillee/NetEffect.
[2] https://arxiv.org/abs/2301.00270.

Table 1. NETEFFECT matches all specs, while baselines miss one or more. '?' and 'N/A' denote unclear and not applicable.

	Property	BP [6,11]	HOLS [4]	General GNNs [9,10]	Het. GNNs [1,3]	NETEFFECT
1. Principled	1.1. Statistical Test	✓	✓			✓
	1.2. Convergence Guarantee	✓	✓			✓
2. Explainable	2.1 Compatibility Matrix Estimation	N/A	N/A			✓
3. General	3.1 Handle Heterophily	?	?	?	✓	✓
	3.2 Handle GNE	?	?			✓
4. Scalable	4.1. Linear Complexity	✓	✓	✓	✓	✓
	4.2. Thrifty	✓	✓	?	?	✓

Analysis by Homophily Statistics. Many studies [14,16,25] utilize homophily ratio to measure how common the labels of the connected node pairs share the same class. Our work focuses on very different aspects, as discussed in the introduction.

Node Classification. GCN [9] and APPNP [10] incorporate neighborhood information to do better predictions and assume homophily. MIXHOP [1], GPR-GNN [3], and H_2GCN [25] make no assumption of homophily. Nevertheless, H_2GCN requires too much memory and thus can not handle large graphs. LINKX [14] introduces multiple large heterophily datasets, but it is not applicable to graphs without node features.

3   Proposed GNE Test

Given a graph with few labels, how can we identify whether the graph has *generalized network-effects* (GNE) or not? In other words, how can we check whether the graph structure is useful for inferring node labels? We propose NETEFFECT_TEST, a statistical approach to identify the presence of GNE in a graph. Applying it to real-world graphs, we show that many popular heterophily graphs exhibit little GNE.

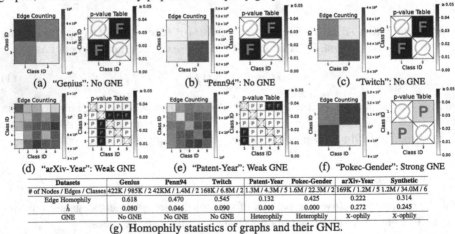

(a) "Genius": No GNE   (b) "Penn94": No GNE   (c) "Twitch": No GNE

(d) "arXiv-Year": Weak GNE   (e) "Patent-Year": Weak GNE   (f) "Pokec-Gender": Strong GNE

Datasets	Genius	Penn94	Twitch	Patent-Year	Pokec-Gender	arXiv-Year	Synthetic
# of Nodes / Edges / Classes	422K / 985K / 2	42KM / 1.4M / 2	168K / 6.8M / 2	1.3M / 4.3M / 5	1.6M / 22.3M / 2	169K / 1.2M / 5	1.2M / 34.0M / 6
Edge Homophily	0.618	0.470	0.545	0.132	0.425	0.222	0.314
$\hat{h}$	0.080	0.046	0.090	0.000	0.000	0.272	0.245
GNE	No GNE	No GNE	No GNE	Heterophily	Heterophily	X-ophily	X-ophily

(g) Homophily statistics of graphs and their GNE.

Fig. 2. NETEFFECT_TEST **works**: It discovers that real-world heterophily graphs do not necessarily have GNE. For each graph, we report the edge counting on the left (not available in practice), and the *p*-value table output from NETEFFECT_TEST on the right, where "P" denotes the presence of GNE, and "F" denotes the absence of GNE.

3.1 NetEffect_Test

We first provide two main definitions regarding GNE:

Definition 1. *If the nodes with class c_i in a graph tend to connect randomly to the nodes with all classes $1, \ldots, c$ (with no specific preference), class c_i has no GNE.*

Definition 2. *If all classes in a graph have no GNE, this graph has no GNE.*

We distinguish heterophily graphs from those with no GNE by the definition. In heterophily graphs, the nodes of a specific class are likely to be connected to the nodes of other classes, such as in bipartite graphs that connect different classes of nodes. In this case, knowing the label of a node gives meaningful information about the labels of its neighbors. On the other hand, if a graph has no GNE, knowing the label of a node gives no useful information about its neighbors. In other words, the structural information of a graph is not useful to infer the unknown labels of nodes.

Next we describe how we propose to determine the existence or absence of GNE. In the inner loop, we need to decide whether class c_i (say, "talkative people"), has statistically more, or fewer edges to class c_j (say, "silent people"). We propose to use Pearson's χ^2 test for that. Specifically, given a class pair (c_i, c_j), the input to the test is a 2×2 contingency table containing the counts of edges that connect pairs of nodes whose labels are in $\{c_i, c_j\}$. The null hypothesis of the test is:

Null Hypothesis 1 *Edges are equally likely to exist between nodes of the same class and those of different classes.*

If the p-value from the test is no less than 0.05, we accept the null hypothesis, which represents that the chosen class pair (c_i, c_j) exhibits no statistically significant GNE in the graph. Then we call them *mutually indistinguishable*:

Definition 3 (Mutually indistinguishable). *Two classes c_i and c_j are mutually indistinguishable if we can not reject the null hypothesis above.*

Novel Implementation Details. The detailed procedure of NETEFFECT_TEST is in Appx. B.1. A practical challenge on the test is that if the numbers in the table are too large, p-value becomes very small and meaningless [15]. Uniform edge sampling can be a natural solution, but sampling for only a single round can be unstable and output very different results. To address this, we combine p-values from different random sampling by Universal Inference [23]. We firstly sample edges to add to the contingency table until the frequency is above a specified threshold, and compute the χ^2 test statistic for each class pair. Next, following Universal Inference, we repeat the procedure for random samples of edges for B rounds and average the statistics. At last, we use the average statistics to compute the p-value table $F_{c \times c}$ of χ^2 tests. Our NETEFFECT_TEST is robust to noisy edges thanks to the sampling process, and works well given either a few or many node labels. Given a few observations, the χ^2 test works well when the frequency in the contingency table is at least 5; given many observations, our sampling trick ensures the correctness and the consistency of the computed p-value.

If a class accepts the null hypotheses with all other classes, this class has little GNE, and satisfies Def. 1. Moreover, if all classes exhibit little GNE, the whole graph satisfies Def. 2. In that case, no label propagation methods will help with node classification.

3.2   Discoveries on Real-World Graphs

We apply NETEFFECT_TEST to 6 real-world graphs and analyze their GNE. For each dataset, we sample 5% of node labels and compute the p-value table using NETEFFECT_TEST. This is because a) only few labels are available in most node classification tasks in practice, and thus it is reasonable to make the same assumption in the analysis, and b) NETEFFECT_TEST can analyze GNE even from partial observations. B is set to 1000 to output stable results. Based on Def. 2 our surprising discoveries are:

Discovery 1 (No GNE) NETEFFECT_TEST *identifies the lack of GNE in "Genius"* [13], *"Penn94"* [21], *and "Twitch"* [18]. They are widely known as heterophily graphs. In "Genius" (Fig. 2a), we see that both classes 1 and 2 tend to connect to class 1, making class 2 indistinguishable by the graph structure. NETEFFECT_TEST thus accepts the null hypothesis, and identifies the lack of GNE. We can observe a similar phenomenon in "Penn94" (Fig. 2b). "Twitch" (Fig. 2c) used to be considered as a heterophily graph because of its weak homophily effect, but NETEFFECT_TEST finds that each of the classes uniformly connects to both classes, and thus it has little GNE.

Discovery 2 (Heterophily and X-ophily) NETEFFECT_TEST *identifies GNE in "Arxiv-Year", "Patent-Year", and "Pokec-Gender".* While "Patent-Year" and "Pokec-Gender" exhibit heterophily (Fig. 2e and 2f), "Arxiv-Year" exhibits X-ophily, i.e., not straight homophily or heterophily (Fig. 2d). They are thus used in our experiments.

Discovery 3 (Weak vs strong GNE) NETEFFECT_TEST *identifies weak, and strong GNE: "Arxiv-Year" and "Patent-Year" exhibit weak GNE; and "Pokec-Gender" exhibits strong GNE.* We consider graphs to have weak GNE if there exists at least one class which is not distinguishable from some other classes. Such graphs limit the accuracy of node classification, compared with graphs with strong GNE (i.e., all classes have GNE), regardless of the specific method used for classification.

Discussion of Homophily Statistics. In Fig. 2g, we report two homophily statistics. Edge homophily [25] is the edge ratio that connect two nodes with the same class, and $\hat{h}$ [14] is an improved metric which is insensitive to the class number and size. We find even using all labels, they are not enough to capture the interrelations of all class pairs in detail, and the graphs with low homophily statistics are not guaranteed to be heterophily. They can only detect the absence of homophily, instead of distinguishing different non-homophily cases, including heterophily, X-ophily, and no GNE. In contrast, our NETEFFECT_TEST identifies whether the graph exhibits GNE or not from only a few labels.

4   Proposed GNE Estimation

Given that a graph exhibits GNE, how can we estimate the all-pair relations between classes? A *compatibility matrix* is a natural strategy to describe the relations, which has been widely used in the literature. We propose NETEFFECT_EST, which turns the compatibility matrix estimation into an optimization problem based on a closed-form formula. NETEFFECT_EST not only overcomes the limitation of naive edge counting, but is also robust to noisy observations even with few observed labels.

4.1   Why NOT Edge Counting

The graph in Fig. 3a exhibits heterophily between class pairs $(1, 2)$ and $(3, 4)$, while it exhibits homophily in classes 5 and 6. A compatibility matrix is commonly used in existing studies, but assumed given by domain experts, instead of being estimated. A naive way to estimate it is via counting labeled edges, but it has two limitations: 1) rare labels are neglected, and 2) it is noisy or biased due to few labeled nodes. The result is even more unreliable if the given labels are imbalanced. In Fig. 3, we upsample the training labels $10\times$ for class 1 using the graph in Fig. 1b. Edge counting in Fig. 3b biases towards the upsampled class and clearly fails to estimate the correct compatibility matrix in Fig. 3a, while our proposed NETEFFECT_EST succeeds in Fig. 3c. This commonly occurs in practice, since we observe only limited labels, and becomes fatal if the observed distribution is different from the true one.

4.2   Closed-Form Formula

We begin the derivation by rewriting Eqn. 1 of BP. The main insight is reminiscent of 'leave-one-out' cross validation. That is, we find $\hat{H}$ that would make the results of the propagation (RHS of Eqn. 2) to the actual values (LHS of Eqn. 2):

$$\underbrace{\hat{E}}_{\text{reality}} \approx \underbrace{A\hat{E}\hat{H}}_{\text{estimate}} \tag{2}$$

Formally, we want to minimize the difference between the reality and the estimate:

$$\min_{\hat{H}} \sum_{i\in\mathcal{P}}\sum_{u=1}^{c} \left\| \hat{E}_{iu} - \sum_{k=1}^{c}\sum_{j\in N(i)\cap\mathcal{P}} \hat{E}_{jk}\hat{H}_{ku} \right\|^2, \tag{3}$$

where $N(i)$ denotes the neighbors of node i. In other words, we aim to minimize the difference between initial belief $\hat{E}$ of each node $i \in \mathcal{P}$ by the ones of its neighbors $N(i) \in \mathcal{P}$, i.e., $N(i) \cap \mathcal{P}$. To estimate the compatibility matrix $\hat{H}$, we solve the optimization problem in Eqn. 3 with the proposed closed-form formula:

Lemma 1 (Network Effect Formula (NEF)). *Given adjacency matrix A and initial beliefs $\hat{E}$, the closed-form solution of vectorized compatibility matrix $vec(\hat{H})$ is:*

$$\boxed{vec(\hat{H}) = (X^T X)^{-1} X^T y} \tag{4}$$

where $X = I_{c\times c} \otimes (A\hat{E})$ and $y = vec(\hat{E})$.

Proof. See Appx. A.1.   ∎

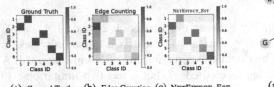

(a) Ground Truth   (b) Edge Counting   (c) NETEFFECT_EST

Fig. 3. NETEFFECT_EST handles imbalanced case well. Labels of class 1 is upsampled.

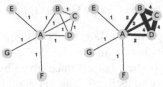

(a) Adj. Matrix   (b) Emphasis Matrix

Fig. 4. Emphasis matrix at work: it prefers well-connected neighbors.

4.3 NETEFFECT_EST

The algorithm is presented in Alg. 1. In practice, we can use any form of adjacency matrix for the estimation. The proposed NEF allows us to estimate the compatibility matrix by solving this optimization problem, but there still exists a practical challenge that need to be addressed. With few labels, it is difficult to properly separate them into training and validation sets for the regression, and the estimation can easily be interfered by the noisy observations. We thus use ridge regression with leave-one-out cross-validation (RidgeCV) instead of the regular linear regression. This allows us to fully utilize the observations without having biases caused by random splits of training and validation sets. Moreover, the regularization effect of RidgeCV makes the compatibility matrix more robust to noisy observations. It is noteworthy that its computational cost is negligible.

Algorithm 1: NETEFFECT_EST

Data: Adjacency matrix A, initial belief $\hat{E}$, and priors $\mathcal{P}$
Result: Estimated compatibility matrix $\hat{H}$
1 $X \leftarrow I_{c \times c} \otimes (A\hat{E})$;   // feature matrix
2 $y \leftarrow \mathrm{vec}(\hat{E})$;   // target vector
3 Extract indices i with nodes in priors $\mathcal{P}$;
4 $\hat{H} \leftarrow RidgeCV(X[i], y[i])$;
5 Return $\hat{H}$;

---

5 Proposed GNE Exploitation

We propose NETEFFECT_EXP to exploit GNE for accurate and fast node classification with few labels. With few labels, it becomes crucial to better utilizing the graph structure. First, we address this by paying attention to influential neighbors by the proposed "emphasis" matrix; and then describe NETEFFECT_EXP with theoretical analysis.

5.1 "Emphasis" Matrix

Rationale and Overview. With few priors, we propose to better utilize the graph structure, by paying attention to only the most important part of it. That is to say, not all neighbors are equally influential: In Fig. 4, best practice shows that well-connected neighbors (i.e., nodes 'B', 'C', and 'D') have more influence on node 'A' than the rest.

Thus, we propose "emphasis" matrix A^*, to pay attention to such neighbors. NET-EFFECT_EST can also benefit from it by replacing A with A^*, where we denote the improved compatibility matrix as $\hat{H}^*$. Alg. 2 shows the details. In short, it has 3 steps:

1. **Favors influential neighbors** by quickly approximating the node-to-node proximity using (non-backtracking) random walks with restarts (lines 2-5);
2. **Touches-up** the new node-proximity by applying a series of transformations (including the best-practice element-wise logarithm) on the proximity matrix (line 6);
3. **Symmetrizes and weighs** the adjacency matrix with structural-aware embedding (lines 7-8), giving higher weights to neighbors with closer embeddings (line 9).

Algorithm 2: "Emphasis" Matrix

Data: Adjacency matrix A, number of trials M, number of steps L, and dimension d
Result: Emphasis matrix A^*

1 $W' \leftarrow O_{n \times n}$;
/* approximate proximity matrix by random walk   */
2 **for** node i in G **do**
3   **for** $m = 1, ..., M$ **do**
4   **for** $j \in \mathcal{W}_m(i, L)$ **do**
5              $W'_{ij} \leftarrow W'_{ij} + 1$;

/* masking, degree normalization and logarithm   */
6 $W_{n \times n} \leftarrow \log(D^{-1}(W' \odot A))$;
7 $U_{n \times d}, \Sigma_{d \times d}, V^T_{d \times n} \leftarrow \text{SVD}(W, d)$;   // embedding
8 $U \leftarrow \sqrt{\Sigma}U$;   // scaling
/* boost weights of close-embedded neighbors   */
9 Weigh $A^*_{n \times n}$, where $A^*_{ij} = \mathcal{S}(U_i, U_j), \forall \{i, j | A_{ij} = 1\}$;
10 Return A^*;

Proximity Matrix Approximation. We propose to utilize random walks to approximate the proximity matrix. The approximated proximity matrix W'_{ij} records the times we visit node j if we start a random walk from node i. Only the well-connected neighbors will be visited more often. We theoretically show that it converges quickly:

Lemma 2 (Convergence of Random Walks). *With probability $1 - \delta$, the error ϵ between the approximated and true distributions for a node walking to its 1-hop neighbor by random walks of length L with M trials is no greater than $\frac{\lceil(L-1)/2\rceil}{L}\sqrt{\frac{\log(2/\delta)}{2LM}}$.*

We can make the convergence even faster by using "non-backtracking" random walks [2]. Given the start node s and walk length L, its function is defined as follows:

$$\mathcal{W}(s, L) = \left\{ (w_0 = s, ..., w_L) \quad \begin{matrix} w_l \in N(w_{l-1}), \forall l \in [1, L] \\ w_{l-1} \neq w_{l+1}, \forall l \in [1, L-1] \end{matrix} \right. \tag{5}$$

Thanks to it, we improve Lemma 2 to have a tighter bound of error ϵ:

Lemma 3 (Convergence of Non-backtracking Random Walks). *With the same condition as in Lemma 2, the error ϵ by non-backtracking random walks is no greater than $\frac{\lceil(L-1)/3\rceil}{L}\sqrt{\frac{\log(2/\delta)}{2LM}}$.*

Proof. See Appx. A.2.   ∎

Structural-Aware Node Representation. Based on W, we apply a series of transformations to generate better and unbiased representations of nodes in a fast way. An element-wise multiplication by A is done to keep the approximation of 1-hop neighbor for each node, which is sparse but supplies sufficient information. We use the inverse of the degree matrix D^{-1} to reduce the influence of nodes with large degrees. This prevents them from dominating the pairwise distance by containing more elements in their rows. The element-wise logarithm rescales the distribution in W, in order to enlarge the difference between smaller structures. We use Singular Value Decomposition (SVD) for efficient rank-d decomposition of sparse W, and multiply the left-singular vectors U by the squared eigenvalues $\sqrt{\Sigma}$ to correct the scale.

"Emphasis" Matrix Construction. Directly measuring the node similarity in the graph is not trivial, or may be time consuming (e.g., by counting motifs). Therefore, we propose to compute the node similarity via the structural-aware node representations, which capture the higher-order information, and construct the "emphasis" matrix A^* by weighing A with the node similarity. The intuition is that the nodes that are closer in the embedding space are better connected with higher-order structures. The similarity function is $\mathcal{S}(U_i, U_j) = e^{-\mathcal{D}(U_{ik}, U_{jk})}$, where e is the Euler's number. It is a universal law [19], which turns the distance into similarity, and bounds it from 0 to 1. While $\mathcal{D}$ can be any distance metric, we use Euclidean as it works well empirically.

5.2 NETEFFECT_EXP

The algorithm of NETEFFECT_EXP is in Appx. B.2. NETEFFECT_EXP takes as input the "emphasis" matrix A^*, the compatibility matrix $\hat{H}^*$ estimated by A^*, and the initial beliefs $\hat{E}$. It computes the beliefs $\hat{B}$ iteratively by aggregating the beliefs of neighbors through A^* until they converge. This reusage of A^* aims to draw attention to the neighbors that are more structurally important. By exploiting GNE with $\hat{H}^*$, NETEFFECT propagates properly in heterophily graphs.

Convergence Guarantee. To ensure the convergence of NETEFFECT_EXP, we introduce a scaling factor f during the iterations. A smaller f leads to a faster convergence but distorts the results, thus we set f to $0.9/\rho(A^*)$. Its exact convergence is:

Lemma 4 (Exact Convergence). *The criterion for the exact convergence of* NETEFFECT_EXP *is* $0 < f < 1/\rho(A^*)$, *where* $\rho(\cdot)$ *denotes the spectral radius of the matrix.*

Proof. See Appx. A.3.   ∎

Complexity Analysis. NETEFFECT_EXP uses sparse matrix representation of graphs and scales linearly. Its complexity is:

Lemma 5. *The time complexity of* NETEFFECT_EXP *is approximately* $O(m)$ *and the space complexity is* $O(\max(m, n \cdot L \cdot M) + n \cdot c^2)$.

Proof. See Appx. A.4.   ∎

6 Experiments

In this section, we aims to answer the following questions:

Q1. **Accuracy**: How well does NETEFFECT work by estimating and exploiting GNE?
Q2. **Scalability**: How does the running time of NETEFFECT scale w.r.t. graph size?
Q3. **Explainability**: How does NETEFFECT explain the real-world graphs?

Datasets. We focus on large graphs and include 8 graphs with at least 20K nodes. For each dataset, we sample only a few node labels for training for five times and report the average. "Synthetic" is the enlarged graph in Fig. 1b, which exhibits X-ophily GNE.

Baselines. We compare NETEFFECT with five baselines and separate them into four groups: *General GNNs:* GCN [9], APPNP [10]. *Heterophily GNNs:* MIXHOP [1], GPR-GNN [3]. *BP-based methods:* HOLS [4]. *Our proposed methods:* NETEFFECT-Hom and NETEFFECT. NETEFFECT-Hom is NETEFFECT using identity matrix as compatibility matrix, which assumes homophily and does not handle GNE.

Experimental Settings. For GNNs, one-hot node degrees are used as the node features, as implemented by PyG [5]. Experiments are run on a server with 3.2GHz Intel Xeon CPU. Details of the experimental setup are in Appx. C.

Table 2. NETEFFECT wins on X-ophily and Heterophily datasets.

Dataset	Synthetic			Pokec-Gender [20]			arXiv-Year [8]			Patent-Year [12]		
# of Nodes / Edges / Classes	1.2M / 34.0M / 6			1.6M / 22.3M / 2			169K / 1.2M / 5			1.3M / 4.3M / 5		
Label Fraction	4%			0.4%			4%			4%		
GNE Strength	Strong X-ophily			Strong Heterophily			Weak X-ophily			Weak Heterophily		
Method	Accuracy	Time (s)	Rel. Time	Accuracy	Time (s)	Rel. Time	Accuracy	Time (s)	Rel. Time	Accuracy	Time (s)	Rel. Time
GCN	16.7±0.0	3456	4.1×	51.8±0.1	2906	3.4×	35.3±0.1	132	2.5×	26.0±0.0	894	2.3×
APPNP	18.6±1.1	7705	9.2×	50.9±0.3	6770	7.8×	33.5±0.2	423	8.1×	27.5±0.2	2050	5.2×
MIXHOP	16.7±0.0	58391	70.0×	53.4±1.2	53871	62.1×	39.6±0.1	2983	57.4×	26.8±0.1	18787	47.6×
GPR-GNN	18.9±1.2	7637	9.1×	50.7±0.2	6699	7.7×	30.1±1.4	400	7.7×	25.3±0.1	2034	5.1×
HOLS	46.1±0.1	1672	2.0×	54.4±0.1	8552	9.9×	34.1±0.3	566	10.9×	23.6±0.0	510	1.3×
NETEFFECT-Hom	45.6±0.1	835	1.0×	56.9±0.2	869	1.0×	37.0±0.3	52	1.0×	24.3±0.0	429	1.1×
NETEFFECT	80.4±0.0	841	1.0×	67.3±0.1	867	1.0×	38.9±0.1	52	1.0×	28.7±0.1	395	1.0×

Table 3. NETEFFECT wins on Homophily datasets.

Dataset	Facebook [17]			GitHub [17]			arXiv-Category [22]			Pokec-Locality [20]		
# of Nodes / Edges / Classes	22.5K / 171K / 4			37.7K / 289K / 2			169K / 1.2M / 40			1.6M / 22.3M / 10		
Label Fraction	4%			4%			4%			0.4%		
Method	Accuracy	Time (s)	Rel. Time	Accuracy	Time (s)	Rel. Time	Accuracy	Time (s)	Rel. Time	Accuracy	Time (s)	Rel. Time
GCN	67.0±0.8	12	2.0×	81.0±0.6	28	2.2×	25.4±0.3	216	2.3×	17.3±0.4	4002	2.9×
APPNP	50.5±2.2	46	7.7×	74.2±0.0	73	5.6×	19.4±0.6	1176	12.3×	16.8±1.7	11885	8.6×
MIXHOP	69.2±0.7	296	49.3×	77.8±1.3	526	40.5×	33.0±0.6	3203	33.4×	16.9±0.3	52139	37.9×
GPR-GNN	51.9±1.5	47	7.8×	74.1±0.1	75	5.8×	19.7±0.3	1174	12.2×	30.0±2.0	11959	8.7×
HOLS	86.0±0.4	934	155.7×	80.8±0.5	126	9.7×	61.4±0.2	627	6.5×	63.7±0.3	8139	5.9×
NETEFFECT-Hom	85.2±0.5	6	1.0×	81.3±0.5	13	1.0×	61.7±0.2	96	1.0×	66.0±0.2	1437	1.0×
NETEFFECT	85.2±0.5	6	1.0×	81.3±0.5	13	1.0×	58.8±0.6	108	1.1×	64.8±0.8	1377	1.0×

6.1 Q1 - Accuracy

In Table 2 and 3, we report the accuracy and running time. We highlight the top three from dark to light by ▬, ▬ and ▬ denoting the first, second and third place. *In summary,* NETEFFECT *wins on* X-*ophily, heterophily and homophily graphs.*

x-ophily and Heterophily. In Table 2, NETEFFECT outperforms all the competitors significantly by more than 34.3% and 12.9% accuracy on "Synthetic" and "Pokec-Gender", respectively. These graphs exhibit strong GNE, thus NETEFFECT boosts the accuracy owing to precise estimations of compatibility matrix. Heterophily GNNs give results close to majority voting when the observed labels are not adequate. With homophily assumption, General GNNs and BP-based methods also not perform well. Both "arXiv-Year" and "Patent-Year" have weak GNE (Sec. 3.2). Even so, NETEFFECT still outperforms the competitors by estimating a reasonable compatibility matrix (Fig. 6c).

Homophily. In Table 3, NETEFFECT-Hom outperforms all the competitors on 3 out of 4 homophily graphs, namely "GitHub", "arXiv-Category" and "Pokec-Locality", and NETEFFECT performs similarly to NETEFFECT-Hom. In addition, NETEFFECT-Hom performs competitively with HOLS on "Facebook", while being 155.7× faster.

Ablation Study. *Our optimizations make a difference.* We evaluate different compatibility matrices – (i) NETEFFECT-EC uses edge counting on the labels of adjacent nodes in the priors, and (ii) NETEFFECT-A uses the adjacency matrix instead of "emphasis" matrix as the input of NETEFFECT_EST. To evaluate the cases when imbalanced labels are given, we upsample 5% labels to the class with the fewest labels in the datasets with weak GNE during the estimation. In Table 4, we find that NETEFFECT outperforms all its variants in all datasets. In the graphs with strong GNE, NETEFFECT shows its robustness to the structural noises and gives better results. In the imbalanced graphs, while NETEFFECT-EC brings its vulnerability to light, NETEFFECT stays with high accuracy. This study highlights the importance of a compatibility matrix estimation, as well as forming it into an optimization problem as shown in Lemma 1.

6.2   Q2 - Scalability

NETEFFECT *is scalable and thrifty.* We vary the edge number in "Pokec-Gender" and plot against the running time, including training and inference. In Fig. 5, NETEFFECT scales linearly as expected (Lemma 5). Table 5 shows the estimated AWS dollar cost in "Pokec-Gender", assuming that we use a CPU machine for NETEFFECT, and a GPU one for GCN. NETEFFECT achieves up to 45× savings. Details in Appx. C.3.

6.3   Q3 - Explainability

Figure 6 shows the compatibility matrices that NETEFFECT recovered. *In a nutshell, the results agree well with the intuition.* For "Synthetic", NETEFFECT matches the answer used for graph generation. For "Pokec-Gender" (Fig. 6a), NETEFFECT report heterophily, where people incline to have more opposite gender interactions [7]. For "arXiv-Year" and "Patent-Year", NETEFFECT find that papers and patents tend to cite the ones published in nearby years, which also agrees with intuition (Fig. 6b and 6c).

Table 4. Ablation Study: Estimating compatibility matrix by proposed "emphasis" matrix is essential.

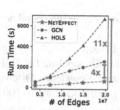

Datasets	GNE Strength	NETEFFECT-Hom	NETEFFECT-EC	NETEFFECT-A	NETEFFECT
Synthetic	Strong	77.7±0.0	68.0±0.1	77.4±0.0	80.5±0.0
Pokec-Gender		56.9±0.1	64.9±0.2	64.8±0.2	67.3±0.1
arXiv-Year (imba.)	Weak	37.0±0.3	36.5±1.0	35.7±0.6	38.4±0.0
Patent-Year (imba.)		24.1±0.0	24.0±0.9	28.7±0.1	28.7±0.0

Table 5. NETEFFECT is thrifty. AWS dollar cost ($) is reported, by t3.small and p3.2xlarge.

Datasets	NETEFFECT	GCN
Pokec-Gender	$ 0.33 (1.0×)	$ 12.61 (45.0×)
Pokec-Locality	$ 0.53 (1.0×)	$ 13.66 (29.1×)

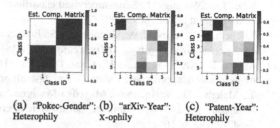

(a) "Pokec-Gender": Heterophily (b) "arXiv-Year": X-ophily (c) "Patent-Year": Heterophily

Fig. 5. NETEFFECT is scalable. It is fast and scales linearly with the edge number.

Fig. 6. NETEFFECT is explainable. Our estimated compatibility matrices are much more robust to noises compared to edge counting (in Fig. 2).

7 Conclusions

We analyze the *generalized network-effects* (GNE) in node classification in the presence of only few labels. Our proposed NETEFFECT has the following desirable properties:

1. *Principled*: NETEFFECT_TEST to statistically identify the presence of GNE,
2. *General* and *Explainable*: NETEFFECT_EST to estimate GNE with derived closed-form solution, if there is any, and
3. *Accurate* and *Scalable*: NETEFFECT_EXP to efficiently exploit GNE for better performance on node classification.

Applied on a real-world graph with 22.3M edges, NETEFFECT only requires 14 *minutes*, and outperforms baselines on both accuracy and speed ($\geq 7\times$).

References

1. Abu-El-Haija, S., Et al. : Mixhop: higher-order graph convolutional architectures via sparsified neighborhood mixing. In: ICML, pp. 21–29 (2019)
2. Alon, N., Benjamini, I., Lubetzky, E., Sodin, S.: Non-backtracking random walks mix faster. Commun. Contemp. Math. **9**(04), 585–603 (2007)
3. Chien, E., Peng, J., Li, P., Milenkovic, O.: Adaptive universal generalized pagerank graph neural network. In: ICLR (2021)
4. Eswaran, D., Kumar, S., Faloutsos, C.: Higher-order label homogeneity and spreading in graphs. In: The Web Conference, pp. 2493–2499 (2020)

5. Fey, M., Lenssen, J.E.: Fast graph representation learning with PyTorch Geometric. In: ICLR Workshop on Representation Learning on Graphs and Manifolds (2019)
6. Gatterbauer, W., Günnemann, S., Koutra, D., Faloutsos, C.: Linearized and single-pass belief propagation. PVLDB **8**(5), 581–592 (2015)
7. Ghosh, A., Monsivais, D., Bhattacharya, K., Dunbar, R.I., Kaski, K.: Quantifying gender preferences in human social interactions using a large cellphone dataset. EPJ Data Sci. **8**(1), 9 (2019)
8. Hu, W., et al.: Open graph benchmark: datasets for machine learning on graphs. NeurIPS **33**, 22118–22133 (2020)
9. Kipf, T.N., Welling, M.: Semi-supervised classification with graph convolutional networks. arXiv preprint arXiv:1609.02907 (2016)
10. Klicpera, J., Bojchevski, A., Günnemann, S.: Predict then propagate: Graph neural networks meet personalized pagerank. arXiv preprint arXiv:1810.05997 (2018)
11. Koutra, D., Ke, T.-Y., Kang, U., Chau, D.H., Pao, H.-K.K., Faloutsos, C.: Unifying guilt-by-association approaches: theorems and fast algorithms. In: Gunopulos, D., Hofmann, T., Malerba, D., Vazirgiannis, M. (eds.) Machine Learning and Knowledge Discovery in Databases, pp. 245–260. Springer, Berlin, Heidelberg (2011). https://doi.org/10.1007/978-3-642-23783-6_16
12. Graphs over time: densification laws, shrinking diameters and possible explanations. In: Proceedings of the Eleventh ACM SIGKDD International Conference on Knowledge Discovery in Data Mining, pp. 177–187. Association for Computing Machinery, New York (2005)
13. Lim, D., Benson, A.R.: Expertise and dynamics within crowdsourced musical knowledge curation: A case study of the genius platform. arXiv preprint arXiv:2006.08108 (2020)
14. Lim, D., et al.: Large scale learning on non-homophilous graphs: New benchmarks and strong simple methods. NeurIPS **34**, 20887–20902 (2021)
15. Lin, M., Lucas, H.C., Jr., Shmueli, G.: Research commentary-too big to fail: large samples and the p-value problem. Inf. Syst. Res. **24**(4), 906–917 (2013)
16. Ma, Y., Liu, X., Shah, N., Tang, J.: Is homophily a necessity for graph neural networks? In: ICLR (2022)
17. Rozemberczki, B., Allen, C., Sarkar, R.: Multi-scale attributed node embedding (2019)
18. Rozemberczki, B., Sarkar, R.: Twitch gamers: a dataset for evaluating proximity preserving and structural role-based node embeddings. arXiv preprint arXiv:2101.03091 (2021)
19. Shepard, R.N.: Toward a universal law of generalization for psychological science. Sci. **237**(4820), 1317–1323 (1987)
20. Takac, L., Zabovsky, M.: Data analysis in public social networks. In: International Scientific Conference and International Workshop Present Day Trends of Innovations. vol. 1 (2012)
21. Traud, A.L., Mucha, P.J., Porter, M.A.: Social structure of facebook networks. Phys. A **391**(16), 4165–4180 (2012)
22. Wang, K., Shen, Z., Huang, C., Wu, C.H., Dong, Y., Kanakia, A.: Microsoft academic graph: when experts are not enough. Quant. Sci. Stud. **1**(1), 396–413 (2020)
23. Wasserman, L., Ramdas, A., Balakrishnan, S.: Universal inference. Proc. Natl. Acad. Sci. **117**(29), 16880–16890 (2020)
24. Wu, F., Souza, A., Zhang, T., Fifty, C., Yu, T., Weinberger, K.: Simplifying graph convolutional networks. In: ICML, pp. 6861–6871. PMLR (2019)
25. Zhu, J., Yan, Y., Zhao, L., Heimann, M., Akoglu, L., Koutra, D.: Large scale learning on non-homophilous graphs: New benchmarks and strong simple methods. NeurIPS **34**, 7793–7804 (2020)

Learning to Rank Based on Choquet Integral: Application to Association Rules

Charles Vernerey[1] , Noureddine Aribi[2] , Samir Loudni[1](✉) , Yahia Lebbah[2] ,
and Nassim Belmecheri[3]

[1] TASC - DAPI, IMT-Atlantique, LS2N - CNRS, 44307 Nantes, France
{Charles.Vernerey,samir.loudni}@imt-atlantique.fr
[2] Lab. LITIO, Université Oran1, 31000 Oran, Algeria
{aribi.noureddine,lebbah.yahia}@univ-oran1.dz
[3] Simula Research Laboratory, Oslo, Norway
nassim@simula.no

Abstract. Discovering relevant patterns for a particular user remains a challenging data mining task. One way to deal with this difficulty is to use interestingness measures to create a ranking. Although these measures allow evaluating patterns from various sights, they may generate different rankings and hence highlight different understandings of what a good pattern is. This paper investigates the potential of learning-to-rank techniques to learn to rank directly. We use the Choquet integral, which belongs to the family of non-linear aggregators, to learn an aggregation function from the user's feedback. We show the interest of our approach on association rules, whose added-value is studied on UCI datasets and a case study related to the analysis of gene expression data.

1 Introduction

The interest in ranking models, paradigms, and functions is not new and is still an important research topic in many fields. In Information Retrieval, where often, documents must be sorted according to their relevance to a given query, ranking is paramount since it directly affects retrieval quality. Several methods have been developed for learning functions to sort objects from example orders [10,13]. This task is referred to as *Learning-to-Rank* (LTR) in machine learning [13].

In pattern mining, when a domain expert is confronted with a set of patterns often counting in thousands or even millions, a common approach relies on using objective interestingness measures to select and rank the result set based on their potential interest [17]. Although these measures allow evaluating patterns from various sights, they may generate different rankings and hence highlight different understandings of what a good pattern is. The difficulty of defining appropriate measures of interest to select the best patterns has led to a trend of works that focus on learning aggregation functions to compare patterns evaluated on multiple criteria by synthesizing their performances into global scores [5]. Authors in [6] have exploited RankingSVM [10] to learn a Weighted sum over a set of patterns, while [2] proposed AHPRank, another approach based on Analytical Hierarchy Process [16]. However, existing approaches have a short-coming:

© The Author(s), under exclusive license to Springer Nature Singapore Pte Ltd. 2024
D.-N. Yang et al. (Eds.): PAKDD 2024, LNAI 14645, pp. 313–326, 2024.
https://doi.org/10.1007/978-981-97-2242-6_25

they are not sufficiently expressive to fit the user's preferences, allowing, for instance, modeling any kind of interaction between measures.

Choquet integral [7] defines a family of non-linear aggregators that performs a Weighted sum of criterion values using a capacity function assigning a weight to any combination of criteria, thus enabling positive and/or negative interactions among them. In this paper, we propose elicitation approaches to learn a ranking function from user feedback and aggregate a set of interestingness measures using the Choquet integral. We therefore describe a new heuristic to select pairs of patterns from an existing database, which the user has to compare. Our approach has shown high effectiveness on UCI datasets and gene expression data, successfully identifying relevant patterns across different configurations with minimal user preferences, while maintaining a favorable trade-off between quality and time.

2   Preliminaries

A) Itemset and Association Rules Mining. Let $\mathcal{I}$ be a set of n *items*, an *itemset* (or pattern) X is a non-empty subset of $\mathcal{I}$. A transactional dataset $\mathcal{D}$ is a multiset of transactions over $\mathcal{I}$, where each *transaction* $t \subseteq \mathcal{I}$. A pattern X *occurs* in a transaction t, iff $X \subseteq t$. The *support* $sup(X)$ of X in $\mathcal{D}$ is the numbers of transactions that contain X. An association rule is an implication $r : X \Rightarrow Y$ where X and Y are itemsets such that $X \cap Y = \emptyset$ and $Y \neq \emptyset$. The problem of mining for association rules involves discovering all the rules that correlate the presence of one itemset with another under minimum support θ and minimum confidence c thresholds.

B) Non-Additive Measures and Choquet Integral. The Choquet Integral is a complex aggregation function [7] that covers a whole range of aggregation functions in which the weights depend not only on the rank of the criteria (as in the Weighted Average or the Ordered Weighted Average [19]), but also on a *fuzzy measure* (called *capacity*). This measure makes it possible to extend the notion of weight to a subset of criteria, to express their *degree of interaction*. Let $N = \{1, ..., n\}$ be a finite set of criteria and $\mu : 2^N \to [0, 1]$ a *fuzzy* measure. For each set $S \subseteq N$, we interpret $\mu(S)$ as the *weight* or, say, the importance of the set of criteria S. Unfortunately, additive measures cannot model any kind of interaction between elements.

Consider, for example, N as a set of criteria relevant for a job, like "English speaking" and "Python programming skill" , and $\mu(S)$ as the evaluation of a candidate satisfying criteria $S \subseteq N$ [18]. Suppose that $S = \{English, Spanish\}$ and $T = \{Python\}$. For instance, S and T can be considered complementary, where meeting a single criterion is insufficient. This is because both language skills and programming skills are crucial for the job. That is $\mu(S \cup T) > \mu(S) + \mu(T)$. Likewise, elements can interact positively: If two sets S and T are partly redundant or competitive, then $\mu(S \cup T) < \mu(S) + \mu(T)$. For instance, $S = \{Python\}$ and $T = \{Java\}$ could be considered redundant, as one programming language is typically sufficient. The above considerations motivate the use of non-additive measures, also called capacities or fuzzy measures, which are normalized and monotone [7]: $\mu(\emptyset) = 0$, $\mu(N) = 1$ and $\mu(S) \leq \mu(T) \ \forall \ S \subseteq T \subseteq N$.

Table 1. Running Example.

S	$\emptyset$	$\{1\}$	$\{2\}$	$\{3\}$	$\{1,2\}$	$\{1,3\}$	$\{2,3\}$	$\{1,2,3\}$
$\mu(S)$	0	0.1	0.2	0.2	0.9	0.3	0.3	1

(a) Example of Choquet capacity.

S	$\{1\}$	$\{2\}$	$\{3\}$	$\{1,2\}$	$\{1,3\}$	$\{2,3\}$
$m_\mu(S)$	0.45	0.5	0.05	-0.4	0.4	0

(b) Möbius transform of a Choquet capacity.

A useful representation of non-additive measures μ is in terms of the Möbius transform [8]. The Möbius transform m_μ of the capacity μ is given for any $S \subseteq N$ by: $m_\mu(S) = \sum_{T \subseteq S}(-1)^{|S|-|T|}\mu(T)$. In a general case, μ can formally be specified by $(2^n - 2)$ values. To reduce this number, k-additive measures are generally used. Let $k \in \{1, ..., n\}$, a capacity μ is said to be k-additive [7], if k is the smallest integer such that $m_\mu(T) = 0$ for all $T \subseteq N$ with $|T| > k$.

Choquet integral. Let $\mathcal{X}$ be the set of *alternatives* or *solutions* that need to be compared. Each alternative $x \in \mathcal{X}$ is evaluated w.r.t. a set of n criteria, and is characterized by a performance vector $(x_1, \ldots, x_n)$ where x_i represents the performance of x w.r.t. criterion i. For simplicity, x will indifferently denote the alternative or its performance vector. An important question, then, is how to aggregate the evaluations of individual criteria, i.e., the values x_i, into an overall evaluation, in which the criteria are properly weighted according to the measure μ. In the case of a non-additive measure, the method of aggregating the overall score is nothing else than the Choquet integral. The Choquet integral C_μ of a vector $x = (x_1, ..., x_n)$ is defined by:

$$C_\mu(x) = \sum_{i=1}^{n}(x_{(i)} - x_{(i-1)})\mu(X(i)) \text{ with } x_{(0)} = 0 \tag{1}$$

where $(.)$ is a permutation of N that sorts the components of x by increasing order (i.e. $x_{(i)} \leq x_{(i+1)}$) and $X(i) = \{(i), \ldots, (n)\}$ (i.e. $X(i)$ is the set of criteria that have a value $\geq x_{(i)}$). $C_\mu(x)$ represents the overall performance of alternative x. In terms of the Möbius transform of μ, the Choquet integral can also be expressed as follows: $C_{m_\mu}(x) = \sum_{T \subseteq N} m_\mu(T) \bigwedge_{i \in T} x_i$, where $\bigwedge$ denotes the min operator.

Example 1 (Choquet capacity). Consider a problem defined on 3 criteria, i.e. $N = \{1, 2, 3\}$, two alternatives $x = (8, 7, 5)$ and $y = (4, 9, 8)$, and Choquet capacities defined in table 1a. We have $C_\mu(x) = 5\,\mu(\{1,2,3\}) + (7-5)\,\mu(\{1,2\}) + (8-7)\,\mu(\{1\}) = 6.9$ and $C_\mu(y) = 4\,\mu(\{1,2,3\}) + (8-4)\,\mu(\{2,3\}) + (9-8)\,\mu(\{2\}) = 5.4$. Hence we have $C_\mu(x) > C_\mu(y)$, meaning that x is strictly preferred to y.

Example 2. Consider the 2-additive Möbius transform m_μ defined in table 1b and the vector $x = (8, 7, 5)$. We have $C_{m_\mu}(x) = 8 \cdot m_\mu(\{1\}) + 7 \cdot m_\mu(\{2\}) + 5 \cdot m_\mu(\{3\}) + min(8,7) \cdot m_\mu(\{1,2\}) + min(8,5) \cdot m_\mu(\{1,3\}) = 6.55$.

C) Interactive Learning of Pattern Rankings The problem of learning a ranking function from user feedback is known as *object ranking* [13]. Here, we introduce the formal definition of the pattern ranking learning task and a generic algorithm for solving it.

Algorithm 1. Template algorithm for interactive pattern ranking

---

Input: Transactional dataset $\mathcal{D}$, set of patterns $\mathcal{X}$, Number of iterations L
Output: Ranking function f
1: $\mathcal{P} \leftarrow \emptyset$, $f^0 \leftarrow$ initial function estimates
2: **for** $t = 1, 2, ..., L$ **do**
3:     select a query $\mathcal{X}^t$ based on f^{t-1} ▷ Mine
4:     ask query $\mathcal{X}^t$ to the user and get feedback $\mathcal{U}^t$ ▷ Interact
5:     $\mathcal{P} \leftarrow \mathcal{P} \cup \mathcal{U}^t$, compute f^t based on f^{t-1} and $\mathcal{P}$ ▷ Learn f
6: **return** f

Learning Pattern Rankings. Let $\mathcal{D}$ be a transactional dataset, and $\mathcal{X}$ a set of patterns from a given language $\mathcal{L}$. Assuming the existence of an unknown, user-specific target ranking R over all patterns of $\mathcal{X}$, where $X \succ Y$ means that the user subjectively finds pattern X more interesting than pattern Y, the problem can be defined as follows:

Problem (Learning-to-Rank). The objective is to collect user preferences $\mathcal{P}$ and use them to learn a subjective pattern interestingness function f that is highly consistent with R. The collected feedback $\mathcal{P}$ consists of rankings of small pattern sets from $\mathcal{X}$, which are referred to as *queries*.

Generic Algorithm. Algorithm 1 presents an algorithmic template of the learning-to-rank framework. The interactive process proceeds iteratively for some reasonable number of iterations L, which is task-dependent. The algorithm maintains an internal estimate f^t of the unobserved function, where $t \in [L]$ is the iteration index. At each iteration, it selects a query $\mathcal{X}^t$ to present to the user. The user's feedback, along with all the feedback received previously, is then used to compute a new estimate f^{t+1} of the true function. We discuss key questions concerning instantiations of this framework.

1) User Interaction & Pattern Representations User feedback w.r.t. patterns takes the form of providing a partial/total order over a (small) set of patterns [6, 15]. Pattern representations determine how the user characterizes patterns of interest to him. In this paper, patterns correspond to alternatives and are represented using their performance vectors $x = (x_1, ..., x_n)$, so that criteria relevance can be learned from the user feedback. Table 2 (see Suppl. Mat.) summarises the different notations used in this paper.

2) Learning from Feedback Acquiring ranking functions from sample orders to evaluate patterns is a very natural way of representing preferences [11]. Common solving techniques involve minimizing pairwise loss, e.g., the number of discordant pairs, as in RankingSVM [10]. In this paper, we propose to use the Choquet integral to learn non-linear models from a set of user feedback w.r.t. some interestingness criteria.

3) Pattern Selection The goal of query selection is to minimize user feedback – thus reducing user effort – while accelerating model learning. As selecting an optimal query is an NP-hard problem [1], heuristics are typically employed. In this paper, we specifically focus on heuristics that involve selecting a query of size two, i.e. $|\mathcal{X}^t| = 2$.

3 Learning a Choquet Ranking Function

We propose an instantiation of the previous framework that uses the Choquet integral to learn a non-linear ranking function from user feedback. We formulate this problem as an optimization problem, and we use the *Generalized Least Squares* (GLS) method [9] to solve it. We describe two strategies to exploit our learning scheme: *passive* and *active*.

GLS for Learning Choquet Capacities. Given a set of user preferences $\mathcal{P}$ corresponding to the ranking of a subset of alternatives $\mathcal{X}$, our objective aims at determining suitable Möbius capacities for the Choquet integral that best fit these preferences. Several approaches have been proposed in the literature to determine the capacity of a Choquet integral. [3] proposed an elicitation process to progressively reduce the set of admissible capacities until a robust recommendation can be made using minimax regret. This approach only computes the bounds of the Choquet capacity and cannot be used to score any alternative. Recently, [12] proposed an outranking approach eliciting the Decision Maker's preferences in the form of a Simple Ranking Method using the Reference Profiles (SRMP) model until a robust recommendation can be made. Our setting is different from outranking. The authors of [4] proposed a machine learning-based approach for the identification of the capacity of a hierarchical 2-additive Choquet integral and the associated utility functions. Learning utility functions is out of the scope of this work. To solve this problem, we use the GLS method. The main idea of the proposed approach is to minimize a quadratic error between Choquet values $C_{m_\mu}(x)$ and desired overall scores $\mathcal{Y}(x)$ for each alternative $x \in \mathcal{X}$. The optimization problem to solve is the following quadratic Mixed Integer Linear Program (MIP), denoted `solveGLS`$(\mathcal{P}, k, \delta_{\mathcal{Y}})$:

$$\text{Minimize} \quad \sum_{x \in \mathcal{X}} [C_{m_\mu}(x) - \mathcal{Y}(x)]^2$$

$$\text{subject to} \quad \begin{cases} \sum_{\substack{T \subseteq S \\ |T| \leq k-1}} m_\mu(T \cup i) \geq 0, \ \forall i \in N, \ \forall S \subseteq N \setminus i & (1) \\ \sum_{\substack{T \subseteq N \\ 1 \leq |T| \leq k}} m_\mu(T) = 1 & (2) \\ C_{m_\mu}(x) - C_{m_\mu}(y) \geq \delta_C, \quad \forall (x, y) \in \mathcal{P} & (3) \end{cases}$$

The variables of the model are presented in Table 2b. Constraints (1) and (2) normalize the Möbius capacities and ensure they satisfy monotonicity. Constraint (3) models the preference between alternatives x and y w.r.t. $\mathcal{P}$, where $\delta_C \geq 0$ is a minimal indifference threshold specified by the user. A solution of this MIP consists of the Möbius capacity m_μ and the overall scores $\mathcal{Y}(x)$ for each $x \in \mathcal{X}$. The main advantage of this approach lies in its ability to find a Choquet integral model that optimizes user preferences. When the objective function's value is 0, it means that the preferences specified by the user are accurately represented by the acquired capacity, resulting in $C_{m_\mu}(x) = \mathcal{Y}(x)$ for all $x \in \mathcal{X}$. Alternatively, two scenarios may arise: (1) The discovery of a Choquet model compatible with $\mathcal{P}$ but exceeding the indifference threshold, (2) The identification of a Choquet model that deviates from the user's preference.

ChoquetRank Algorithm. We introduce `ChoquetRank`, a Choquet ranking learning function that operates in two modes: *passive* or *active learning*. It takes as input a

Algorithm 2. ChoquetRank $(\mathcal{X}, \mathcal{P}, k, \delta_C, L)$

In: : Set of user preferences $\mathcal{P}$, Set of alternatives $\mathcal{X}$, Number of iterations L, Integer k, Indifference threshold δ_C;
Out: Möbius capacity m_μ;
1: **begin**
2:    **if** $(\mathcal{P} \neq \emptyset)$ **then**
3:        **return** solveGLS$(\mathcal{P}, k, \delta_C)$ ▷ Passive learning
4:    **else**
5:        **for** $T \subseteq N$ **do**   ▷ Active learning
6:            **if** $(|T| == 1)$ **then** $m_\mu^0(T) = \frac{1}{|N|}$
7:            **else** $m_\mu^0(T) = 0$
8:        **for** $t = 1, \dots, L$ **do**
9:            $\mathcal{U}^t \leftarrow$ AskPreference$(\mathcal{H}_{sbh}(\mathcal{X}, m_\mu^{t-1}))$
10:           $\mathcal{P} \leftarrow \mathcal{P} \cup \mathcal{U}^t$
11:           $m_\mu^t \leftarrow$ solveGLS$(\mathcal{P}, k, \delta_C)$
12:       **return** m_μ^L
13: **end**

set of patterns $\mathcal{X}$, a set user feedback $\mathcal{P}$ and a list of parameters $\langle \delta_C, k, L \rangle$ representing respectively the indifference threshold, desired k-additivity of the Choquet integral and the number of iterations (only used in the active learning setting). ChoquetRank begins by assessing if $\mathcal{P}$ is not empty. If so, it acts as a passive learning procedure and employs the solveGLS algorithm on $\mathcal{P}$ for learning (lines 2-3). In the absence of feedback, it transitions to the active mode, prompting a series of queries to the user.

a) *Passive learning*: Initially, patterns are generated from $\mathcal{D}$ and divided into training and test sets. The user establishes a complete pre-order over the training set patterns, which is then used as a batch input for the solveGLS learning algorithm.

b) *Active learning*: In this setting, the algorithm begins with a predefined set of patterns. It continuously presents pairwise pattern comparisons to a user until a stopping criterion, such as a fixed number of iterations L, is met. At the beginning, capacity values m_μ related to singleton criteria are initialized using an *uniform capacity* (cf. lines 5-7). During each iteration, a heuristic selects a pair of patterns from $\mathcal{X}$ for user evaluation. User preferences for this pair of patterns are added into $\mathcal{P}$, accumulating comparisons from the previous iterations. These accumulated preferences are then used to learn new Möbius capacities m_μ. Further details on pattern pair selection are given bellow.

Sensitivity-Based Heuristic. The proposed heuristic, $\mathcal{H}_{sbh}$, uses learned Möbius capacity values to select similar pattern pairs. It prompts the user to compare alternatives based on the overall predicted score. The similarity between pattern pairs is determined using an adapted L_1 norm, denoted as d. $\mathcal{H}sbh$ outputs a pair $(x, y)_{sbh}$ that minimizes $d(x, y)$, where $d(x, y) = \frac{|C_{m_\mu}(x) - C_{m_\mu}(y)|}{\sum_{i=1}^n |x_i - y_i|}$. Here, x_i represents the performance of pattern x w.r.t. criterion i.

4   Experiments

Rule Interestingness Measures. [17] identified seven independent families of measures having similar properties. We selected five criteria to evaluate the rules: Yule's Q,

Table 2. Characteristics of datasets used in the experiments and variables of our MIP model. $|\mathcal{T}|$ is the number of transactions of the dataset, $|\mathcal{I}|$ its number of items and ρ its density.

Dataset	$|\mathcal{T}|$	$|\mathcal{I}|$	ρ (%)	$|\mathcal{X}|$
Mushroom	8,124	119	18.75	7,363
Connect	67,557	129	33.33	312,033
Hepatitis	137	68	50.00	1,764
Retail	88,162	16,470	0.06	1,609

(a) Characteristics of the datasets.

Notations	Significance
$m_\mu(T)$	*Möbius variables*: represent the searched Möbius capacity for each set of criteria T ($T \subseteq N |T| \le k$)
$\mathcal{Y}(x)$	*Alt. choquet variables*: represent the ideal choquet value of the alternative x ($\forall x \in \mathcal{X}$)
$\mathcal{P}$	The set of user preferences
δ_C	The minimal indifference threshold used in equation (3)
m_μ^L	Learned Möbius function

(b) Variables of the MIP model .

Cosine, Goodman Kruskal's, Added Value, and Certainty Factor (see Suppl. Mat. A1). To ensure commensurability between criteria, we normalize them in the interval $[0, 1]$.

User Ranking Functions. We use various hidden ranking functions ψ to simulate user feedback as the target ranking. (1) LINEAR: The user ranking follows a random Weighted sum, where weights w_i are randomly generated in the interval $[0, 1]$ and sum up to 1; (2) CHISQUARED: The user ranking is defined by χ^2, a statistical measure assessing independence between a rule's antecedent and consequent; (3) CHOQUET-PEARSON: The user ranking is derived from a capacity function inspired by [14]. The singleton capacities are randomly generated, while the remaining capacities are computed based on correlations between two criteria using the Pearson coefficient.

Evaluation Metrics. Let R_T (resp. R_L) denotes the target (resp. learned) ranking for a given set of rules $\mathcal{X}$, i.e. $R_T(i)$ represents the i-th element of the target ranking. Let $topK(R)$ be the top-k rules w.r.t. R, i.e. $topK(R) = \{R(i) \mid i \in [1, k]\}$. To evaluate the effectiveness of our approach, we consider two well-known metrics: (i) the recall at k, defined as the proportion of rules in common between the top-k of the target and the learned ranking ($rec@k = \frac{|topK(R_L) \cap topK(R_T)|}{k}$), and the average precision $AP@k$ which evaluates how recall values change as k changes ($AP@k = \frac{1}{k}\sum_{i=1}^{k} rec@i$).

Experimental Protocol. We evaluate our approach on four real datasets from the FIMI repository (see Table 2a). The set of valid rules $\mathcal{X}$ is mined using a standard association rules algorithm without any knowledge about the user. Our ChoquetRank approach is implemented in Java[1]. We use the KAPPALAB package [9], implemented in R language, to solve our MIP model. We compare our approach to Learning-to-Rank techniques ListNet, RankNet, RankBoost [13], and with RankingSVM [10] and AHPRank [2]. Implementation of AHPRank was made available to us by the authors. All learning-to-rank techniques are from the open-source RANKLIB package. We used the standard implementation of RankingSVM. All experiments were conducted as single-threaded runs on Intel core i7, 1.80GHz × 8 with 31Gb of RAM using a timeout of **one hour**.

---

[1] Available (with all Supplementary Material) at https://gitlab.com/chaver/choquet-rank.

Table 3. Comparison of LTR algorithms (passive mode) in terms of $rec@k$ and $AP@k$ for different ranking functions: ψ_1: CHISQUARED, ψ_2: CHOQUETPEARSON, ψ_3: LINEAR. The size of the training set is fixed to $n = 50$ rules. (1) : AHPRank, (2) : RankingSVM, (3) : ChoquetRank.

	Learning algorithm	CONNECT Rec@ 1%	Rec@ 10%	AP@ 1%	AP@ 10%	RETAIL Rec@ 1%	Rec@ 10%	AP@ 1%	AP@ 10%	HEPATITIS Rec@ 1%	Rec@ 10%	AP@ 1%	AP@ 10%	MUSHROOM Rec@ 1%	Rec@ 10%	AP@ 1%	AP@ 10%
ψ_1	(1)	0.73	0.73	0.84	0.86	0.73	0.71	0.81	0.86	**0.78**	**0.93**	**0.87**	**0.98**	0.00	**0.47**	0.01	0.51
	(2)	0.74	0.71	0.84	0.84	0.33	0.62	0.39	0.71	0.71	0.90	0.80	0.97	0.00	**0.47**	0.01	**0.52**
	(3)	**0.83**	**0.89**	**0.91**	**0.96**	**0.99**	**0.95**	**1.00**	**0.99**	0.72	0.91	0.80	0.97	0.00	**0.47**	0.01	0.50
ψ_2	(1)	0.70	0.72	0.76	0.82	0.59	0.93	0.73	**0.99**	0.76	0.87	0.86	0.96	**0.96**	0.93	**0.99**	0.98
	(2)	0.61	0.70	0.67	0.80	0.40	0.83	0.37	0.93	0.74	0.87	0.85	0.95	**0.96**	0.92	**0.99**	0.98
	(3)	**0.91**	**0.97**	**0.97**	**1.00**	**0.80**	**0.96**	**0.86**	**0.99**	**0.96**	**0.99**	**1.00**	**1.00**	0.95	**0.98**	**0.99**	**1.00**
ψ_3	(1)	0.73	0.81	0.81	0.90	0.77	0.96	0.89	**0.99**	0.74	0.84	0.82	0.92	0.95	0.92	**0.99**	0.97
	(2)	0.64	0.78	0.68	0.87	0.61	0.85	0.60	0.95	0.80	0.86	0.87	0.94	0.96	0.93	**0.99**	0.98
	(3)	**0.89**	**0.97**	**0.96**	**1.00**	**0.84**	**0.94**	**0.86**	0.98	**0.93**	**0.99**	**0.98**	**1.00**	**0.97**	**0.99**	**0.99**	**1.00**

CONNECT and RETAIL.   HEPATITIS and MUSHROOM.

We use default parameter values in the implementations of learning-to-rank methods. We address the following questions: **Q1** How (in terms of $rec@k$, $AP@k$ and running time) does ChoquetRank compare with AHPRank, RankingSVM and with learning-to-rank methods in **passive learning**? **Q2** How does the size of the training data impacts the quality of the learning? **Q3** How does ChoquetRank compare with AHPRank and RankingSVM from an **active learning** perspective? **Q4** How robust the ChoquetRank is in the presence of user preference inconsistencies?

1) Passive Learning Results. To evaluate the efficiency of our approach regarding the research questions **Q1-Q2**, for each dataset, we perform a 5-fold cross-validation on $\mathcal{X}$ with different sample sizes as follows: in each fold, $n \in \{10, 20, 50, 100\}$ rules are selected uniformly at random to form the training data and we use the remaining rules (i.e., $(|\mathcal{X}| - n)$) for validation (testing data). Rules of the training data are ranked according to the (hidden) user preference function. All algorithms use the same training data. This procedure is repeated 5 times for each dataset; average metric values are reported using the test set to evaluate how well the results generalize to new data.

Q1-A) Comparing ChoquetRank with AHPRank and RankingSVM. In this experiment, we fix $n = 50$. Table 3 reports for each dataset and each approach, the average values of $rec@k$ and $AP@k$ ($k \in \{1\%, 10\%\}$), according to the different user-specific rankings. The results clearly show that ChoquetRank performs significantly better than AHPRank and RankingSVM in terms of recall ($rec@1\%$, $rec@10\%$) for all user ranking functions. On CONNECT dataset, we may notice that $rec@1\%$ values using ChoquetRank range approximately between 0.83 and 0.91, while for RankingSVM (resp. AHPRank) the values range between 0.61 and 0.74 (resp. between 0.7 and 0.73). Interestingly, ChoquetRank ensures an average precision of 1 at the 10% top of the ranking on almost all tested datasets. This indicates that our approach can identify the properties of the target ranking from ordered lists of rules. However, on MUSHROOM dataset, the three algorithms perform very similarly and none of them enabled to learn accurate rankings at the 1% top of the ranking with CHISQUARED.

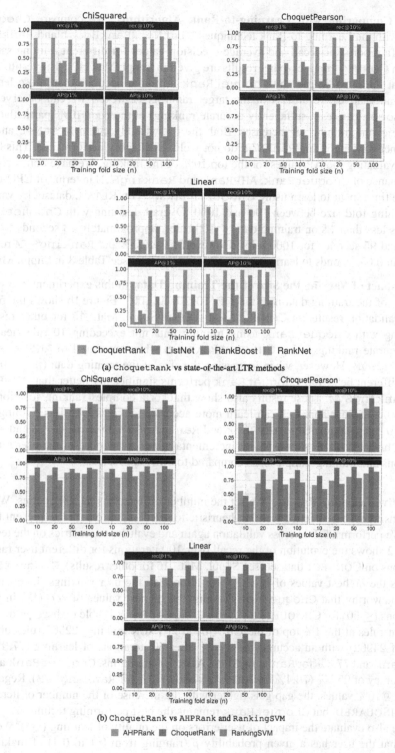

Fig. 1. Accuracy comparison on CONNECT dataset (passive learning): different training fold size.

Q1-B) Comparing with Learning-to-Rank Algorithms. We compare Choquet Rank with the Learning-to-Rank techniques ListNet, RankBoost and RankNet. Figure 1a reports the recall and average precision values for different training sample sizes on CONNECT dataset (other results are given in Suppl. Mat. $A2$). The results indicate that the ranks learned by ChoquetRank are more accurate than those learned by the alternatives, regardless of the target ranking. However, we can observe that RankBoost can learn sufficiently accurate rankings at the $rec@10\%$, particularly as the size of the training set increases. Both the list-wise algorithm ListNet and the neural network algorithm RankNet do not yield satisfactory performance. This holds when evaluating the AP values for the top 10% of the ranking. We have compared the performance of ChoquetRank, AHPRank and RankingSVM in terms of CPU-times (i.e. the time spent to learn using CHOQUETPEARSON) on RETAIL dataset by varying the training fold size between 100 and 1000. Overall, learning with ChoquetRank requires less than 1 s on training data ≤ 100 rules, approximately 13 seconds for 500 rules, and 95 seconds for 1000 rules. However, AHPRank and RankingSVM require less than 0.05 seconds to learn on the same training data (see Table 3 in Suppl. Mat.).

Q2: Impact of Varying the Size of the Training Data. In this experiment, we varied the size of the training data n in the set $\{10, 20, 50, 100\}$. Figure 1b shows the 5-fold cross-validation results on CONNECT test set (see Suppl. Mat. $A5$ for other results). Learning with ChoquetRank using training data not exceeding 10 rules leads to less accurate rankings using CHOQUETPEARSON when compared to AHPRank and RankingSVM. However, when increasing the size of the training data ($n \geq 20$), we see a different behavior: ChoquetRank performs significantly better than AHPRank and RankingSVM. These results also show that for a complex ranking function like CHISQUARED, our approach can learn more accurate rankings with less training data (only 10 rules) compared to AHPRank and RankingSVM. In terms of rec and AP at 10%, ChoquetRank shows higher incremental gains for almost all the user ranking functions and training sample sizes compared to other approaches.

(2) Active Learning Results. We set the number of iterations to 100 queries. We use our Sensitivity-based active learning heuristic to select a pair of rules to present to the user. We perform a 5-fold cross validation again and evaluate our metrics on the test set. Figure 2 shows the evolution of the recall over 100 iterations for different user ranking functions on CONNECT dataset (see Suppl. Mat. $A6$ for other results). ChoquetRank ensures the highest values of $rec@1\%$ for almost all the user rankings. Interestingly, it is noteworthy that ChoquetRank attains the highest values of $rec@1\%$ in a few iterations (≤ 50) for CHOQUETPEARSON: ChoquetRank is able to discover the most relevant rules at the 1% top of the CHOQUETPEARSON ranking (22565 rules under a total of 24962) with an accuracy of 90.4% after 50 iterations of learning (77.9% for AHPRank and 77.3% for RankingSVM). After 100 iterations, ChoquetRank attains an accuracy of 95.5% (76.4% for AHPRank and 79.5% for RankingSVM). Regarding the $rec@10\%$ values, the gap gets tighter with the increase of the number of iterations for CHISQUARED, but ChoquetRank remains the best performing technique.

We also evaluate the impact of user errors on the quality of learning (**Q4**). We suppose that the user has a given probability p (ranging from 0.1 to 0.4) of making an

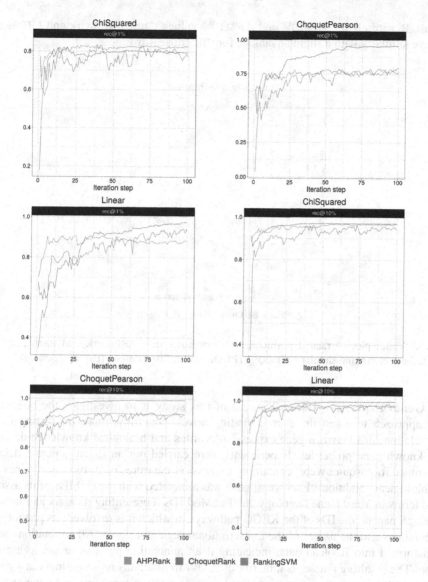

Fig. 2. Active learning results on CONNECT.

error, i.e. if she prefers a rule r^i to a rule r^j, she says the opposite. Figures 6–8 (see Suppl. Mat. $A7$) show the evolution of the recall over 100 learning iterations of CHO-QUETPEARSON on different datasets. ChoquetRank remains quite stable with 10% of errors: $rec@1\%$ drops from 95.5% (without errors) to 93.5% (with $p = 0.1$). When p increases to 0.2, the performance of ChoquetRank is highly impacted: $rec@1\%$ decreases from 93.5% to 81.5%. The decline in the performance of RankingSVM is even more pronounced particularly with 40% of errors, while AHPRank remains quite robust. Regarding the $rec@10\%$ and $AP@10\%$ values, ChoquetRank and AHPRank behave similarly with a slight advantage for ChoquetRank.

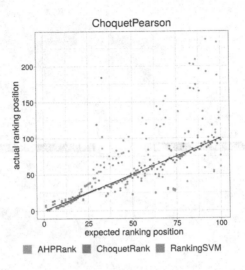

Fig. 3. Scatter plot of ranked positions for ChoquetRank, AHPRank and RankingSVM w.r.t. the user ranking simulator CHOQUETPEARSON on Eisen dataset (passive learning).

(3) Genes Expression Data. The aim of this study is to investigate the benefit of our approach to assist the user in finding novel, hopefully, more interesting rules, i.e. relationships between gene expression profiles and biological knowledge describing known gene properties. Experiments were carried out on Eisen genomic dataset (i3s.unice.fr/ pasquier/web/) containing expression measures of $2,465$ Yeast genes for 79 biological conditions. Each yeast gene was annotated with the GO IDs of its associated terms in Yeast Gene Ontology, the PubMed IDs representing its associations with research papers, the IDs of the KEGG pathways in which it is involved, its phenotypes annotations and the names of the transcriptional regulator genes. All annotations were transformed into Boolean data, indicating if an annotation pertains or not to a given gene. The resulting dataset is a matrix of $2,465$ transactions representing yeast genes and $9,634$ items representing discretized gene expression measures and gene annotations. We first extracted a set of association rules with $\theta = 1\%$ and $c = 80\%$, leading to $20,658$ rules. Then, we selected uniformly at random 100 rules as a training set and used the remaining rules for evaluation (test set). After training the model, we selected the 100 top rules of the ranking of each user-specific ranking function on the test set and compared their ranking position with the one given by the learned model. For instance, a point with the coordinates (x,y) means that the rule was ranked at position x by the user-specific ranking function and at position y by the learned model. Figure 3 depicts a scatter plot of ranked positions for ChoquetRank, AHPRank and RankingSVM w.r.t. the user ranking CHOQUETPEARSON. The line $y = x$ draws the case where both the learned and user-specific ranking functions obtain the same ranking for the 100 top rules. We can observe that the ranking given by `ChoquetRank` is almost perfect (i.e. all points are close to the line $y = x$). However, for `RankingSVM` and `AHPRank` most of the points are more dispersed on both sides of the line $y = x$, which means that the quality of the learned model is not as good as the one of `ChoquetRank`. Table 4 (see Suppl. Mat. $A4$) shows examples of rules that are mostly well ranked by `ChoquetRank`. For example, rule R#9 is ranked $56th$ by `ChoquetRank` while `AHPRank` ranks it at $83rd$ position. Interestingly, rules 7–8 and 10 do not appear in the top 100 rules produced by `AHPRank`. Rules 7–9 relate specific morphological phenotypes (defects) of essential genes in yeast with common GO annotation(s). Rule 10 states that genes responsible for utilizing non-fermentable carbon sources are involved in protein synthesis.

5 Conclusions

In this paper, we have proposed a framework for learning a non-linear ranking function based on the Choquet integral using user feedback. We formulated this problem as an optimization task, specifically using the GLS method and described two learning strategies. We focus our study on association rules discovery, in which the learned Choquet integral is used for ranking a set of rules. Experiments on UCI datasets have shown that our method, `ChoquetRank`, compares favorably in terms of accuracy with LTR methods and with `AHPRank` and `RankingSVM`. Scaling our approach toward large-size training data is an important topic of ongoing and future work.

References

1. Ailon, N.: An active learning algorithm for ranking from pairwise preferences with an almost optimal query complexity. J. Mach. Learn. Res. **13**, 137–164 (2012)
2. Belmecheri, N., Aribi, N., Lazaar, N., Lebbah, Y., Loudni, S.: Boosting the learning for ranking patterns. Algorithms **16**(5), 218 (2023)
3. Benabbou, N., Perny, P., Viappiani, P.: Incremental elicitation of Choquet capacities for multicriteria choice, ranking and sorting problems. Artif. Intell. **246**, 152–180 (2017)
4. Bresson, R., Cohen, J., Hüllermeier, E., Labreuche, C., Sebag, M.: Neural representation and learning of hierarchical 2-additive Choquet integrals. In: IJCAI 2020, pp. 1984–1991 (2020)
5. Choi, D.H., Ahn, B.S., Kim, S.H.: Prioritization of association rules in data mining: multiple criteria decision approach. Exp. Syst. Appl. **29**(4), 867–878 (2005)
6. Dzyuba, V., van Leeuwen, M., Nijssen, S., Raedt, L.D.: Interactive learning of pattern rankings. Int. J. Artif. Intell. Tools **23**(6), 1460026 (2014)
7. Grabisch, M., Roubens, M.: Application of the Choquet integral in multicriteria decision making. In: Grabisch, M., Murofushi, T., Sugeno, M. (eds.) Fuzzy Measures and Integrals - Theory and Applications, pp. 348–374. Physica Verlag (2000)
8. Grabisch, M.: The möbius transform on symmetric ordered structures and its application to capacities on finite sets. Discret. Math. **287**(1–3), 17–34 (2004)
9. Grabisch, M., Kojadinovic, I., Meyer, P.: A review of methods for capacity identification in Choquet integral based multi-attribute utility theory: applications of the Kappalab R package. Eur. J. Oper. Res. **186**(2), 766–785 (2008)
10. Joachims, T.: Optimizing search engines using clickthrough data. In: KDD 2002, July 2002, pp. 133–142. Association for Computing Machinery, New York (2002)

11. Kamishima, T., Kazawa, H., Akaho, S.: A survey and empirical comparison of object rank-ing methods. In: Fürnkranz, J., Hüllermeier, E. (eds.) Preference Learning, pp. 181–201. Springer, Heidelberg (2010). https://doi.org/10.1007/978-3-642-14125-6_9
12. Khannoussi, A., Olteanu, A., Labreuche, C., Meyer, P.: Simple ranking method using ref-erence profiles: incremental elicitation of the preference parameters. 4OR-Q J. Oper. Res. **20**(3), 499–530 (2022). https://doi.org/10.1007/s10288-021-00487-w
13. Li, H.: Learning to rank for information retrieval and natural language processing. In: Syn-thesis Lectures on Human Language Technologies. Morgan & Claypool (2011)
14. Nguyen Le, T.T., Huynh, H.X., Guillet, F.: Finding the most interesting association rules by aggregating objective interestingness measures. In: Richards, D., Kang, B.-H. (eds.) PKAW 2008. LNCS (LNAI), vol. 5465, pp. 40–49. Springer, Heidelberg (2009). https://doi.org/10.1007/978-3-642-01715-5_4
15. Rüping, S.: Ranking interesting subgroups. In: Danyluk, A.P., Bottou, L., Littman, M.L. (eds.) Proceedings of ICML 2009, vol. 382, pp. 913–920 (2009)
16. Saaty, T.L.: What is the analytic hierarchy process? In: Mitra, G., Greenberg, H.J., Lootsma, F.A., Rijkaert, M.J., Zimmermann, H.J. (eds.) Mathematical Models for Decision Sup-port, vol. 48, pp. 109–121. Springer, Heidelberg (1988). https://doi.org/10.1007/978-3-642-83555-1_5
17. Tan, P.N., Kumar, V., Srivastava, J.: Selecting the right objective measure for association analysis. Inf. Syst. **29**(4), 293–313 (2004)
18. Tehrani, A.F., Cheng, W., Dembczynski, K., Hüllermeier, E.: Learning monotone nonlinear models using the Choquet integral. Mach. Learn. **89**(1–2), 183–211 (2012)
19. Yager, R.R.: On ordered weighted averaging aggregation operators in multicriteria decision-making. IEEE Trans. Syst. Man Cybern. **18**(1), 183–190 (1988)

Saliency-Aware Time Series Anomaly Detection for Space Applications

Sangyup Lee [ID] and Simon S. Woo[✉] [ID]

Department of Computer Science and Engineering, Sungkyunkwan University, Suwon, South Korea
{sangyup.lee,swoo}@g.skku.edu

Abstract. Detecting anomalies in real-world multivariate time series data is challenging due to the deviation between the distributions of normal and anomalous data. Previous studies focused on capturing time and spatial features but lacked an effective criterion to measure differentiation from normal data. Our proposed method utilizes saliency detection, similar to anomaly detection, to identify the most significant region and effectively detect abnormal data. In this work, We propose a novel framework, Saliency-aware Anomaly Detection (SalAD), for detecting anomalies in multivariate time series data. SalAD comprises three main components: 1) a saliency detection module to remove redundant data, 2) an unsupervised saliency-aware forecasting model, and 3) a saliency-aware anomaly score to differentiate anomalies. We evaluate our model using the real-world Korea Aerospace Research Institute (KARI) orbital element dataset, which includes six orbital elements and unexpected disturbances from satellites, as well as conducting extensive experiments on four benchmark datasets to demonstrate its effectiveness and superiority over other baselines. The SalAD framework has been deployed on the K3A and K5 satellites.

Keywords: Satellite Orbit Maneuver Detection · Multivariate Anomaly Detection · Time Series Analysis

1 Introduction

Over the past few decades, numerous artificial objects, including satellites, spacecraft, telescopes, and probes, have been launched to outer space, resulting in a crowded Earth orbit [14]. Occasionally, overcrowding on the Earth's outer orbit with artificial objects and space debris forces the satellite's operating system to alter the planned orbital path in consideration of unforeseen factors, e.g., changes in air density and collisions with space debris. Furthermore, such unexpected disturbances that alter the satellite's orbital path may force the satellite to fall back to Earth or lose its capacity to carry out its mission. Therefore, detecting unexpected disturbances for appropriate avoidance maneuvers is crucial and essential to prevent collisions for a long-term satellite mission objective.

© The Author(s), under exclusive license to Springer Nature Singapore Pte Ltd. 2024
D.-N. Yang et al. (Eds.): PAKDD 2024, LNAI 14645, pp. 327–339, 2024.
https://doi.org/10.1007/978-981-97-2242-6_26

In particular, anomalous behavior, i.e., unexpected disturbances, can be detected through multivariate time series anomaly detection by utilizing six orbital elements with additional altitude data. However, similar to multivariate time series anomaly detection, many difficult challenges arise in detecting unanticipated disruptions with real-world data. Since exploring historical patterns and determining the correlations between orbital components are important, one of the major challenges of multivariate time series comes in designing a model to capture both time series correlations and temporal dependencies. Also, as enormous amounts of data are generated in real-time, manual labeling becomes impracticable due to the associated high cost. To alleviate such challenges, many proposed unsupervised anomaly detection models utilize the normal patterns during the training process to capture both correlations and temporal dependencies of normal instances, allowing the model to differentiate from abnormal instances. On the other hand, since visual saliency detection is essentially a similar task, a more practical approach using saliency detection for time-series anomaly detection has been proposed [13] for real-world application. Visual saliency is widely used in computer vision tasks to identify the most "stand out" region of an image. By utilizing the Spectral Residual (SR) [5] method, Ren et al. [13] first proposed SR-CNN to bring the visual saliency detection to anomaly detection in the time series domain. The suggested model seemed promising for time series anomaly detection as a saliency map can provide generalized detection capability to various time series patterns. However, since SR-CNN is designed for univariate anomaly detection, this approach is unsuitable for multivariate time series anomalous behaviors, which is collectively caused by a plethora of connected sensors.

To address the aforementioned problems, we propose a novel and effective multivariate saliency-aware anomaly detection model, SalAD, a framework that focuses on the most salient part of the multivariate time series data. The main technical contributions of our SalAD include three major components: 1) saliency detection module, 2) saliency-aware forecasting model, and 3) saliency-aware anomaly score. In this paper, we utilize the orbital data from Korea Aerospace Research Institute (KARI), a national research center in South Korea that researches and develops aerospace scientific technologies. To effectively perform prior collision avoidance maneuvers and predict the new KARI satellites' unexpected orbital disturbances and changes, we apply the proposed SalAD for evaluation.

Our main contributions are summarized as follows:

- We propose SalAD, a novel Saliency-aware Anomaly Detection framework for multivariate time series that does not rely on saliency labels, ensuring a realistic training process, generating the saliency-aware anomaly score to improve performance.
- Extensive experiments on several benchmark datasets and real-world satellite precision orbit elements (ELE) dataset from the KARI database demonstrate that our SalAD architecture achieves state-of-the-art performance.

– The SalAD framework has been deployed on the K3A and K5 satellites, with plans to extend its application to the K3 and K6 missions in 2024. Our code is available here[1].

2   Background and Related Work

Background on Satellite Orbital Elements. KARI has successfully launched multipurpose satellites, Korea Multi-Purpose Satellite 5 (K5) and 3A (K3A), in 2013 and 2015, respectively [7]. The main objectives of K5 and K3A are ground observations on the Korean Peninsula for national safety, resource management, and environmental monitoring by providing high-resolution overhead imagery. While satellites serve numerous crucial missions, satellite missions are highly vulnerable to system failures from various unpredictable orbital changes. Thus, it is essential to take proactive actions by anomaly detection for spacecraft health monitoring systems. In this work, to detect the root causes of the orbital disturbances in satellites, we utilize precise orbital elements of K5 and K3A represented as six variables collected from Jan. 2016 to Dec. 2020. Moreover, we calculate the altitude from the semi-major axis for additional data. The orbital elements and the calculated altitude consist of the following seven variables: **1) Semi-major axis** (α). The sum of periapsis and apoapsis distances is divided by two for the altitude calculation. **2) Eccentricity** (ϵ). The degree to which the conical curve deviates from the circle for the shape and size of the orbit. **3) Inclination** (ι). The angle between the equatorial plane and the orbital plane indicates the degree to which the satellite's orbit is tilted toward the Earth. **4) Right ascension of the ascending node** (Ω). Right ascension of the ascending node (RAAN) represents the angle between the first point of Aries and the ascending node of the ellipse to define the orientation of the orbital plane. **5) Argument of perigee** (ω). Argument of perigee defines the ellipse's orientation in the orbital plane as an angle measured from the ascending node to the perigee. **6) Mean anomaly** (v). The angular distance between a satellite and the perigee represents the satellite's position in orbit. **7) Altitude** (δ). Altitude represents the distance between the Earth's surface and the satellite. The altitude values are calculated as follows, where 6378.137 represents the diameter of Earth in kilometers: $\delta = \frac{\alpha}{1000} - 6378.137$.

Anomaly Detection (AD) Methods. Generally, there are reconstruction and forecasting-based anomaly detection methods. First, reconstruction-based methods presume that anomalies are substantially more difficult to reconstruct properly than normal samples. To mitigate this problem, various methods attempt to employ long-term-based time series information and features. Past reconstruction-based anomaly detection utilizes autoencoders [4] and Long Short-Term Memory (LSTM)-Encoder-Decoder [4] for reconstruction-based anomaly identification. An et al. [1] used Variational Autoencoders (VAE), which outperformed autoencoder-based and principle component-based algorithms in

---

[1] https://github.com/Clench/SalAD.

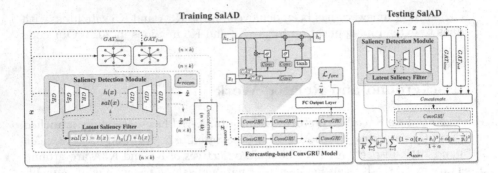

Fig. 1. Overall architecture of SalAD divided into training and testing phases.

evaluations; however, this method has a limitation that it cannot explicitly identify the relationship between different time series variables. On the other hand, forecasting-based methods typically utilize historical and timestamped data to forecast or predict future anomaly behaviors at any point in time. To solve the problem, models such as Recurrent Neural Network (RNN)-based [13], LSTM-based [6], Convolutional LSTM-based, and graph attention network is used to predict future time based on given past time-series data [23]. TranAD [18] adopts transformer architecture to time series anomaly detection that utilizes self-conditioning and adversarial training. These proposed methods support the model to gain training stability and amplify prediction errors, respectively. Zhao et al. [23] utilized the time and feature-oriented graph attention layer to model spatio-temporal dependencies to forecast future values in time.

Saliency Detection in Time Series. Despite the aforementioned approaches focusing on capturing feature correlations and temporal dependencies, the saliency detection method that makes the model focus on the crucial part of the data has not been widely used in multivariate time series. Visual saliency detection is a computer vision technique that seeks to identify the most important objects in an image. Ren et al. [13] first introduced the application of saliency detection to the time series field, where anomaly detection is essentially similar to visual saliency detection. SR-CNN [13] applied Spectral Residuals [5] to calculate the saliency map, and the 1D-CNN captured the features of normal patterns to differentiate the abnormal behaviors. However, Spectral Residual approach [13] can only be applied to univariate time series anomaly detection, making it challenging to model the feature correlations of multivariate time series.

3   Methodology

In this section, we describe our proposed Saliency-aware Anomaly Detection (SalAD) framework, consisting of three major components: 1) saliency detection module, 2) saliency-aware forecasting model, and 3) saliency-aware anomaly score. Both 1) and 2) components are trained simultaneously in an unsupervised fashion, using datasets divided into training and testing datasets.

Saliency Detection Module. For vision saliency detection, the majority of the outperforming state-of-the-art methods [19,22] require labels indicating the salient part of the image during the training process. However, labeling tasks are not practical for the real-world time series data, which brought various alternative methods such as unsupervised methods for time series anomaly detection. Thus, this work utilizes a similar approach from the Spectral Residual-based vision saliency detection [5]. As the SR-based detection method does not require any salient part labeling of the dataset, it is highly efficient to be adopted in the time series domain [13]. The main function of the SR method is to reduce the redundant information of the given sample by embedding the sample to the frequency domain to form a log spectrum using Fourier Transform. Afterward, the spectral residuals can be calculated with the following equation: $\mathcal{R}(f) = L(f) - AL(f)$, where $\mathcal{R}(f)$ is the spectral residuals, $L(f)$ is the log spectrum of the sample, and $AL(f)$ is the average log spectrum. The average log spectrum is treated as the statistical invariant properties, leaving the innovative part of the sample in the spectral residuals [5]. However, since the SR method can only be applied to univariate time series [13], we expand the residual concept by converting the Fourier transform and inverse Fourier transform to Gated Recurrent Unit (GRU)-based encoder and decoder, respectively. Then, we are able to use multivariate time series datasets as inputs to the model and calculate the new saliency map. Furthermore, we introduce a latent saliency filter to calculate the latent residuals, as shown in Fig. 1. Both the encoders and decoders in our saliency detection module are composed of l GRU layers. The recurrent encoder reduces the number of features in the input x and maps it to the latent space, resulting in a vector $h(x)$. Input x has the shape $n \times K$, where n is the length of the time series sequence, and K is the number of features. Similar to the previous reconstruction-based anomaly detectors [4], we reconstruct $h(x)$ to acquire the reconstructed $\hat{x}$. Furthermore, we train our autoencoder in the saliency detection module to minimize the reconstruction error $\mathcal{L}_{recon} = \frac{1}{K}\sum_{i=1}^{K}\|x_i - \hat{x}_i\|^2$, where we utilize the mean square error loss function. Not only can we reconstruct input x for training the autoencoder, but we can also generate a saliency map by building a latent saliency filter, as shown in the red box in Fig. 1. To minimize redundant and invariant information using the latent space vector $h(x)$, we use the latent saliency filter that outputs latent saliency vector $sal(x)$. Then, we pass $sal(x)$ to the decoder to generate the saliency map $\hat{x}^{sal}$:

$$\hat{x}_i = GD(h(x)), \tag{1}$$

$$\hat{x}_i^{sal} = GD(sal(x)), \tag{2}$$

$$A(x) = h_q(f) * h(x), \tag{3}$$

$$sal(x) = h(x) - A(x), \tag{4}$$

where GD indicates the GRU decoder. And, $h_q(f)$ is the average filter, $h_q(f) = \frac{1}{q^2}[I]$, where the filter is a kernel with a size of $q \times q$, and $[I]$ is a matrix of ones. We calculate the latent saliency vector $sal(x)$ by subtracting the average latent vector $A(x)$ which is approximated by convoluting ($*$) the latent vector $h(x)$ with the average filter $h_q(f)$.

Saliency-Aware ConvGRU Net. In real-world time series anomaly detection, such as satellite orbital disturbance detection, training a model with a supervised technique is challenging due to the lack of anomalies in the dataset. Therefore, we follow an unsupervised anomaly detection process where the model is trained on a dataset without anomalies and is evaluated with the data that includes abnormal behaviors; thus, we utilize a multi-step forecasting-based model in SalAD to predict future time series, and calculate the error to generate anomaly scores. The main objective of a forecasting-based model is to minimize the forecasting error during training and to maximize the error during testing phase where abnormal data is present. In this work, our SalAD implements the forecasting error in the anomaly score. For training the forecasting model, we utilize the mean square error as a loss function. Another essential task when detecting unexpected anomalies in real-world time series data is capturing spatio-temporal dependencies among multivariate time series data. To achieve this, we use Convolutional GRU (ConvGRU) [3] cells in our saliency-aware forecasting model to jointly capture spatio-temporal dependencies as shown in Fig. 1. Moreover, we employ feature-oriented and time-oriented graph attention layers (GAT) [23] to explicitly capture spatio-temporal features and to improve the effectiveness of the anomaly detection model while training both reconstruction (saliency detection module) and forecasting-based models (ConvGRU). Our ConvGRU forecasting model is comprised of two ConvGRU layers and a fully connected layer that outputs the number of features in x. The model forecasts future sequences $\overrightarrow{y}$ of length m. To create a model that concentrates on the salient portion of the time series, we execute a feature-wise concatenation of x and $\hat{x}^{sal}$. Additionally, to jointly train the captured spatio-temporal features, we concatenate the feature and time-oriented GAT outputs to create a vector x^{concat} that is passed to the ConvGRU model: $x^{concat} = x \oplus \hat{x}^{sal} \oplus GAT_{feat}(x) \oplus GAT_{time}(x)$, where $\oplus$ indicates feature-wise vector concatenation. GAT_{feat} and GAT_{time} indicates the features and time-oriented graph attention layers respectively. We set the final objective function of SalAD as: $\mathcal{L}_{total} = \mathcal{L}_{recon} + \mathcal{L}_{fore}$.

Saliency-Aware Anomaly Score. When previously unseen data points are observed, we pass through to the pretrained SalAD model, which generates the observed sample's reconstruct (through the saliency detection module) and forecast (by the saliency-aware ConvGRU model) error values. The saliency detection module reconstructs the predicted value $\hat{x}$ and compares it to the original input x. A second criterion comes from the saliency-aware ConvGRU model, which forecasts y given the input x^{concat}. Then, $\overrightarrow{y}$ is compared to the observed sequence value y, with equivalent sequence length of m. The errors for each feature are calculated using the squared error ($(x_i - \hat{x}_i)^2$ and $(y_i - \overrightarrow{\hat{y}}_i)^2$), and then we sum the errors for all features. As both models try to predict values within the distribution of the training dataset, models will generate high errors when abnormal data is observed. To compensate for the high errors, we use the anomaly score $\mathcal{A}_{score}$ as follows:

$$\mathcal{A}_{score} = \frac{1}{K}\sum_{i=1}^{K}\hat{x}_i^{sal} \cdot \sum_{i=1}^{K}\frac{(1-\alpha)(x_i-\hat{x}_i)^2+\alpha(y_i-\hat{\overline{y}}_i)^2}{1+\alpha}, \tag{5}$$

where α is an adjustable parameter for balancing the saliency detection module and saliency-aware ConvGRU model. Since the saliency map can capture the holistic view and the innovative part of the multivariate time series, as shown as an example in Fig. 1, we also apply our saliency map $\hat{x}^{sal}$ to the anomaly score. We take the mean of $\hat{x}^{sal}$ for all features and multiply it with the balanced error to finalize our anomaly score $\mathcal{A}_{score}$. We conduct an ablation study on α value, and the effectiveness of utilizing $\hat{x}^{sal}$.

4   Experiments

Datasets.1) KARI-ELE Dataset. We evaluate SalAD and the baseline models with the KARI satellite precision orbit Elements (ELE) dataset consisting of 6 orbital elements and altitude collected by K3A and K5 operations from Jan 2016 to Dec 2020. Specifically, the ELE dataset includes 2,191,681 samples at a minute interval. We split the ELE dataset into training and testing sets. The training set consists of 439,201 samples collected from Jan 2016 to Dec 2016, and the testing set consists of 1,752,480 samples collected from Jan 2017 to Dec 2020. Since the primary purpose of SalAD is to detect anomalies by training with normal samples only, we utilize the samples collected in 2016 as a training set as fewer unexpected disturbances are observed during the period. We sample data from 2017 to 2020 to evaluate the anomaly detection performance for the testing set.

2) Benchmark Datasets. We use four multivariate time series anomaly detection benchmark datasets, namely MSL, SMAP [12], SMD [6], and SWaT [11], to comparatively analyze the effectiveness of our model, SalAD, with baseline models.

Baseline Methods. To verify the effectiveness of our model, we compare SalAD with state-of-the-art models for multivariate time-series anomaly detection on the benchmark datasets and the ELE dataset. The baseline models include Tele-mAnom [6], TelemAnom with GRU cells, OmniAnomaly [17], USAD [2], MTAD-GAT [23], and MSCRED [21]. Note that we do not compare with the original saliency detection in time series model SR-CNN [13] because it is inapplicable to multivariate time series. We also assess the multivariate anomaly detection performance with classical approaches, IF, OCSVM, and Vector Auto Regression (VAR) [16].

SalAD Implementation. For the saliency detection module, we set the number of GRU encoder and decoder layer l as 2, and the number of hidden dimensions as $\lfloor K/2 \rfloor$ where K is the number of features of the given multivariate time series dataset. Latent saliency filter kernel size q is set to 3. Output shapes of saliency

Table 1. Performance comparison on multivariate time series anomaly detection benchmark datasets. PA F1 and AUC denote F1-score with point adjustments and AUROC, respectively.

Methods	MSL (%)				SMAP (%)				SMD (%)			
	Pre	Re	PA F1	AUC	Pre	Re	PA F1	AUC	Pre	Re	PA F1	AUC
Isolation Forest [8]	75.37	30.11	43.02	50.69	99.58	49.53	66.15	48.07	68.93	87.53	68.13	72.04
OCSVM [10]	53.65	96.87	69.06	52.50	87.90	56.34	68.66	39.27	64.72	79.34	69.02	77.18
VAR [16]	91.32	51.20	65.61	60.59	98.82	25.79	40.91	44.52	68.82	79.34	74.24	73.51
TelemAnom (LSTM) [6]	59.00	77.97	67.17	59.83	98.48	24.70	39.49	41.96	39.48	84.88	51.19	72.59
TelemAnom (GRU) [6]	30.80	93.75	46.36	60.01	85.61	56.37	67.98	41.17	45.67	84.90	53.66	72.85
OmniAnomaly [17]	52.92	96.87	68.45	53.10	98.30	52.98	68.85	45.11	95.14	63.01	70.21	76.26
USAD [2]	31.23	75.25	44.14	65.62	63.83	47.39	54.39	39.73	20.83	73.30	32.15	67.71
MTAD-GAT [23]	46.36	99.34	63.21	66.00	9.85	52.83	69.10	48.13	62.08	61.71	60.82	75.91
MSCRED [21]	47.36	96.87	63.62	62.99	98.36	53.56	69.35	39.45	71.94	71.24	69.42	81.77
SalAD (Ours)	85.64	58.13	69.25	67.30	92.08	55.77	69.47	48.23	58.83	84.88	67.34	76.58

Methods	SWaT (%)				KARI K3A (%)				KARI K5 (%)			
	Pre	Re	PA F1	AUC	Pre	Re	PA F1	AUC	Pre	Re	PA F1	AUC
Isolation Forest [8]	99.99	65.73	79.32	71.37	10.85	33.33	16.38	49.14	97.88	56.45	71.61	51.29
OCSVM [10]	99.84	66.27	79.66	79.26	43.05	16.67	24.03	52.44	40.85	60.75	48.85	52.54
VAR [16]	99.29	71.47	79.66	70.46	4.15	100.0	7.97	50.42	28.86	100.0	44.79	51.26
TelemAnom (LSTM) [6]	81.43	76.15	78.70	79.01	34.14	16.67	22.40	55.54	41.27	45.16	43.13	55.18
TelemAnom (GRU) [6]	87.82	74.82	80.80	79.21	1.06	83.33	2.10	55.54	55.63	45.97	50.34	55.13
OmniAnomaly [17]	45.89	88.82	60.51	74.56	83.48	16.67	27.79	56.28	78.46	76.64	77.54	53.36
USAD [2]	100.0	65.73	79.32	78.86	3.47	16.67	5.75	47.08	20.97	31.78	25.27	54.55
MTAD-GAT [23]	99.16	73.48	84.41	35.19	11.91	99.99	21.28	61.97	97.62	31.73	47.89	55.14
MSCRED [21]	20.19	78.48	32.12	50.37	29.00	16.67	21.17	59.93	26.39	77.57	39.39	53.23
SalAD (Ours)	95.22	71.40	81.61	79.21	38.55	33.33	35.75	63.56	59.86	89.52	71.74	55.22

Table 2. Ablation study on the effectiveness of saliency module (SM), graph attention modules (GATs), and anomaly score $\mathcal{A}_{score}$ with $\hat{x}_{sal}$ (PA F1).

Methods	MSL	SMAP	SMD	SWaT	K3A	K5
SalAD	**69.25**	**69.47**	**67.34**	**81.61**	**35.75**	**71.74**
w/o SM	60.18	64.72	57.21	80.38	23.15	70.16
w/o SM, GATs	61.36	61.89	52.56	78.56	22.14	62.56
w/o $\hat{x}_{sal}$	61.22	67.84	62.46	80.33	25.00	71.65

detection module, feature-oriented GAT, and time-oriented GAT have the equal size to x, $n \times K$, where n is the length of the input sequence. This forms a vector with size of $n \times 4K$ for x^{concat}. The ConvGRU model consists of two consecutive ConvGRU layers with the hidden dimension of 32 and 64. The last linear layer after the ConvGRU cells reduces the number of features to K. Finally, we set the output sequence length m of the saliency-aware ConvGRU model as 10.

Evaluation Metrics. Conventionally, many state-of-the-art models [2] assess the anomaly detection performance using precision, recall, and F1-score with a predefined threshold (e.g., Peaks Over Threshold (POT) [15] or brute force threshold search). However, such methods can be biased since the performances are highly volatile depending on the parameter value of POT. In addition, model performance is evaluated [2] using latency and sparsity-aware evaluation methods such as Point adjustment (PA) [20]. PA considers the entire segment as true positives if at least one observation is detected as an anomaly in the segment, which can also be highly biased since PA can switch false positives to true positives. Likewise, both evaluation techniques have a great possibility of overestimating the models' performances, leading to misguided baseline comparison. Therefore, in addition to precision, recall, and F1-score with PA, we also evaluate the performance of SalAD and baseline models with Area Under the Receiver Operating Characteristic (AUROC) [9], which is calculated at various threshold settings. We utilize each model's inference or anomaly scores and test possible anomaly thresholds with POT using the validation dataset (30% of the training dataset) to calculate the highest PA F1-score for the fair evaluation.

5 Results

Anomaly Detection Performance. The performance of anomaly detection on baselines and SalAD on various datasets are reported in Table 1, where the best performance in terms of AUROC (AUC) and point adjustment F1-score (PA F1) are indicated in bold. The second-best performance is underlined. In Table 1, SalAD outperforms most of the datasets compared to the baseline methods, demonstrating its effectiveness in distinguishing abnormal behaviors from normal patterns by explicitly training with saliency-aware architecture. Specifically, SalAD achieves state-of-the-art results in the space-related domain, outperforming the best baseline on both PA F1-score and AUROC on MSL and SMAP datasets from NASA. Furthermore, on the real-world satellite orbital disturbance detection dataset, KARI-ELE K3A, SalAD outperforms other models with 63.56% AUROC and 35.75% PA F1-score. However, most of the models perform poorly on this dataset compared to other datasets due to the seasonality effect caused by the "Inclination" of the K3A dataset, where the data values increase over time. Since all models adopt the unsupervised anomaly detection training strategy where the model is trained with a normal operation period, we only utilize the first year of the ELE data where there are no orbital disturbances. As a result, anomaly scores from the models gradually increase throughout the K3A test dataset, making it challenging to find the optimal static threshold. For the KARI-ELE K5 dataset, SalAD shows on-par performance with 55.22% AUROC. Although OmniAnomaly reports the best performance in the K5 dataset in terms of PA F1-score with 77.54%, our SalAD model demonstrates the second-best performance and outperformed all other baseline models. Another important observation is that the result from the KARI-ELE K5 dataset shows relatively lower AUROC than results from the K3A dataset. This phenomenon occurs because the training set in the K5 dataset includes more unexpected orbital disturbances, making the model difficult to differentiate between normal satellite orbital paths and disturbances.

Effectiveness of the Framework. To verify the effectiveness of the saliency detection module and the saliency-aware forecasting model in improving the

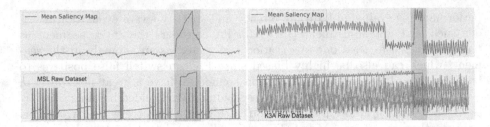

Fig. 2. Examples of mean saliency map of $\hat{x}_{sal}$, from different datasets. The blue lines of the upper graphs illustrate the $\hat{x}_{sal}$, and the bottom is the raw multivariate time series in the same period. (Color figure online)

anomaly detection performance, we conduct an ablation study with the vanilla forecasting-based backbone model, ConvGRU, as shown in Table 2. We demonstrate the effectiveness of utilizing the saliency module as the most significant improvement of 13.61% PA F1-score compared to the vanilla backbone model ("w/o SM, GATs") on a real-world dataset, KARI-ELE K3A. In addition, we observed that utilizing both the saliency module and GATs effectively improve the overall performance and reported the best performance from all the datasets.

Saliency Detection in Time Series. We illustrate the effectiveness of the saliency module in capturing the most important regions in the time series domain by plotting mean saliency maps, $\frac{1}{K}\sum_{i=1}^{K}\hat{x}_i^{sal}$, in the blue line, as shown in Fig. 2. To visualize the location of anomalies, we highlight the anomaly period as red boxes. The bottom graphs of the five examples are the raw SMD, MSL, and KARI-ELE K3A datasets, and the red box indicates the abnormal period. From Fig. 2, we can clearly observe that the saliency maps can capture an innovative and essential part of multivariate time series. As shown in the first sample of Fig. 2, the dynamic part of the saliency map in the blue line demonstrates the significance of anomalies compared to the period where there are no fluctuations in the data. Furthermore, for the regions of anomalies, we can observe high peaks in the mean saliency map, which reflects the differentiating capabilities between the area of anomalies and the normal time period of the saliency map.

Saliency-Aware Anomaly Score. To validate the effectiveness of the saliency map in the enhancement of detection performance at inference time, we conduct an ablation study on utilizing the saliency-aware anomaly score, where we multiply the mean saliency map $\frac{1}{k}\sum_{k}^{i=1}\hat{x}_i^{sal}$ to the anomaly score $\mathcal{A}_{score}$ in Eq. 5. In Table 2, we report the PA F1-score performances from the SalAD's anomaly score $\mathcal{A}_{score}$ with and without $\hat{x}_i^{sal}$. The results show an increase in PA F1-score on all the datasets. Thus, we believe that utilizing the mean saliency map improves anomaly detection performance and introduces a holistic view of abnormal behaviors. The biggest PA F1-score gain from all datasets with the mean saliency map was 10.75% on the KARI-ELE K3A dataset. Additionally, we perform an ablation study on all benchmark and KARI-ELE datasets with different α values in $\mathcal{A}_{score}$, [0.0, 0.1, 0.3, 0.5, 0.7, 0.9, 1.0]. On datasets SMAP,

SMD, SWaT, KARI-ELE K3A, and K5, SalAD achieved higher detection performance when the α value was high (from 0.7 to 1.0) in terms of PA F1-score. This indicates that errors generated from the forecasting-based ConvGRU model $((y_i - \hat{\overrightarrow{y}}_i)^2)$ are much more effective in differentiating the anomalies compared to errors from the reconstruction-based GRU model $((x_i - \hat{x}_i)^2)$. However, for the MSL dataset, a combination of both reconstruction and forecasting-based showed relatively higher performance. To conclude, we observe that the α value varies and is sensitive to dataset selection. Thus, we plan to optimize the anomaly score for more generalizability in future work.

6   Conclusion

In the space environment, detecting the abnormal orbital disturbances of the satellite is one of the critical tasks to carry out for long-term operation and satellite health monitoring. In this paper, we proposed a novel multivariate time series anomaly detection model, SalAD, a saliency detection-based technique with the ConvGRU forecasting model as the backbone. We applied our SalAD in a real-world multivariate time series dataset, KARI-ELE, from KARI satellite orbital data to verify the model's effectiveness and four benchmark datasets to demonstrate the generalization capabilities of the proposed model. Through extensive experiments, we demonstrated the effectiveness of SalAD on diverse multivariate time series anomaly datasets, and further it is planning to be deployed to support the real-world space operation.

Acknowledgements. This work was partly supported by Institute for Information & communication Technology Planning & evaluation (IITP) grants funded by the Korean government MSIT: (No. 2022-0-01199, Graduate School of Convergence Security at Sungkyunkwan University) (No. 2022-0-01045, Self-directed Multi-Modal Intelligence for solving unknown, open domain problems) (No. 2022-0-00688, AI Platform to Fully Adapt and Reflect Privacy-Policy Changes) (No. 2021-0-02068, Artificial Intelligence Innovation Hub) (No. 2019-0-00421, AI Graduate School Support Program at Sungkyunkwan University), and (No. RS-2023-00230337, Advanced and Proactive AI Platform Research and Development Against Malicious Deepfakes). Lastly, this work was supported by Korea Internet & Security Agency (KISA) grant funded by the Korea government (PIPC) (No.RS-2023-00231200, Development of personal video information privacy protection technology capable of AI learning in an autonomous driving environment).

References

1. An, J., Cho, S.: Variational autoencoder based anomaly detection using reconstruction probability. Spec. lect. IE **2**(1), 1–18 (2015)
2. Audibert, J., Michiardi, P., Guyard, F., Marti, S., Zuluaga, M.A.: USAD: unsupervised anomaly detection on multivariate time series. In: Proceedings of the 26th ACM SIGKDD International Conference on Knowledge Discovery & Data Mining, pp. 3395–3404 (2020)

3. Ballas, N., Yao, L., Pal, C., Courville, A.: Delving deeper into convolutional networks for learning video representations. arXiv preprint arXiv:1511.06432 (2015)
4. Hagemann, T., Katsarou, K.: Reconstruction-based anomaly detection for the cloud: a comparison on the yahoo! webscope S5 dataset. In: Proceedings of the 2020 4th International Conference on Cloud and Big Data Computing, pp. 68–75. ICCBDC 2020, Association for Computing Machinery, New York, NY, USA (2020). https://doi.org/10.1145/3416921.3416934
5. Hou, X., Zhang, L.: Saliency detection: a spectral residual approach. In: 2007 IEEE Conference on Computer Vision and Pattern Recognition, pp. 1–8. IEEE (2007)
6. Hundman, K., Constantinou, V., Laporte, C., Colwell, I., Soderstrom, T.: Detecting spacecraft anomalies using LSTMs and nonparametric dynamic thresholding. In: Proceedings of the 24th ACM SIGKDD International Conference On Knowledge Discovery & Data Mining, pp. 387–395 (2018)
7. KARI: Satellite information database. https://ksatdb.kari.re.kr/satIntro.do (2015). Accessed 21 Jan 2022
8. Liu, F.T., Ting, K.M., Zhou, Z.H.: Isolation forest. In: 2008 Eighth IEEE International Conference on Data Mining, pp. 413–422. IEEE (2008)
9. Mandrekar, J.N.: Receiver operating characteristic curve in diagnostic test assessment. J. Thorac. Oncol. 5(9), 1315–1316 (2010)
10. Manevitz, L.M., Yousef, M.: One-class svms for document classification. J. mach. Learn. res. 2(Dec), 139–154 (2001)
11. Mathur, A.P., Tippenhauer, N.O.: SWaT: a water treatment testbed for research and training on ICS security. In: 2016 International Workshop on Cyber-physical Systems for Smart Water Networks (CySWater), pp. 31–36. IEEE (2016)
12. O'Neill, P., Entekhabi, D., Njoku, E., Kellogg, K.: The NASA soil moisture active passive (SMAP) mission: Overview. In: 2010 IEEE International Geoscience and Remote Sensing Symposium, pp. 3236–3239. IEEE (2010)
13. Ren, H., et al.: Time-series anomaly detection service at Microsoft. In: Proceedings of the 25th ACM SIGKDD International Conference on Knowledge Discovery & Data Mining, pp. 3009–3017 (2019)
14. Rigby, M.: A chunk of satellite almost hit the ISS, requiring an 'urgent change of orbit' (2021). https://www.sciencealert.com/a-chunk-of-Chinese-satellite-almost-hit-the-international-space-station Accessed on 25 Nov 2021
15. Siffer, A., Fouque, P.A., Termier, A., Largouet, C.: Anomaly detection in streams with extreme value theory. In: Proceedings of the 23rd ACM SIGKDD International Conference on Knowledge Discovery and Data Mining, pp. 1067–1075 (2017)
16. Stock, J.H., Watson, M.W.: Vector autoregressions. J. Econ. perspect. 15(4), 101–115 (2001)
17. Su, Y., Zhao, Y., Niu, C., Liu, R., Sun, W., Pei, D.: Robust anomaly detection for multivariate time series through stochastic recurrent neural network. In: Proceedings of the 25th ACM SIGKDD International Conference on Knowledge Discovery & Data Mining, pp. 2828–2837 (2019)
18. Tuli, S., Casale, G., Jennings, N.R.: TranAD: Deep transformer networks for anomaly detection in multivariate time series data. arXiv preprint arXiv:2201.07284 (2022)
19. Wang, L., et al.: Learning to detect salient objects with image-level supervision. In: Proceedings of the IEEE Conference on Computer Vision and Pattern Recognition, pp. 136–145 (2017)
20. Xu, H., et al.: Unsupervised anomaly detection via variational auto-encoder for seasonal KPIs in web applications. In: Proceedings of the 2018 World Wide Web Conference, pp. 187–196 (2018)

21. Zhang, C., et al.: A deep neural network for unsupervised anomaly detection and diagnosis in multivariate time series data. In: Proceedings of the AAAI Conference on Artificial Intelligence, vol. 33, pp. 1409–1416 (2019)
22. Zhang, J., Zhang, T., Dai, Y., Harandi, M., Hartley, R.: Deep unsupervised saliency detection: a multiple noisy labeling perspective. In: Proceedings of the IEEE Conference on Computer Vision and Pattern Recognition, pp. 9029–9038 (2018)
23. Zhao, H., et al.: Multivariate time-series anomaly detection via graph attention network. In: 2020 IEEE International Conference on Data Mining (ICDM), pp. 841–850. IEEE (2020)

A Model for Retrieving High-Utility Itemsets with Complementary and Substitute Goods

Raghav Mittal[1]($\boxtimes$), Anirban Mondal[1], P. Krishna Reddy[2],
and Mukesh Mohania[3]

[1] Ashoka University, Sonipat, India
raghav.mittal@alumni.ashoka.edu.in
[2] IIIT Hyderabad, Hyderabad, India
pkreddy@iiit.ac.in
[3] IIIT Delhi, Delhi, India
mukesh@iiitd.ac.in

Abstract. Given a retail transactional database, the objective of high-utility pattern mining is to discover high-utility itemsets (HUIs), i.e., itemsets that satisfy a user-specified utility threshold. In retail applications, when purchasing a set of items (i.e., itemsets), consumers seek to replace or *substitute* items with each other to suit their individual preferences (e.g., Coke with Pepsi, tea with coffee). In practice, retailers, too, require substitutes to address operational issues like stockouts, expiration, and other supply chain constraints. The implication is that items that are *interchangeably purchased*, i.e., substitute goods, are critical to ensuring both user satisfaction and sustained retailer profits. In this regard, this work presents (i) an efficient model to identify HUIs containing substitute goods *in place of* items that require substitution, (ii) the **S**ubsti**T**ution-based **I**temset inde**X** (STIX) to retrieve HUIs containing substitutes, and (iii) an experimental study to depict the benefits of the proposed approach w.r.t. a baseline method.

Keywords: Utility Mining · Pattern Mining · Itemsets · Substitutes · Indexing

1 Introduction

In e-commerce, as well as brick and mortar retail environments, consumers prefer to purchase products that complement each other (i.e., complementary goods) to satisfy their need for one-stop-shopping [1,2,6]. In this regard, utility mining approaches, which exploit the complementary nature of items (estimated by evaluating their support) and their utility (i.e., price/profits), have been investigated [11,23,24]. In practice, when users purchase a set of items (itemsets), they

---

A. Mondal—With grief, this work reports the passing of Dr. Anirban Mondal in 2022. The authors of this work owe him a debt of gratitude for his guidance in preparing this manuscript.

© The Author(s), under exclusive license to Springer Nature Singapore Pte Ltd. 2024
D.-N. Yang et al. (Eds.): PAKDD 2024, LNAI 14645, pp. 340–352, 2024.
https://doi.org/10.1007/978-981-97-2242-6_27

meticulously select products that suit their niche preferences. For example, some users may prefer to buy {tea, bread} for breakfast, while others may purchase {coffee, bread}. Here, note that tea and coffee are interchangeably purchased *in place of each other*. In economics, such items are referred to as *substitute goods* (Fig. 1).

Fig. 1. Examples Of Substitute Goods and Complementary Goods

Substitute goods are essential products stocked by retailers to cater to the individual needs and preferences of their loyal consumer base, thereby also *significantly driving consumer demand*. Hence, if retailers greedily recommended *only* itemsets containing higher-priced complementary items to its consumers, they would fail to capture the importance of substitute goods in attracting consumer demand, which in turn would decrease the sales of complementary items. Consequently, this would lead to a significant deduction in retailer revenue.

In retail applications, phenomena such as product expiration [17], i.e., when items expire, and stockouts [18], i.e., when item stock is depleted, pose practical limitations to improving retail profits. A viable solution to these challenges may involve *substituting* non-expiry goods with those nearing expiration or replacing stocked-out products with in-stock alternatives. Substitutes may also help retailers balance user demand, mitigate operational risks, and incorporate seasonal items into HUIs. These reasons, along with other retailing needs (refer to Fig. 2) underscore the importance of substitute goods in addressing consumer needs, as well as the challenges faced by retailers in everyday operations. In the absence of substitutes for items in HUIs, retailers may face significant losses due to the decreased demand for HUIs, as well as lost sales due to operational inefficiencies.

High-Utility Itemset	USER NEED	SELLER NEED	COMMON
	Budget Limit	Expiry Control	Out-Of-Stock
	Health, Safety	Risk Mitigation	Global Events
	Quality, Brand	Diversification	Item Upgrade
	Occasion	Regulation	Legal Ban
	Fashion Need	Marketing Tactic	Customizing
	Lifestyle	Branding Strategy	Cultural Change
	Seasonality	Competition	'Need' v/s 'Want'
SUBSTITUTES	Personal Taste	Inventory Control	Sustainability

Fig. 2. Substitutions In High-Utility Itemsets

Adopting a naive approach, if retailers impulsively discard or phase out items in HUIs that need substitution - say, due to expiry - they incur the risk of losing valuable insights into historic user purchase patterns. Consider an itemset: {tea, bread, jam}. Suppose that items {tea, jam} need to be substituted. If a retailer decides to reduce the set {tea, bread, jam} to just {bread}, it diminishes the *significance* of items {tea, jam} in attracting user demand. Essentially, the retailer must substitute {tea, jam} in {tea, bread, jam} with suitable alternatives, like {coffee, butter}, to create a fresh itemset ({coffee, bread, butter}). This process is critical to preserve valuable knowledge about user purchase behavior in HUIs.

While substitutes can be formulated manually across a small number of HUIs, it becomes practically impossible in large-scale retail applications, especially due to the combinatorial explosion of itemsets. In certain cases, retailers may wish to substitute an individual item (e.g., a ball) with an itemset (e.g., {boxing gloves, punching bag}). Conversely, retailers may wish to substitute a subset of items (in an itemset) with a single item. In other cases, retailers may wish to replace all items contained in HUIs. This issue is exacerbated by the fact that user preferences for substitutes may be dynamic in nature and evolve based on advertisements. Hence, *strategic indexing* of data about HUIs along with profitable and relevant substitute goods becomes a necessity for service providers.

Existing works have focused on (a) product substitution [7,8], (b) HUI mining [11], and (c) applications of HUI mining [10,13,22]. Efforts in (b,c) propose methods for the discovery and utilization of HUIs, while (a) addresses substitutions for single items. In essence, none of the existing works *efficiently* perform substitutions across a given set of HUIs, s.t. their utility is improved. This substantially limits their applicability in practical use, e.g., in e-commerce services.

This work introduces *product affinity* as a measure of homogeneity or substitutability between different pairs of itemsets (or items, i.e., itemsets of size 1). For e.g., Coke and Pepsi may be considered homogeneous as they have a similar purpose. Hence, they may have higher *affinity*. Conversely, Coke and a potato may be considered heterogeneous, and thus may have lower affinity. Affinity scores among distinct sets of items can be computed by examining concept hierarchies, descriptions, reviews, and so on (as per application) and provided as input to the proposed model. Other inputs to the model include set $s=\{s_1...s_n\}$ of n itemsets (or items) that need substitutes, and no. N_s of substitutes needed for all $s_i \in s$.

In the context of the discussed problem, the challenge is to effectively substitute item(sets) in s while formulating HUIs, such that the utility of the retrieved HUIs is improved. A naive approach to this would be to replace all items in s_i with top-utility items in the database. However, these items may not serve a similar purpose in relation to the substituted item(set), and hence, may not be purchased by users. Consider a HUI {tea, butter, jam}; here, if tea is replaced with a pair of shoes, it may not generate sales as tea and a pair of shoes have dissimilar purposes. In a similar vein, consider a HUI: {beer, wafers}. Here, if beer is replaced with an expensive drink, it may also not yield sales as consumers may not purchase expensive drinks in the face of their budget limitations.

Based on this discussion, it can be concluded that both (a) price similarity and (b) affinity scores are critical to efficiently performing substitution between items contained in HUIs. In this regard, this work develops the concept of the substitution score (weighted sum of (a) and (b)) to discover the top-substitutes for item(sets) that need substitution. In order to retrieve HUIs augmented with the top-substitutes, this work further proposes the SubstiTution-based Itemset indeX (STIX): a two-phased index that stores (a) top-substitutes for item(sets) in s, and (b) top HUIs containing these substitutes (as derived from (a)). Given a number T_I of the total top-HUIs that are needed by an application, STIX is exploited to retrieve T_I HUIs. The contributions of this work are three-fold:

1. This work presents a model to efficiently discover HUIs with substitute goods.
2. Based on the model, this work presents the STIX index and a scheme that exploits STIX to efficiently retrieve itemsets containing top substitute items.
3. This work performs a performance study to show the effectiveness of STIX.

To the best of found knowledge, this is the first work to address substitution in HUIs, s.t. the utility of the retrieved HUIs is improved. This paper is organized as follows. Section 2 discusses related works. Section 3 describes a model of HUIs with substitutes. Section 4 presents STIX and a HUI retrieval scheme. Section 5 reports the performance study. This work concludes with research issues in Sect. 6.

2   Related Work

Existing approaches can be broadly categorized into approaches of three types:

(I) Product Substitution: The concepts of complementarity and substitutability have been extensively researched in retail as well as business marketing [7,8]. Among popular bodies of work, [4] investigated retail assortment based on complements and substitutes, while [26] examined the impact of price promotions on complements and substitutes. [25] examined the impact of complements and substitutes on user satisfaction. The work in [21] conducted experiments to study users' willingness to pay for distinct types of complements and substitutes.

(II) Utility Mining: The HUI-Miner [11] stores heuristic information about HUIs by employing *utility-lists*. The Utility Pattern Growth (UP-Growth) [24] employs the Utility Pattern Tree (UP-Tree) for discovering HUIs, while [23] uses the NVUV-list data structure to identify HUIs with high average utility scores. The MinFHM [5] algorithm discovers minimal HUIs, while the CHUI-Miner [20] retrieves closed HUIs. Incremental HUIM approaches are considered in [9,27]. TKHUIM-GA, which is a genetic algorithm for HUIM is proposed in [12].

(III) Applications Of HUI-Mining: In the context of retail stores, HUIs have been used to address business challenges such as urgency in sales [15], market segmentation [14], diversification [16], and item-type-aware [13] as well as slot-premiumness-based placement [19]. In addition, [22] presented a technique for retrieving high-utility coverage patterns for banner ads, while [10] examined use of HUIs in IoT scenarios. [3] examined the aspect of composite items.

In essence, none of these works considered the issue of efficiently discovering and/or retrieving high-utility itemsets with a provision for substitute goods.

3   Model of High-Utility Itemsets with Substitute Items

Consider a set $D = \{t_1, t_2...t_{|D|}\}$ of transactions over a set A with m items. Each item $i \in A$ has a (i) utility (price): $util_i$ and (ii) support: σ_i, i.e., no. of times i appears in D. Moreover, consider a set $s = \{s_1...s_n\}$ with n itemsets (or items, i.e., itemsets of size one) that need to be substituted with N_s substitutes each.

This work assumes that each item(set) in s has an equal degree of urgency, or priority for substitution. Moreover, in the context of substitution, this work considers that *all* instances of itemsets in s need to be substituted. Consistent with practice, this work also assumes that item(sets) in s that are not substituted will not be sold. In retail environments, users purchase sets of items (i.e., itemsets) to satisfy their need for one-stop shopping. If certain items in an itemset are not available, users may be unable to benefit from one-stop-shopping, hence, they may not purchase the itemset. In a similar vein, this work assumes that retailers cannot sell itemsets with expired or stocked-out items that need substitution.

Given D, A, and s, this section discusses the following concepts: (i) expected utility, (ii) affinity scores, (iii) price similarity scores, and (iv) substitution scores.

Definition 1 (Expected Utility of Item). *Let i be an item s.t. $i \in A$. Let $util_i$ be the utility (price) of i. Let σ_i be the no. of times i appears in D. The expected utility EU_i of i is defined as the product of $util_i$ and σ_i, i.e., $EU_i = util_i \times \sigma_i$.*

Definition 2 (Expected Utility of an Itemset). *Let z be an itemset s.t. all items in z are contained in A. Let $util_z$ be the sum of utility (price) of items in z. Let σ_z be the no. of times items in z together occur in D. The expected utility EU_z of itemset z is defined as the product of $util_z$ and σ_z, i.e., $EU_z = util_z \times \sigma_z$.*

Note that the definition of EU_z may be biased towards itemsets of larger size. Hence, given expected utility value EU_z, the normalized expected utility NEU_z of an itemset z is computed as EU_z divided by total no. of items in z.

Definition 3 (Normalized Expected Utility of an Itemset). *Let z be an itemset s.t. each item in z is contained in A, and EUz be the expected utility of z. The normalized expected utility NEU_z of an itemset z is defined as the expected utility EU_z of z divided by the total number of items in z, i.e., $NEU_z = EU_z/|z|$.*

Recall from Sect. 1, that in order to efficiently perform item substitutions, retailers need to consider the product affinity scores among pairs of items(ets).

Definition 4 (Affinity Score α of Two Itemsets). *Let z_1 and z_2 be two itemsets such that each of the items in z_1 and z_2 are contained in A. The affinity score $\alpha(z1, z2)$ of itemsets z_1 and z_2 is defined as their degree of substitutability.*

In retail applications, the computation of product affinity between pairs of items and/or itemsets is subject to a variety of factors, which may vary across different categories and sub-categories of products, as well as contexts. Given domain knowledge, affinity values among items and/or itemsets can be computed by retailers according to the need of their application. This work assumes that affinity scores between items(ets) are provided as input by the retailers.

Definition 5 (Price Similarity Score P of Two Itemsets). *Let z_1 and z_2 be two sets s.t. items in z_1 and z_2 belong to A. Let util(z1) and util(z2) be price values of z_1 and z_2 respectively. The price similarity P(z1,z2) between z_1 and z_2 is computed as normalized price value difference between util(z1) and util(z2).*

In Defn. 5, alternate scores, such as the cosine similarity score, may be used (with an appropriate normalization method), in conjunction with this model to compute the price similarity score P, depending on the nature of the application.

Note that if items are substituted only on the basis of α it may not yield sales as substitutes may have highly contrasting price values. Conversely, if items are substituted only on the basis of P, it may not yield sales, as substitutes may not serve a similar purpose. Hence, both P and α need to be considered in tandem.

Definition 6 (Substitution Score of Two Itemsets). *Let z_1 and z_2 be two itemsets such that items in z_1 and z_2 are contained in set A. Let P(z1,z2) and $\alpha(z1, z2)$ be price similarity and affinity score between z_1 and z_2. The substitution score S(z1,z2) between z_1 and z_2 is computed as the weighted sum of P(z1,z2) and $\alpha(z1, z2)$, i.e., $S(z1, z2) = [w_1 \times \alpha(z1, z2)] + [w_2 \times P(z1, z2)] | s.t.(w_1 + w_2 = 1)$*

Note that S considers both affinity and price similarity scores among items(ets) in s and H. Also note that weighted scores give retailers the flexibility to prioritize P or α based on application, operational needs, and/or domain knowledge.

Problem Statement: Consider a set D of transactions over set A with m items. Each item in A has utility score $util_i$ and support σ_i (i.e., frequency). Consider a set s = $\{s_1, s_2, s_3, ..., s_n\}$ of n items(ets) that need N_s substitutes each. Also consider a matrix of affinity scores between pairs of items(ets) in set H (mined using any HUI approach) and s. Given total number T_I of HUIs that are queried, the problem is to substitute items(ets) provided in s while retrieving the top-T_I HUIs, such that the total utility of retrieved HUIs is improved.

4 Description of HUI Retrieval Scheme

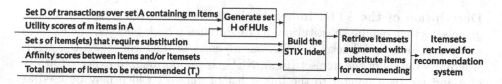

Fig. 3. Schematic diagram of the HUI retrieval scheme

Basic Idea: Existing utility mining approaches examine the utility scores of itemsets to determine the top-HUIs. During this process, HUIs with lower-priced, relevant substitutes often get eliminated. As discussed in Sect. 1, in practical

applications, certain important items contained within HUIs often require substitution. Removing these items from HUIs poses the risk of losing valuable
insights into user purchase patterns. Instead, these items are required to be substituted with relevant alternatives to (a) preserve insights into user purchase
patterns, (b) provide a diverse range of products to consumers, and (c) improve
the efficiency of retail operations (e.g., relating to expiry, and/or stockouts).

In this regard, Sect. 3 presents a model for computing the substitution score
(S), which helps in determining the top-substitutes for items in HUIs that require
substitution. The substitution score (S) may be used in conjunction with existing utility mining approaches for identifying substitutes on-the-go during the
HUI mining process. Conversely, these may be used to determine substitutes for
items in HUIs after HUIs have been mined. As such, this work is agnostic to the
underlying method used to mine HUIs. In the context of this paper, the challenge is to *efficiently* retrieve HUIs containing top-substitutes in response to user
queries. Hence, this work presents the STIX index to index and retrieve HUIs
with top-substitutes. Figure 3 presents an overview of the developed approach.

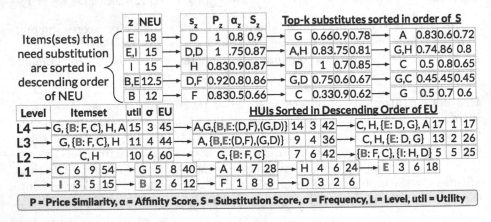

Fig. 4. Illustration Of The Two-Phased STIX Index

Description of the STIX index: in the first phase, given n items(ets) in a
given set s that require substitution, STIX initiates n hash buckets, one for each
entry in s. Each hash bucket in STIX has an entry of the form $\{z, NEU_z, ptr_z\}$,
where z is the ID of the itemset (in s), NEU_z is the normalized expected utility
of z and ptr_z is a pointer to the linked list of itemsets which may be used as
substitutes for z. Each entry in the linked list is of the form $\{s_z, P_z, \alpha_z\ S_z\}$,
where s_z refers to the unique identifier of substitute itemset, P_z, α_z, and S_z are
the price similarity, affinity score, and substitution score, respectively, between
s_z and itemset z depicted in the hash bucket. The linked-list entries are sorted in
descending order of S. Note that each linked list stores only the top-k HUIs for
an itemset in s. Here, the value of k is application-dependent, and can be decided
by the retailer. For *efficiently* retrieving substitutes for an itemset z (in s), STIX

can directly traverse to the zth hash bucket and traverse its linked list to retrieve relevant itemsets, till all itemsets in the given set s have been substituted.

Note that the hash buckets in STIX are sorted in descending order of NEU of the items(ets) so that it can prioritize substitution of items(ets) that can yield higher profits. Such prioritizing is critical as there may be an overlap between items(ets) that need substitution (e.g., {E} and {E,I}). Figure 4 depicts an instance of first phase of STIX with five itemsets (or items, i.e., itemsets of size one). Note how the linked list entries for each itemset in s contains itemsets (or items) that constitute its substitutes. In addition, note that the linked lists of substitute itemsets are sorted in descending order of their substitution scores S.

The second phase of STIX contains L_{max} levels, each associated with a different itemset size. Here, each level has a hash bucket that points to the linked list of itemsets, which are sorted in descending order of their EU. Moreover, HUIs at each level that require substitution are augmented with their substitutes in descending order of their substitution scores. In this way, STIX is able to efficiently store data about HUIs of different sizes, along with their substitutes. Figure 4 shows an instance of second phase of STIX. Note that each level contains HUIs of the corresponding size. Further note how itemsets of different sizes, which require substitution (denoted in red), are augmented with their substitutes.

Inputs: (a) Item Utility (b) Affinity Scores

Item	Util		A	B	C	D	E	F	G	H	I
A	4	A	1	0.1	0.2	0.1	0.6	0.2	0.3	0.2	0.2
B	2	B	0.1	1	0.9	0.1	0.1	0.5	0.7	0.1	0.1
C	6	C	0.2	0.9	1	0.2	0.1	0.2	0.1	0.1	0.8
D	3	D	0.1	0.1	0.2	1	0.8	0.1	0.1	0.1	0.7
E	3	E	0.6	0.1	0.1	0.8	1	0.2	0.9	0.1	0.1
F	1	F	0.2	0.5	0.2	0.1	0.2	1	0.2	0.1	0.2
G	5	G	0.3	0.7	0.1	0.1	0.9	0.2	1	0.1	0.2
H	4	H	0.2	0.1	0.1	0.1	0.1	0.1	0.1	1	0.9
I	3	I	0.2	0.1	0.8	0.7	0.1	0.2	0.2	0.9	1

HUIs Retrieved from STIX

HUIs With Substitutes	EU	NEU
C, H	60	30
G, {B: F, C}	42	21
G, {B: F, C}, H	44	14.6
{B: F, C}, {I: H, D}	25	12.5
A,{(B,E): (D,F), (G,D)}	36	12
G, {B: F, C}, H, A	45	11.2
A,G,{(B,E): (D,F), (G,D)}	42	10.5
C, H, {E: D, G}	26	8.6
C, H, {E: D, G}, A	17	4.2

Sample Query

N_s	T_I
1	5

Top-HUIs
C, H
G, F
G, F, H
F, H
A, D, F

Fig. 5. Illustrative Example of the HUI Retrieval Scheme

HUI retrieval scheme: this work considers a set D of transactions and a set s of items(ets) that need substitution (with N_s substitutes each). Moreover, a set H of HUIs, which may be mined from D using any HUI method, is provided. By examining itemsets in H, it is possible to find substitutes for entries in set s by computing their substitution scores (S). Alternatively, the retailer may directly provide us with items(ets) that constitute viable substitutes, which may then be stored in STIX. In case the items(ets) given by the retailer do not appear in D, transactions in D may be enriched to mark their presence. Given total no. T_I of HUIs that are needed by the retailer, the scheme extracts HUIs from STIX, and sorts them in descending order of their NEU. Next, for each HUI, if it does not contain any augmented substitute items(ets), HUIs are directly added to the output. If it contains items(ets) with potential substitutes, the list of potential substitutes is *iteratively* traversed to select top-S N_S substitute items(ets).

Illustrative Example: Figure 5 depicts the HUI retrieval scheme. Figure 5 considers items A to I with utility values and affinity scores for different pairs of items. Observe how EU and NEU are computed for HUIs retrieved from STIX. Given set s of items(ets) that require substitution, top-substitutes may be discovered using S scores, or directly provided by the retailer. The scheme exploits STIX to retrieve top-NEU HUIs, which have augmented substitutes for items(ets) in s. In the sample query, N_s and T_I depict no. of substitutes needed for items(ets) in s, and total HUIs needed. The retrieval scheme first selects the top-HUIs sorted in order of NEU, and then iterates through their potential substitutes to select the top-1 (since $N_s = 1$) substitute for HUI that needs substitution. For e.g., for HUIs that contain B, the scheme replaces instances of B with F, which is B's top substitute. Similarly, item I is replaced with H, itemset {B,E} with {D,F} and so on. In this way, the scheme substitutes items in HUIs. Algorithm 1 presents the pseudocode of the STIX index and HUI retrieval scheme.

Algorithm 1: Model for retrieving HUIs with substitute goods

Input: D: set of transactions; A: list of m tuples of form <i,util> (i is an item and util is its utility); s: list of tuples of form <z,N_s> (z is an itemset and N_s is the number of substitutes needed); M: matrix with affinity scores; T_I: queried no. of itemsets; L_{max}: no. of levels in STIX; k: no. of max. substitutes in STIX

Output: Out: List of itemsets augmented with top-substitutes

Variables: H, J, C: list of tuples of form <z,util> (z: set, util: real value)

1 Using D and A, compute set H of HUIs using any HUI mining algorithm
2 Compute NEU for each itemset z ∈ H ∪ s; sort on basis of NEU and store in J
  Populating the STIX index
3 **foreach** entry <z,N_s> in s: initialize a hash bucket /*Phase-1 of STIX*/
4   Using J,A,M, compute substitution score (S) between z & itemsets in H
5   Insert k top-S itemsets from H into linked list attached to the hash bucket
6 Using J, arrange the hash buckets of each z ∈ s in descending order of NEU
7 **for** L in range 1 to L_{max} **do** initialize a hash bucket for L /*Phase-2 of STIX*/
8   **foreach** entry <HUI,NEU>∈ J where HUI is of size L
9   **foreach** itemset z ∈ s /*To check if HUI has item(s) needing subst.*/
10   **if** z ⊆ HUI: replace z with linked list of k substitutes from Phase-1
11   Add HUI augmented with linked-list of substitutes in hash bucket of L
  Retrieving HUIs with Substitute Items
12 Retrieve itemsets from the hash bucket of each level (1 to L_{max}) of STIX
13 Using J, store itemsets from STIX in list C in descending order of their NEU
14 **while** T_I>0
15   **foreach** entry <z,NEU> ∈ C; T_I -= 1
16   **if** z contains a linked list of k substitute items(ets)
17   Using s, add z with top-N_s (of the total k) substitutes to Out
18   **else** directly add z to Out

---

5 Performance Evaluation

This section reports the results of the performance study. For the study, a 13th Gen Intel(R) Core(TM) i7-13700H processor running on Ubuntu 22.04.3 LTS with 8 GB RAM is used. The code is implemented in Python. Two real-world datasets are considered: *Chainstore* (T: 1million, I: 46k) and *E-Commerce* (T: 14.9k, I: 3.4k). The datasets and their details are available on the SPMF[1] site.

Recall from Sect. 4 that STIX needs a set H of HUIs as input. Although any approach can be used to generate H, this work uses the kUI-index [16] to generate HUIs. The kUI index has multiple levels, each associated with a unique itemset size. At each level of the kUI index, the top-λ HUIs are stored in descending order of their utility. This enables the kUI index to swiftly retrieve HUIs of any size. The kUI index was implemented with six levels with $\lambda = 3000$. *It is critical to note that the kUI index is oblivious to the substitution needs of retailers.*

In this study, total no. T_I of HUIs needed was varied from 500–2500 (default 1500) and no. n of substitute itemsets was varied from 200–1000 (default 600). The datasets were divided into training and test sets (80% and 20% of transactions). The results were evaluated in the test set. Performance metrics include: (a) total utility T_U and (b) execution time ET required for the discovery and retrieval of HUIs. In the study, the transactions in the test set are iterated and itemset utility (price) values are added to T_U *only* if items in the transaction have been retrieved as HUIs in the training phase. The utility values of items(ets) that require substitution (as indicated by values in s) are not computed.

Set s of length n was generated by *randomly* selecting n HUIs in H. For each $s_i \in s$, 20 random substitutes were created, which are denoted using their ID. For each occurrence of $s_i \in s$ in the training and test set, the transactions were enriched with the IDs of their substitutes. For each substitute ID, a random value in the range [0.75, 1] was generated which denoted their substitution score (S). Consistent with practice, this study assumed that profits earned through these substitutes would be proportional to S. For e.g., for a substitute with S = 0.8, it was assumed that its utility would be 80% of the original HUI. These scores can be developed using Defn. 6 or directly provided by retailers. Of the 20 IDs generated, only the top ten substitutes were stored in STIX (i.e., with k=10).

As a baseline, this work considered a reference scheme designated as the **S**ubstitution **O**blivious **I**temset **R**etrieval scheme (SOIR), which greedily sorts HUIs in H on the basis of NEU, but is oblivious to retailer substitution needs.

---

[1] www.philippe-fournier-viger.com/spmf/index.php?link=datasets.php.

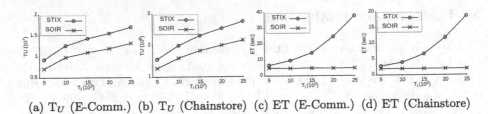

(a) T_U (E-Comm.) (b) T_U (Chainstore) (c) ET (E-Comm.) (d) ET (Chainstore)

Fig. 6. Effect of varying total number T_I of required HUIs

Effect of variations in the no. T_I of required HUIs: The results in Fig. 6 depict the effect of varying no. T_I of queried HUIs. As T_I increases, an increase in T_U is observed for both schemes. This occurs because retrieval of a higher no. of HUIs implies more sales for retailers, leading to a higher T_U. STIX outperforms SOIR in terms of T_U as it meticulously selects HUIs with substitutes, which allows users to substitute items with each other. These substitutions result in a higher sales, which results in higher T_U. SOIR is oblivious to substitutes for entries in s, hence, it ignores HUIs with profitable substitutes.

The results in Fig. 6(c-d) indicate that ET increases for STIX and SOIR (albeit slightly) with increase in T_I as more HUIs need to be examined. STIX incurs higher ET than SOIR since it meticulously selects HUIs with profitable substitutes. It can be noted that this is a small price to pay for increased profits.

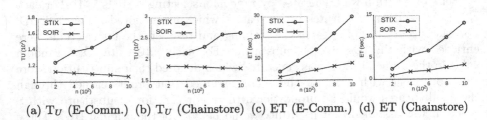

(a) T_U (E-Comm.) (b) T_U (Chainstore) (c) ET (E-Comm.) (d) ET (Chainstore)

Fig. 7. Effect of varying number (n) of substitute items(ets)

Effect of variations in the no. (n) of substitute items(ets): The results in Fig. 7 depict the effect of varying no. n of items(sets) that need substitution. As n increases, T_U increases for STIX but decreases for SOIR. This occurs because with a higher no. of substitutes, STIX is able to provide more opportunities for substitutions to users, while allowing retailers to manage operational needs in tandem. Since SOIR is incapable of performing substitutions, it incurs losses due to a lower number of sales in the test sets, resulting in lower profits (T_U).

The results in Fig. 7(c-d) depict that SOIR incurs lower execution time (ET) than STIX, which is consistent with the rationale provided for the results for Fig. 6(c-d). STIX depicts an increase in ET as it determines more HUIs comprising profitable substitutes. Similar trends are observed for both datasets.

6   Conclusion

While purchasing sets of items (itemsets), users seek to replace, or *substitute* items with each other to suit their needs. In practice, retailers, too, need substitutes to address operational challenges. Therefore, this work presents a model and the STIX index to efficiently discover and retrieve HUIs containing substitute goods. The performance study shows the benefits of STIX. In the future, this work will be extended in multiple directions: (i) the proposed model will be augmented with data on specific needs of users and retailers (as listed in Fig. 2). Moreover, (ii) specific impacts of phenomena such as expiry, inventory, diversification of substitutes, and the like will be incorporated into the developed model. The authors also plan to (iii) investigate the cost-effective integration of the proposed STIX model into existing retail-based recommendation systems.

References

1. Agrawal, R., Srikant, R.: Fast algorithms for mining association rules. In: VLDB, pp. 487–499 (1994)
2. Ahn, K.I.: Effective product assignment based on association rule mining in retail. Expert Syst. Appl. **39**, 12551–12556 (2012)
3. Basu Roy, S., Amer-Yahia, S., Chawla, A., Das, G., Yu, C.: Constructing and exploring composite items. In: ACM SIGMOD, pp. 843–854 (2010)
4. Diehl, K., Van Herpen, E., Lamberton, C.: Organizing products with complements versus substitutes: effects on store preferences as a function of effort and assortment perceptions. J. Retail. **91**(1), 1–18 (2015)
5. Fournier-Viger, P., Lin, J.C., Wu, C., Tseng, V.S., Faghihi, U.: Mining minimal high-utility itemsets. In: DEXA, pp. 88–101 (2016)
6. Han, J., Pei, J., Yin, Y.: Mining frequent patterns without candidate generation. In: ACM SIGMOD, pp. 1–12 (2000)
7. Hicks, J.: Value and Capital (1939)
8. Lange, O.: Complementarity and interrelations of shifts in demand. Rev. Econ. Stud. **8**(1), 58–63 (1940)
9. Lee, J., Yun, U., Lee, G., Yoon, E.: Efficient incremental high utility pattern mining based on pre-large concept. EAAI **72**, 111–123 (2018)
10. Lin, J.C.W., Djenouri, Y., Srivastava, G., F.-Viger, P., Xue, X.: Mining profitable and concise patterns in large-scale Internet of Things environments. WCMC (2021)
11. Liu, M., Qu, J.: Mining high utility itemsets without candidate generation. In: CIKM, pp. 55–64 (2012)
12. Luna, J.M., Kiran, R.U., Fournier-Viger, P., Ventura, S.: Efficient mining of top-k high utility itemsets through genetic algorithms. Inf. Sci. **624**, 529–553 (2023)
13. Mittal, R., Mondal, A., Reddy, P.K.: A consumer-good-type aware itemset placement framework for retail businesses. In: PAKDD, pp. 276–288 (2023)
14. Mittal, R., Mondal, A., Reddy, P.K.: A market segmentation aware retail itemset placement framework. In: DEXA, pp. 273–286 (2022)
15. Mittal, R., et al.: An urgency-aware and revenue-based itemset placement framework for retail stores. In: DEXA, pp. 51–57 (2021)
16. Mondal, A., Mittal, R., Chaudhary, P., Reddy, P.K.: A framework for itemset placement with diversification for retail businesses. Appl. Int. 1–19 (2022)

17. Mondal, A., Mittal, R., Khandelwal, V., Chaudhary, P., Reddy, P.K.: PEAR: a product expiry-aware and revenue-conscious itemset placement scheme. In: IEEE DSAA, pp. 1–10 (2021)
18. Mondal, A., Mittal, R., Saurabh, S., Chaudhary, P., Reddy, P.K.: An inventory-aware and revenue-based itemset placement framework for retail stores. Expert Syst. Appl. **216**, 119404 (2023)
19. Mondal, A., Saurabh, S., Chaudhary, P., Mittal, R., Reddy, P.K.: A retail itemset placement framework based on premiumness of slots and utility mining. IEEE Access **9**, 155207–155223 (2021)
20. Nguyen, L.T., et al.: An efficient method for mining high utility closed itemsets. Inf. Sci. **495**, 78–99 (2019)
21. Rousu, M.C., Beach, R.H., Corrigan, J.R.: The effects of selling complements and substitutes on consumer willingness to pay: evidence from a laboratory experiment. Can. J. Agric. Econ. **56**(2), 179–194 (2008)
22. Srinivas, P.G., et al.: Discovering coverage patterns for banner advertisement placement. In: PAKDD, pp. 133–144 (2012)
23. Truong, T., Duong, H., Le, B., F.-Viger, P., Yun, U.: Efficient high average-utility itemset mining using novel vertical weak upper-bounds. KBS **183**, 104847 (2019)
24. Tseng, V.S., Wu, C., Shie, B., Yu, P.S.: UP-Growth: an efficient algorithm for high utility itemset mining. In: ACM SIGKDD, pp. 253–262 (2010)
25. Voss, G.B., Godfrey, A., Seiders, K.: How complementarity and substitution alter the customer satisfaction-repurchase link. J. Marketing **74**(6), 111–127 (2010)
26. Walters, R.G.: Assessing the impact of retail price promotions on product substitution, complementary purchase, and interstore sales displacement. J. Mark. **55**(2), 17–28 (1991)
27. Wu, J.M.T., Teng, Q., Lin, J.C.W., Yun, U., Chen, H.C.: Updating high average-utility itemsets with pre-large concept. JIFS **38**, 5831–5840 (2020)

LPSD: Low-Rank Plus Sparse Decomposition for Highly Compressed CNN Models

Kuei-Hsiang Huang, Cheng-Yu Sie, Jhong-En Lin, and Che-Rung Lee[✉] [iD]

National Tsing Hua University, HsinChu, Taiwan
cherung@cs.nthu.edu.tw

Abstract. Low-rank decomposition that explores and eliminates the linear dependency within a tensor is often used as a structured model pruning method for deep convolutional neural networks. However, the model accuracy declines rapidly as the compression ratio increases over a threshold. We have observed that with a small amount of sparse elements, the model accuracy can be recovered significantly for the highly compressed CNN models. Based on this premise, we developed a novel method, called LPSD (Low-rank Plus Sparse Decomposition), that decomposes a CNN weight tensor into a combination of a low-rank and a sparse components, which can better maintain the accuracy for the high compression ratio. For a pretrained model, the network structure of each layer is split into two branches: one for low-rank part and one for sparse part. LPSD adapts the alternating approximation algorithm to minimize the global error and the local error alternatively. An exhausted search method with pruning is designed to search the optimal group number, ranks, and sparsity. Experimental results demonstrate that in most scenarios, LPSD achieves better accuracy compared to the state-of-the-art methods when the model is highly compressed.

1 Introduction

Deep learning models have demonstrated excellent performance in natural language processing, image recognition, and many other fields. However, the model size is increasing for more accurate and more powerful models. On the other hand, with the popularity of IoT and edge computing, model compression, such as pruning, quantization, and low-rank approximation, is an essential technique to deploy those models on the devices with limited computational power and memory capacity.

Among different model compression methods, low-rank decomposition has the ability to explore the latent dependency in the tensor weights, which can be used to remove the redundant parameters while preserving important information. Although determining the rank for each layer is an NP-complete problem, there have been many studies that have provided approximate solutions [7,11,14], enabling low rank decomposition methods to achieve higher accuracy with fewer parameters.

© The Author(s), under exclusive license to Springer Nature Singapore Pte Ltd. 2024
D.-N. Yang et al. (Eds.): PAKDD 2024, LNAI 14645, pp. 353–364, 2024.
https://doi.org/10.1007/978-981-97-2242-6_28

However, when a model is compressed with an extremely low rank, its accuracy starts to drop dramatically. This is not a surprising result since more and more important information are removed. Our observation found that if we add a small amount of sparse elements in the extremely low rank cases, the model accuracy can be recovered substantially. This finding matches the empirical results of recent studies [3,15], which show that the CNN weight tensor not only exhibits low rank properties but also sparsity. Using only low rank approximation on pre-trained models may lead to a loss of accuracy under high compression ratios. On the other hand, the low rank plus sparse decomposition preserves both coarse and fine grained structures of models, and therefore can achieve superior compression rates with good model accuracy.

Decomposing a matrix into a low rank part plus a sparse structure is not a new problem. [1,6,13,16] However, they have different goals, such as separating the objects in images or video, other than model compression. For deep learning model compression, there are only few studies on how to apply the low-rank plus sparse methods [3,15]. However, their compression ratio and accuracy are not as good as pure low-rank decomposition methods.

In this paper, we propose a new low-rank plus sparse decomposition, LPSD, for extreme model compression. LPSD first splits a network layer into two branches: low rank and sparse. It is known that the decision of optimal rank and the sparsity to minimize the model size with accuracy constraints is an NP-complete problem. LPSD adapts the alternating minimization algorithm [11] to determine the budgets of each layer, and uses an novel algorithm to search the ranks and sparsity of each layer. We designed a faster method to estimate the error, and showed the proposed sparse element selection method is optimal. Experimental results show that under the same compression ratio, LPSD outperforms pure low-rank decomposition methods in terms of model accuracy in most cases, especially for extremely low rank compression scenarios.

The rest of this paper is organized as follows. In Sect. 2, we illustrate the related work. In Sect. 3, we introduce the LPSD algorithm. In Sect. 4, experimental results are presented. The conclusion and future work are given in the last section.

2 Related Work

Low-rank compression involves decomposing the convolution layer as a multidimensional tensor or transforming the weight tensor into a two-dimensional matrix for matrix low-rank decomposition. Tucker decomposition or SVD decomposition are commonly used methods for low-rank decomposition. In [9], Tucker decomposition is applied to compress the convolution layer, and the VBMF algorithm is used to determine the rank of each layer. In [7], authors used SVD decomposition and alternated the weight updates and rank selection process. In [14], authors utilized BudgetAware Tucker decomposition to minimize both the loss and the nuclear norm of the weight tensor under budget constraints, while determining the rank during the training process of the model. The method proposed in [11] divides the weight matrix into multiple subsets and uses SVD decomposition to minimize the maximum relative error in each layer to determine the rank and the number of subsets. In [2], authors used optimization methods to select the proper ranks of decomposed network layers.

Sparsity involves setting certain parameters in the weight to zero, typically achieved by overlaying a sparse mask on the weight. In [4], authors employed a dense-sparse-dense training flow, pruning smaller weights, and retraining the network given the sparsity constraint. In [12], authors dynamically allocated sparse parameters during the training process, incorporating feedback information to reactivate weights that were pruned prematurely. When subsequent steps result in significant cumulative gradient updates that substantially change specific weights, those weights can be reactivated.

In [3], authors initialized a network structure with a combination of low-rank and sparse components, trained it from scratch using L1 regularization, and then sparsified the trained model instead of approximating a pretrained model. In [15], authors utilized the QR decomposition to combine the low-rank and sparse decomposition of weight matrices with feature map reconstruction, alternately optimizing the low-rank and sparse components. The paper visualizes the filters of the original layer and the compressed low-rank and sparse matrices, revealing that the Conv layer exhibits both low-rank property and sparsity. In [5], authors considered a structure that incorporates both low-rank and sparse components, where the low-rank and sparse parts do not simultaneously contribute to an entry. The entries of neural network weights are taken exclusively from either the sparse component or the low-rank component. In [8], authors combined the CP decomposition with sparsity, minimizing the approximation error under constraints on rank and sparsity budgets, and developed an efficient CPU implementation.

3   LPSD Method

Our algorithm is based on the ALDS algorithm [11]. Although it is an state-of-the-art method, when the compression ratio is higher than 60%, the model accuracy drops significantly. We have found that if adding a small amount of sparse elements in the decomposition, the model accuracy can be recovered significantly. Based on this observation, we developed the Low-rank Plus Sparsity Decomposition (LPSD) algorithm, which decomposes each layer of weight tensors W into the sum of a low-rank approximation UV^T and a sparse tensor S, $W \rightarrow UV^T + S$.

Algorithm 1 sketches the procedure of LPSD. Similar to ALDS, LPSD utilizes the alternative optimization method that minimizes the global errors in the entire network and the local errors in each layer alternatively. However, unlike ALDS that only considers the optimal number of groups and optimal ranks, LPSD needs to search the optimal sparsity of each layer and balance remaining size of the low-rank and the sparsity, which adds another dimension of complexity.

Algorithm 1. LPSD (r, W, ρ, N_S)

Input: r: compression ratio, W: parameters, ρ: maximum sparse ratio, N_S: a set of random seeds
Output: J: ranks, K: group numbers, S: sparsity.

1: **for** $i \in N_S$ **do**
2:     $K \leftarrow$ **RandomGroup**(i)
3:     **while** not converged **do**
4:         $B \leftarrow$ **LayerBudget**(W, r, K)   ▷ Global search step
5:         **for** $l = 1, \ldots, L$ **do**
6:             $j_l, k_l, s_l \leftarrow$ **MinimizeError**(W^l, b_l, ρ)   ▷ Local search step
7:         **end for**
8:     **end while**
9:     $R \leftarrow$ **Record**($J = [j_1, \ldots, j_L]$, $K = [k_1, \ldots, k_L]$, $S = [s_1, \ldots, s_L]$)
10: **end for**
11: **return** Optimal J, K, S from R

3.1   Mathematical Formulation

The input of LPSD is a convolution tensor $\mathcal{W}$ with f filters, c channels and $d_1 \times d_2$ kernel. To better compress the weight tensor $\mathcal{W}$, LPSD performs grouping in each layer, which divides the tensor $\mathcal{W}$ into k groups evenly based on the dimension of channels, resulting in k convolution $\{\mathcal{W}_i\}_{i=1}^{k}$. Each group tensor $\mathcal{W}_i$ has at most $\lceil c/k \rceil$ channels. $\mathcal{W}_i$ is a $f \times c_i \times d_1 \times d_2$ tensor with f filters, c_i channels and $d_1 \times d_2$ kernel, where $\sum_{i=1}^{k} c_i = c$, in most cases, $c_i = c/k$. After that, LPSD transforms each group tensor $\{\mathcal{W}_i\}_{i=1}^{k}$ into a group of matrices $\{W_i\}_{i=1}^{k}$.

The output of LPSD are three vectors of length k: J, K, and S, where J records the optimal rank for each layer; K keeps tracking the group number for each layer; and S stores the number of sparse elements in layers.

In the article, let W be a matrix or a tensor. We use $|W|$ to represent the number of elements in W. For a matrix A, we use $\|A\|$ as the 2-norm of A and $\|A\|_F$ as the Frobenius norm of A.

3.2   Alternative Optimization Method

The alternative optimization method used in LPSD has two steps: global and local. Initially, LPSD guesses random group partition numbers, and runs the alternative optimization to obtain a solution based on this initial guess. Such process runs several times with different random seeds.

In the global search step (**LayerBudget**), LPSD utilizes the current group values $[k^1, \ldots, k^L]$ to partition each layer into a group of sub-tensors, and matricizes each sub-tensor into a matrix. After that, LPSD computes the SVD of each matrix. To minimizes the maxima error, LPSD solves the following min-max problem,

Algorithm 2. MinimizeError (W, b, ρ)

Input: W: parameters in a layer, b: budget for a layer, ρ: maximum sparse ratio
Output: j: ranks, k: group numbers, s: sparsity.

1: $Z \leftarrow$ All possible (j, k) that makes $(1 - \rho)|W| \leq |W'_{(j,k)}| \leq b$.
2: **for** $(j_t, k_t) \in Z$ **do**
3:     $s_t \leftarrow |W| - |W'_{(j_t, k_t)}|$.
4:     $\epsilon_t \leftarrow$ Error Estimation using (2)
5:     $R \leftarrow$ **Record**$(t, \epsilon_t, j_t, k_t, s_t)$
6: **end for**
7: $j, k, s = \arg\min_{t \in R} \epsilon_t$
8: **return** j, k, s

$$\min_{j^1, \dots, j^L} \max_{l=1, \dots L} \frac{\|W^l - U_{j^l}^l (V_{j^l}^l)^T\|}{\|W^l\|} \tag{1}$$

$$\text{s.t.} \sum_{l=1, \dots L} \text{sizeof}(U_{j^l}^l, V_{j^l}^l) \leq r$$

where the compression ratio $r = |W'|/|W|$ for the compressed model W' is a number in $[0, 1]$, $U_{j^l}^l (V_{j^l}^l)^T$ is the low-rank approximation obtained from the SVD of W^l with rank j^l (the singular values have been multiplied with $V_{j^l}^l$), and the function sizeof computes the storage size for $U_{j^l}^l, \Sigma_{j^l}^l$, and $V_{j^l}^T$. The layer budget $B = [b_1, b_2, \dots, b_L]$ specifies the allowable model size for each layer, in which b_j equals to sizeof$(U_{j^l}^l, V_{j^l}^l)$ for $j = 1, 2, \dots, L$.

In the local step, based on the layer budget b^l, LPSD searches the best allocation of k^l, j^l, and s^l for layer l. This part distinguishes ALDS and LPSD, because ALDS only searches for the best k^l. The details of the local search step will be presented in the next subsection.

Once the optimal k^l, j^l, and s^l for each layer l have been found, LPSD will use the new group partition $[k^1, k^2, \dots, k^L]$ to perform the global search again, and repeats those two steps alternatively until the values are converged. LPSD will record the current results in R, and tries different random seeds to run the entire process several times. The best result stored in R is returned in the end.

3.3 Local Search Step (MinimizeError)

The local search step searches the optimal j, k, s for each layer so that the size of layer after compression is less than or equals to b_l. In addition, we require the total number of sparse elements in a network is less than $\rho|W^l|$. This is because we want the compressed model can still enjoy the hardware acceleration for structured pruning mostly, and only allows few skip links.

Algorithm 2 shows the pseudo code of the local search step. Basically, it does an exhausted search for all possible (j, k). However, to accelerate the searching, we prune the impossible configurations. Let $W'_{j,k}$ be the compressed model using grouping number k and low-rank approximation with rank j. It should satisfy two constrains. First, its size should be less than the budget b. Second, the remaining sparse elements should be less than $\rho|W|$, which is $|W| - |W'_{(j,k)}| \le \rho|W|$, or $(1 - \rho)|W| \le |W'_{(j,k)}|$.

For the sparse element selection, we only need to know how many sparse elements to be removed. Modern deep learning framework, such as PyTorch, can perform the model pruning based the given number of pruning elements. In step 3, we assign the remaining budgets $|W| - |W'|$ all to the sparse elements. PyTorch or other framework will retain the s_t largest elements from $W - W'$, and that is exactly what we need. The optimality of our sparse element selection can be proven, as shown in Lemma 1 in Appendix A.

The next step is to calculate the error under the configuration j_t, k_t, s_t. However, computing SVDs for the grouped W, its residual, and remove top s_t sparse elements is a time consuming task. Therefore, we use the following formula to estimate the error.

$$mn\frac{-2\sigma^2}{\sigma\sqrt{2\pi}}\left[-te^{\frac{-t^2}{2\sigma^2}} - \sigma\sqrt{2\pi} + \sigma\sqrt{2\pi}F_X(t)\right] \tag{2}$$

where $t = F_X^{-1}\left(1 - \frac{k}{2mn}\right)$, $F_X(t)$ is the cumulative distribution function of $f_X(t) = \frac{1}{\sigma\sqrt{2\pi}}e^{\frac{-t^2}{2\sigma^2}}$, the probability density function of normal distribution with mean 0. The derivation of (2) is given in Appendix B.

3.4   Complexity Analysis

There are lots of computations of SVD to obtain the relative errors in the algorithm in the implementation. Since these relative errors might be used several times, they are stored in a lookup table to make the implementation efficient.

In the global search step (**LayerBudget**), ALDS has an efficient method to implement it efficiently, based on two assumptions and binary search or other root finding algorithms.

In the local search step (**MinimizeError**), the time complexity is determined by the size of Z, the set of all possible combinations of (j_t, k_t). In the implementation, the possible k_t is set to make k_t divisible by c. Therefore, the number of possible combinations of (j_t, k_t) is $r_td(c)$, where r_T is the rank of weight matrix of t-th layer, $d(x)$ is the number of positive factors of x and c is the number of input channels of t-layer. Moreover, estimating the error using equation (2) takes only $O(1)$. Therefore, the time complexity of the local search step is $O(Rd(c) \times 1) = O(cd(c)) = O(c^2)$.

4   Experiments

In this section, we presented the experimental results to evaluate the performance of our algorithm, and compare it with other methods. There are three sets of

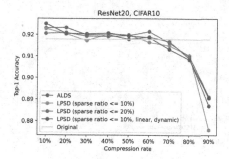

Fig. 1. Comparison of ALDS and LPSD for ResNet 20 on CIFAR-10.

experiments. The first one runs ResNet20 on CIFAR-10, and the second one evaluates the algorithm for ResNet-50 on CIFAR-100, and the last one runs ResNet-50 on ImageNet.

4.1   Experimental Setting

Once determined the low-rank and sparse components, we obtained the compressed model. We finetuned the models to restore its accuracy. For the ResNet20 experiments, We retrained it for 182 epochs, with an initial learning rate of 0.1. After the 91th epoch, the learning rate is reduced to 0.01, and after the 136th epoch, it is further reduced to 0.001. For the ResNet50 experiments on ImageNet, We fine-tuned it for 90 epochs, with an initial learning rate of 0.1. After the 30th epoch, the learning rate is reduced to 0.01. After the 60th epoch, it is further reduced to 0.01. For ResNet50 on CIFAR-100, we fine-tuned it for 200 epochs, with an initial learning rate of 0.1. After the 60th, 120th, 160th epoch, the learning rate is reduced to 0.02, 0.004, 0.0008, respectively. We used cross-validation to perform early stopping based on the validation loss, selecting the model with the lowest validation loss. We evaluate the accuracy of this model on the test set and repeat the experiment three times to obtain the average results.

4.2   Experimental Results for ResNet20 Model on CIFAR-10

For the ResNet20 model, we evaluated the model accuracy of ALDS and LPSD on CIFAR-10 with several different compression ratio, from 10% to 90%. For LPSD, we have set the sparse ratio (ρ) to 10% and 20%. In addition, we have included an experimental result of $\rho = 10\%$ with sparsity constrains (linear, dynamic). The linear method works as follows. Instead of given a global sparse ratio ρ to all the layers, we allow ρ_i varied for different layers. In this experiment, we used $\rho_1 = 2\%, \rho_2 = 12\%, \dots \%$. The sparsity of the entire model will be the same as 10%. The dynamic method is the method of pruning the sparse part during the fine-tuning that works as follows. Instead of retaining the same

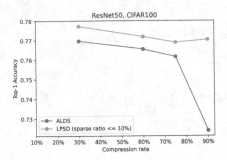

Fig. 2. Comparison of ALDS and LPSD for ResNet 50 on CIFAR-100.

elements of the sparse matrix, the dynamic method retains the largest s_t elements at that time, which are not necessarily the same as the positions at the beginning.

All the experimental results are presented in Fig. 1. As can be seen, for the compression ratio less than or equal to 80%, all three compression methods have similar model accuracy. However, when the compression ratio reaches 90%, the model accuracy of LPSD for $\rho = 20\%$ and $\rho = 10\%$ with linear dynamic is better than that of ALDS. We have also observed that for LPSD with $\rho = 10\%$, the model accuracy drops more than ALDS. But with some adjustment of sparsity for different layers, the model accuracy restored.

The reason why the linear dynamic method for 10% is better than the method of evenly distributed sparsity is the behavior of different layers in CNN is not the same. Such kind of phenomenon has been observed, such as in [10]. For ResNet 20, the former layers are more structured, so the sparsity does not help; the later layers are more unstructured, so pruning them using more sparse elements is more effective.

4.3   Experimental Results for ResNet50 Model on CIFAR-100

We also evaluated LPSD and ALDS on dataset CIFAR-100. The sparse ratio is set to be less than or equal to 10%. The experimental results are in Fig. 2.

As shown in the Fig. 2, the LPSD performs well when the compression ratio is high. When compression ratio is 90%, the accuracy of model compressed by LPSD is 77.03%, while the accuracy of model compressed by ALDS is 72.40%. This gives a concrete evidence that the model accuracy can be well-maintained by including a small amount of sparse components.

4.4   Experimental Results for ResNet50 Model on ImageNet

We evaluate LPSD and ALDS on ImageNet dataset using ResNet50. We used 10% sparse ratio constrain to measure the efficacy of the existence of both low rank and sparse layer. The experimental results are shown in Fig. 3.

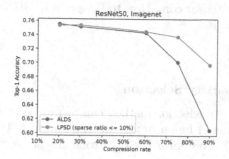

Fig. 3. Comparison of ALDS and LPSD for ResNet 50 on ImageNet.

As can be seen, LPSD shows its superior when the compression ratio ρ is larger than 60%. The ALDS is a pure low-rank approximation method. When we includes a small amount of sparse components, the model accuracy can be well-maintained. The gap of model accuracy between ALDS and LPSD increases as the compression ration grows. For the 90% compression ratio, the model compressed by LPSD can still have above 70% model accuracy; while the model compressed by ALDS drops its accuracy to nearly 60%.

5   Conclusion and Future Work

We have observed that the low-rank approximation for model compression has it limits, especially for highly compressed models. Based on this observation, we developed the Low-rank Plus Sparse Decomposition (LPSD) algorithm that splits the a network layer into a low-rank part and a sparse part. LPSD algorithm utilizes both global and local approaches to allocate sparsity, and employs the alternative optimization method to determine the low rank and sparse components. We conducted experiments to determine the appropriate allocation ratio of low rank parameters and sparsity for two algorithms. The results show that LPSD outperforms ALDS in terms of model accuracy especially for highly compressed models.

There are several future directions to explore. First, in this paper, we only evaluate LPSD for the image classification task. But CNN has a wide range of applications, such as segmentation or object detection. More experiments on various models for different tasks are required to shows the usefulness of LPSD. Second, in our experiments, we found different layers of the network have various requirements of sparsity. The structure of some layer seems more sparse than others. Therefore, how to design a method to detect the sparsity for each layer in a network requires further study. Last, our experimental results show for highly compressed models, sparsity is more important than the low-rank approximation. How to accelerate the sparse structure will become more and more important for the smaller models. Currently, some hardware vendors have proposed solutions, such as Nvidia's sparse tensor core. How to integrate LPSD with their solutions is an important task.

Appendix

A. Optimality of Sparsity Selection

The optimality of sparsity selection method can be proven by the Eckart-Young-Mirsky theorem [11]. Let A be an $m \times n$ matrix, and $nnz(A)$ be the number of nonzero elements pf A. The norm used is Frobenius norm, whose definition is $\|A\|_F = \sqrt{\sum_{i=1}^m \sum_{j=1}^n A_{i,j}^2}$, where $A_{i,j}$ is the (i,j)th element of A. The following lemma shows how to find the optimal sparse matrix S to minimize $\|A - S\|_F$.

Lemma 1. *Let A be an $m \times n$ matrix. The solution to minimize $\|A-S\|_F$ such that $nnz(S)=s$ is the matrix T that contains only the largest s $|A_{i,j}|$ elements at the same indices, and other elements are zeros.*

The proof is straightforward, since $\|A - S\|_F^2 = \sum_{i=1}^m \sum_{j=1}^n (A_{i,j} - S_{i,j})^2$. It minimal value can be obtained by removing the s largest $|A_{i,j}|$ elements, which is equivalent to make $(A_{i,j} - S_{i,j}) = 0$ for those largest elements in magnitude.

B. Error Estimation

Theorem 1 *If a collection of data with size n is in normal distribution with mean 0, then top-k squares sum can be estimated by the formula:*

$$n\frac{-2\sigma^2}{\sigma\sqrt{2\pi}}\left[-te^{\frac{-t^2}{2\sigma^2}} - \sigma\sqrt{2\pi} + \sigma\sqrt{2\pi}F_X(t)\right],$$

where $t = F_X^{-1}\left(1-\frac{k}{2n}\right)$, $F_X(t)$ is the cumulative distribution function of $f_X(t) = \frac{1}{\sigma\sqrt{2\pi}}e^{\frac{-t^2}{2\sigma^2}}$, the probability density function of normal distribution with mean 0.

Proof. Let $X \sim N(\mu = 0, \sigma^2)$ be the random variable of the data, the probability density function of X is $f_X(t) = \frac{1}{\sigma\sqrt{2\pi}}e^{\frac{-t^2}{2\sigma^2}}$. Now consider another random variable $Y = X^2$. We can find out the probability density function of Y by:

$$f_Y(y) = \frac{d}{dy}Pr(Y \leq y) = \frac{d}{dy}Pr(-\sqrt{y} \leq X \leq \sqrt{y}) = \frac{d}{dy}\int_{-\sqrt{y}}^{\sqrt{y}} f_X(x)\,dx$$

We can rewrite $f_Y(y)$ as:

$$f_Y(y) = \frac{d}{dy}\int_{-\sqrt{y}}^{\sqrt{y}} f_X(x)\,dx = \frac{d}{dy}F_X(x)\Big|_{-\sqrt{y}}^{\sqrt{y}}$$
$$= \frac{d}{dy}\left(F_X(\sqrt{y}) - F_X(-\sqrt{y})\right) = f_X(\sqrt{y})\frac{1}{2\sqrt{y}} + f_X(-\sqrt{y})\frac{1}{2\sqrt{y}} = \frac{1}{\sqrt{y}}f_X(\sqrt{y})$$

After obtaining the probability density function of $Y = X^2$, the kth largest square value in data can be found. Assume that the number is t^2 $(t > 0)$, then

$$Pr(Y \le t^2) = 1 - \frac{k}{n}$$

$$\Rightarrow Pr(-t \le X \le t) = 1 - \frac{k}{n}$$

$$\Rightarrow Pr(X > t) = \frac{k}{2n} \Rightarrow Pr(X \le t) = 1 - \frac{k}{2n}$$

$$\Rightarrow t = F_X^{-1}\left(1 - \frac{k}{2n}\right)$$

After obtaining the kth largest square value, t^2, the average of top-k squares can be found by expected value:

$$E[Y|Y \ge t^2] = \frac{\int_{t^2}^{\infty} y f_Y(y)\, dy}{\int_{t^2}^{\infty} f_Y(y)\, dy} = \frac{1}{\frac{k}{n}} \int_{t^2}^{\infty} \frac{y}{\sqrt{y}} f_X(\sqrt{y})\, dy = \frac{1}{\frac{k}{n}} \int_{t^2}^{\infty} \sqrt{y}\, \frac{1}{\sigma\sqrt{2\pi}} e^{\frac{-y}{2\sigma^2}}\, dy$$

Focus on the integral part:

$$\int_{t^2}^{\infty} \sqrt{y}\, \frac{1}{\sigma\sqrt{2\pi}} e^{\frac{-y}{2\sigma^2}}\, dy = \int_{t^2}^{\infty} \sqrt{y}\, \frac{-2\sigma^2}{\sigma\sqrt{2\pi}} e^{\frac{-y}{2\sigma^2}}\, d\left(\frac{-y}{2\sigma^2}\right) = \frac{-2\sigma^2}{\sigma\sqrt{2\pi}} \int_{t^2}^{\infty} \sqrt{y} e^{\frac{-y}{2\sigma^2}}\, d\left(\frac{-y}{2\sigma^2}\right)$$

$$= \frac{-2\sigma^2}{\sigma\sqrt{2\pi}} \left[\sqrt{y} e^{\frac{-y}{2\sigma^2}} \Big|_{t^2}^{\infty} - \int_{t^2}^{\infty} e^{\frac{-y}{2\sigma^2}}\, d(\sqrt{y}) \right] = \frac{-2\sigma^2}{\sigma\sqrt{2\pi}} \left[-te^{\frac{-t^2}{2\sigma^2}} - \sigma\sqrt{2\pi} F_X(\sqrt{y}) \Big|_{t^2}^{\infty} \right]$$

$$= \frac{-2\sigma^2}{\sigma\sqrt{2\pi}} \left[-te^{\frac{-t^2}{2\sigma^2}} - \sigma\sqrt{2\pi} + \sigma\sqrt{2\pi} F_X(t) \right]$$

The top-k squares sum can be estimated by:

$$kE[Y \mid Y \ge t^2] = k \frac{1}{\frac{k}{n}} \frac{-2\sigma^2}{\sigma\sqrt{2\pi}} \left[-te^{\frac{-t^2}{2\sigma^2}} - \sigma\sqrt{2\pi} + \sigma\sqrt{2\pi} F_X(t) \right]$$

$$= n \frac{-2\sigma^2}{\sigma\sqrt{2\pi}} \left[-te^{\frac{-t^2}{2\sigma^2}} - \sigma\sqrt{2\pi} + \sigma\sqrt{2\pi} F_X(t) \right]$$

Corollary 1. *If values of a $a \times b$ matrix W are in normal distribution with mean 0, we can estimate the top-k squares sum by Theorem 1.*

$$n \frac{-2\sigma^2}{\sigma\sqrt{2\pi}} \left[-te^{\frac{-t^2}{2\sigma^2}} - \sigma\sqrt{2\pi} + \sigma\sqrt{2\pi} F_X(t) \right]$$

σ can be estimated by the Frobenius Norm divided by matrix size:

$$\sigma = E[X^2] = \frac{\|W\|_F^2}{n}$$

and $n = ab$

References

1. Cai, J.F., Li, J., Xia, D.: Generalized low-rank plus sparse tensor estimation by fast Riemannian optimization (2022)
2. Chu, B.S., Lee, C.R.: Low-rank tensor decomposition for compression of convolutional neural networks using funnel regularization (2021)
3. Guo, K., Xie, X., Xu, X., Xing, X.: Compressing by learning in a low-rank and sparse decomposition form. IEEE Access **7**, 150823–150832 (2019). https://doi.org/10.1109/ACCESS.2019.2947846
4. Han, S., et al.: DSD: Dense-sparse-dense training for deep neural networks (2017)
5. Hawkins, C., Yang, H., Li, M., Lai, L., Chandra, V.: Low-rank+sparse tensor compression for neural networks (2021)
6. Huang, W., et al.: Deep low-rank plus sparse network for dynamic MR imaging (2021)
7. Idelbayev, Y., Carreira-Perpinan, M.A.: Low-rank compression of neural nets: learning the rank of each layer. In: 2020 IEEE/CVF Conference on Computer Vision and Pattern Recognition (CVPR), pp. 8046–8056 (2020). https://doi.org/10.1109/CVPR42600.2020.00807
8. Kaloshin, P.: Convolutional neural networks compression with low rank and sparse tensor decompositions (2020)
9. Kim, Y.D., Park, E., Yoo, S., Choi, T., Yang, L., Shin, D.: Compression of deep convolutional neural networks for fast and low power mobile applications (2016)
10. Liang, C.C., Lee, C.R.: Automatic selection of tensor decomposition for compressing convolutional neural networks a case study on VGG-type networks. In: 2021 IEEE International Parallel and Distributed Processing Symposium Workshops (IPDPSW), pp. 770–778 (2021). https://doi.org/10.1109/IPDPSW52791.2021.00115
11. Liebenwein, L., Maalouf, A., Gal, O., Feldman, D., Rus, D.: Compressing neural networks: Towards determining the optimal layer-wise decomposition (2021). CoRR **abs/2107.11442**, https://arxiv.org/abs/2107.11442
12. Lin, T., Stich, S.U., Barba, L., Dmitriev, D., Jaggi, M.: Dynamic model pruning with feedback (2020)
13. Otazo, R., Candès, E., Sodickson, D.: Low-rank plus sparse matrix decomposition for accelerated dynamic MRI with separation of background and dynamic components. Magn. Reson. Med. **73**, 1125–1136 (2014). https://doi.org/10.1002/mrm.25240
14. Yin, M., Phan, H., Zang, X., Liao, S., Yuan, B.: BATUDE: budget-aware neural network compression based on tucker decomposition. Proc. AAAI Conf. Artif. Intell. **36**, 8874–8882 (2022). https://doi.org/10.1609/aaai.v36i8.20869
15. Yu, X., Liu, T., Wang, X., Tao, D.: On compressing deep models by low rank and sparse decomposition. In: 2017 IEEE Conference on Computer Vision and Pattern Recognition (CVPR), pp. 67–76 (2017). https://doi.org/10.1109/CVPR.2017.15
16. Zhang, X., Wang, L., Gu, Q.: A unified framework for low-rank plus sparse matrix recovery (2018)

Modeling Treatment Effect with Cross-Domain Data

Bin Han, Ya-Lin Zhang, Lu Yu, Biying Chen, Longfei Li, and Jun Zhou[(✉)]

Ant Group, Hangzhou, China
{binlin.hb,lyn.zyl,bruceyu.yl,biying.cby,longyao.llf,
jun.zhoujun}@antgroup.com

Abstract. Treatment effect estimation has received increasing attention recently. However, the issue of *data sparsity* often poses a significant challenge, limiting the feasibility of modeling. This paper aims to leverage cross-domain data to mitigate the *data sparsity* issue, and presents a framework called **TEC**. **TEC** incorporates a collaborative and adversarial generalization module to enhance information sharing and transferability across domains. This module encourages the learned representations of different domains to be more cohesive, thereby improving the generalizability of the models. Furthermore, we address the issue of poor performance for few-shot samples in each domain, and propose a pattern augmentation module that explicitly borrows samples from other domains and applies the self-teaching philosophy to them. Extensive experiments are conducted on both synthetic and benchmark datasets to demonstrate the superiority of the proposed framework.

Keywords: Treatment Effect Estimation · Cross-Domain Modeling · Representation Learning

1 Introduction

Treatment effect estimation has received increasing attention due to its wide application [1,14,26], and numerous methods have been proposed, leading to significant advancements [9,21]. Broadly speaking, these methods primarily address two widely concerned issues, i.e., *missing counterfactual* and *selection bias* [23]. However, in many scenarios, obtaining sufficient samples is challenging due to cost constraints and other factors, leading to the problem of *data sparsity*, which can significantly degrade the performance [8]. For example, when developing a personalized medical effect estimator for a new anticancer using data collected from a general hospital, the data may be insufficient for effective modeling. As a result, the learned model becomes overly biased towards the limited samples and exhibits poor generalizability to unseen samples. Furthermore, the collected data may contain certain types of patients with very few records, referred to as few-shot samples. The model may perform unsatisfactorily for such cases. These phenomena highlight the importance of addressing the *data sparsity* issue.

B. Han and Y. -L. Zhang—Equal contribution.

© The Author(s), under exclusive license to Springer Nature Singapore Pte Ltd. 2024
D.-N. Yang et al. (Eds.): PAKDD 2024, LNAI 14645, pp. 365–377, 2024.
https://doi.org/10.1007/978-981-97-2242-6_29

Previous studies may provide potential solutions for the *data sparsity* issue. For instance, efforts have been dedicated to combining randomized controlled trials (RCTs) and observational trials [4,7], while RCTs are not always available due to high cost. Other works generalized the problem to unsupervised domain adaptation setting, where only unlabeled samples are available for the target domain [12,20]. However, the lack of supervision severely limits the performance. A recent work [3] studied the heterogeneous transfer learning problem for CATE estimation, where the feature space of different domains is heterogeneous, this is not our primary focus. This paper aims to leverage cross-domain data to mitigate the *data sparsity* issue in treatment effect estimation.

Estimating treatment effects using cross-domain data is a viable choice in various fields. Take the healthcare example again, besides the general hospital, data may be also available from a tumor hospital, where there is a need for a medical effect estimator as well. Constructing the model using data from different domains is a natural choice. Intuitively, samples from different domains are likely to exhibit a high degree of overlap, providing mutual complementation and enhancing generalizability. Furthermore, the underlying decision functions of different domains may be closely interconnected, allowing for information sharing and transferability between domains, which is feasible and beneficial.

However, several challenges should be addressed when modeling with cross-domain data. One is the potential existence of distribution shifts and variations in outcome functions across domains, which hinders information sharing and transferability between domains. Additionally, poor performance may arise for few-shot samples, as limited supervision is available for them.

In this paper, we present a novel framework named **TEC**. Specifically, to enhance the information sharing and transferability across domains, we propose a collaborative and adversarial generalization module that encourages the learned representations of different domains to be more cohesive, thereby improving the generalizability of the models. To address the issue of poor performance with few-shot samples, we propose a pattern augmentation module. This module explicitly borrows samples from other domains to generate augmented patterns, and applies the self-teaching strategy to the augmented samples, leading to more stable and improved performance. We conduct extensive experiments and provide comprehensive results to demonstrate the superiority of **TEC**.

2   Preliminaries

Problem Setup. Let $D = \{(\boldsymbol{x}_i, t_i, y_i, s_i)\}_{i=1}^{N}$ be observational data consisting of N independent samples. Here, $\boldsymbol{x} \in \mathcal{X} \subseteq \mathbb{R}^d$ represents the d-dimensional covariate vector, $t \in \mathcal{T} = \{0,1\}$ represents the binary treatment, $y \in \mathcal{Y}$ is the continuous outcome, $s \in \{0,1\}$ is the domain indicator with two domains. We denote $X = \{\boldsymbol{x}_i\}_{i=1}^{N}$ as the collection of the covariates. Further, we denote $D_0 = \{(\boldsymbol{x}_i, t_i, y_i, s_i = 0)\}_{i=1}^{N_0}$ ($D_1 = \{(\boldsymbol{x}_i, t_i, y_i, s_i = 1)\}_{i=1}^{N_1}$) as the data from domain 0 (domain 1) with N_0 (N_1) samples. Note that we have $N = N_0 + N_1$, $D = D_0 \cup D_1$, and we can similarly denote $X = X_0 \cup X_1$. Each $\boldsymbol{x}$ of D_0 (D_1) is drawn from the distribution $P_0(x)$ $(P_1(x))$. Note that $P_0(x)$ is not necessarily equal to $P_1(x)$. Moreover, the outcome function may vary across domains, meaning that $P_0(y|x)$ is not necessarily equal to $P_1(y|x)$. We assume the treatments are homogeneous across domains, as the basis for information transfer. We aim to improve the performance for both domains.

Following the Neyman-Rubin potential outcomes framework [17], we denote Y_1 (Y_0) as the potential outcome under treatment (control). Our main interest is to estimate the Conditional Average Treatment Effect (CATE), i.e., $\tau(x) = \mathbb{E}[Y_1 - Y_0|X = x] = \mu_1(x) - \mu_0(x)$, which is the expected treatment effect for a unit with the covariate $X = x$. Here, $\mu_1(x) = \mathbb{E}[Y_1|X = x]$ $(\mu_0(x) = \mathbb{E}[Y_0|X = x])$ is the treated (controlled) response function. Building upon the standard assumptions [24] (*SUTVA, Consistency, Strong ignorability*) in the potential outcomes framework, we can express the conditional expectation of potential outcomes as $\mathbb{E}[Y_t|X = x] = \mathbb{E}[Y|X = x, T = t]$ [13], and it can be estimated from the observational data. We extend these assumptions to our setting.

The Fundamental Framework. Representation learning based methods have been extensively developed to solve the treatment effect estimation problem, with the multi-head architecture being widely used [18,19], as simplified in Fig. 1. The overall loss can be roughly denoted as $\mathcal{L} = \mathcal{L}_{FL} + \alpha\mathcal{L}_{DL}$, where $\mathcal{L}_{FL}$ is the factual loss used to encourage the estimated outcomes to approach the factual outcomes, $\mathcal{L}_{DL}$ is the debias loss used to address the selection bias issue, and α is a balance hyperparameter. For simplicity, we omit other regularization terms.

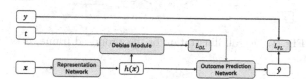

Fig. 1. The fundamental architecture for treatment effect estimation models.

3 Methodology

Overview. We design **TEC** based on representation learning. **TEC** incorporates a collaborative and adversarial generalization (CA) module, which encourages the learned representations of different domains to be more unified, promoting information sharing and transferability across domains. Furthermore, we address the issue of unsatisfactory performance for few-shot samples, and introduce a pattern augmentation (PA) module that utilizes samples from neighboring domains to generate augmented patterns and leverages them with a self-teaching strategy, leading to improved performance. Our whole framework is illustrated in Fig. 2, and the overall loss can be concluded as:

$$\mathcal{L} = \mathcal{L}_{FL} + \alpha\mathcal{L}_{DL} + \beta\mathcal{L}_{CA} + \delta\mathcal{L}_{PA}, \tag{1}$$

In which $\mathcal{L}_{FL}$ and $\mathcal{L}_{DL}$ are the factual and debias loss respectively, $\mathcal{L}_{CA}$ and $\mathcal{L}_{PA}$ are the loss of the CA and PA module, and α, β, δ are used to balance the losses.

We adopt the Inverse Propensity Weighting (IPW) based strategy [16], which includes a classification sub-task to predict the propensity score for each sample and assigns a weight accordingly. The propensity score $e(\boldsymbol{x}, s)$ is defined as the conditional probability of receiving treatment given the sample $\boldsymbol{x}$ and its domain indicator s, i.e., $e(\boldsymbol{x}, s) = P(t = 1|\boldsymbol{x}, s)$, and its weight w can be calculated as $w = \frac{t}{e(\boldsymbol{x},s)} + \frac{1-t}{1-e(\boldsymbol{x},s)}$. The debias loss term can be interpreted as $\mathcal{L}_{DL} = \frac{1}{N}\sum_{i=1}^{N}(\hat{t}_i - t_i)^2$, where t_i represents the factual treatment assignment, and $\hat{t}_i$ is the predicted probability of being treated, obtained using a Multi-Layer Perceptron (MLP). The factual loss term is as $\mathcal{L}_{FL} = \frac{1}{N}\sum_{i=1}^{N} w_i(\hat{y}_i - y_i)^2$, where $\hat{y}_i = M(\boldsymbol{x}_i, t_i, s_i)$ is the predicted outcome of sample $\boldsymbol{x}_i$ from s_i with treatment t_i, and $M(\cdot)$ denotes the outcome function implemented with a MLP, and y_i is the associated factual outcome. Note that in our setting, multiple domains are involved, while the above loss terms remain tenable although the domain information is not explicitly written out. We will elaborate on the CA and PA modules subsequently.

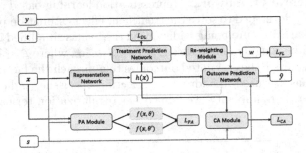

Fig. 2. Overall illustration of the proposed framework.

3.1   Collaborative and Adversarial Generalization

While the complementation of samples and the similarity of outcome functions offer the possibility of improving performance by using cross-domain data, challenges arise due to distribution shifts among domains and discrepancies in outcome functions. Absent a carefully calibrated approach, these discrepancies can lead to inferior performance. Within the scope of representation learning, these issues are intrinsically linked to the quality of the learned embeddings. If the embeddings of samples from different domains exhibit significant variation, it hampers information sharing and transferability, resulting in unsatisfactory performance. The proposed Collaborative and Adversarial Generalization (CA) module aims to encourage more unified latent representations across domains. This facilitates enhanced information sharing and transferability. Specifically, a collaborative module is introduced to identify representative samples, and then a metric is optimized to minimize the discrepancy among these representatives in the embedding space.

Concretely, we introduce a sub-network to predict the domain information from the latent representation $h(\boldsymbol{x})$ of $\boldsymbol{x}$. We denote $\hat{P}(s|h(\boldsymbol{x}))$ as the predicted possibility of $\boldsymbol{x}$ belonging to source s, which can be simply achieved using a MLP. The underlying idea is to reduce the discriminability of the samples' latent representations concerning the domain information. To this end, we define the *distinguishableness score* $DS(\boldsymbol{x}_i)$ for each sample $\boldsymbol{x}_i$ from domain s_i as follows:

$$DS(\boldsymbol{x}_i) = \hat{P}(s_i|h(\boldsymbol{x}_i)) - \hat{P}(1 - s_i|h(\boldsymbol{x}_i)). \tag{2}$$

Intuitively, higher values of $DS(\boldsymbol{x}_i)$ indicate that $\boldsymbol{x}_i$ is more representative of domain s_i. For each domain, we select samples with higher distinguishableness scores to form the distinguishable sample set, i.e.,

$$\hat{D}_0 = \{\boldsymbol{x}_i|\boldsymbol{x}_i \in X_0, DS(\boldsymbol{x}_i) \geq \alpha_0\}, \hat{D}_1 = \{\boldsymbol{x}_i|\boldsymbol{x}_i \in X_1, DS(\boldsymbol{x}_i) \geq \alpha_1\}. \tag{3}$$

where $\hat{D}_j$ represents the set of the selected samples from domain j, and α_j is the ad-hoc threshold to select the top k samples ($j \in \{0, 1\}$). Additionally, we select the sample $\hat{\boldsymbol{x}}_{s_j}$ with the lowest $|DS(\boldsymbol{x}_i)|$ as the most *indistinguishable* sample for each domain[1]. To mitigate the overbiased representation problem, we aim to eliminate the domain-specific factors in the latent space. Concretely, using the selected samples, we minimize the distance between the *distinguishable* samples from domain j and the *indistinguishable* sample from domain $(1 - j)$, where $j \in \{0, 1\}$. The loss is as:

$$\mathcal{L}_{DA} = \frac{1}{k}(\sum_{\boldsymbol{x}_i \in \hat{D}_0} d(h(\boldsymbol{x}_i), h(\hat{\boldsymbol{x}}_{s_1})) + \sum_{\boldsymbol{x}_j \in \hat{D}_1} d(h(\boldsymbol{x}_j), h(\hat{\boldsymbol{x}}_{s_0}))), \tag{4}$$

where $d(\cdot)$ is the distance function (we use the L_2 distance in our implementation, and other options are also possible). Note that a collaborative classification task is involved to predict the domain information, and its loss is denoted as:

$$\mathcal{L}_{CC} = -\frac{1}{N} \sum_{i=1}^{N} [s_i * \log(\hat{P}(s_i|h(\boldsymbol{x}_i))) + (1 - s_i) * \log(\hat{P}(1 - s_i|h(\boldsymbol{x}_i)))]. \tag{5}$$

In summary, the CA module involves a collaborative sub-task to identify representative samples, and the latent representations are adversarially driven by $\mathcal{L}_{DA}$. We refer to this approach as the Collaborative and Adversarial Generalization strategy, and the loss for this module can be summarized as:

$$\mathcal{L}_{CA} = \mathcal{L}_{DA} + \lambda\mathcal{L}_{CC}, \tag{6}$$

where λ is a parameter to balance the two terms.

3.2   Pattern Augmentation

Due to the *data sparsity* issue, few-shot samples exist extensively in each domain, which are always with unsatisfactory predictions. As shown in the upper figures in Fig. 3a, for both domains, there may exist underrepresented regions (the shadow regions) with extremely limited samples, and the learned models will exhibit unstable and unsatisfactory predictions for these regions. For example, when examining the predictions for sample P in Fig. 3a, the prediction may resemble the blue line in Fig. 3b as the training progresses, which is undesirable.

---

[1] We use a classification threshold of 0.5 for explanation purposes.

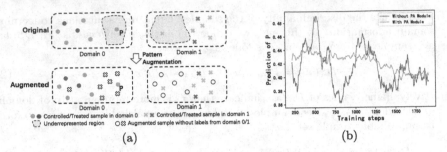

Fig. 3. Illustration of the pattern augmentation module. (a): Samples before and after the augmentation process in PA module. (b): The prediction of P with/without PA module as training progresses.

The multi-domain situation presents an opportunity to mitigate this obstacle, as complementary data can be accessed for each domain. For ease of explanation, we define a pattern $(\boldsymbol{x}, t, s)$ as the combination of a covariate $\boldsymbol{x}$, a treatment assignment t, and a domain assignment s. The few-shot samples can be then regarded as the corresponding few-shot patterns. Initially, only the observed patterns $(\boldsymbol{x}_i, t_i, s_i)$ in the original dataset are encountered for the models, while the models may yield unsatisfactory predictions for the unobserved patterns. In this module, we propose a pattern augmentation strategy to obtain an augmented dataset $\tilde{D} = \tilde{D}_0 \cup \tilde{D}_1$, which can be detailed as:

$$\tilde{D}_0 = \{(\boldsymbol{x}_i, t_i, s_i = 0)|\boldsymbol{x}_i \in X_1, t_i \in \mathcal{T}\}, \tilde{D}_1 = \{(\boldsymbol{x}_i, t_i, s_i = 1)|\boldsymbol{x}_i \in X_0, t_i \in \mathcal{T}\}. \quad (7)$$

Note that when constructing $\tilde{D}$, we only borrow the covariates $\boldsymbol{x}_i$ from X, while the outcome information y_i is not used due to the potential outcome function discrepancy. Furthermore, to make full use of the augmented dataset $\tilde{D}$, we propose to take a self-teaching strategy [25]. Inspired by [22,25], we also take the predictions in the early stage as the guidance for the subsequent training stage. Additionally, we extend to use the running average of the predictions as the targets. Specifically, we denote the model's predicted outcome for pattern $(\boldsymbol{x}_i, t_i, s_i)$ as $p_i = M(\boldsymbol{x}_i, t_i, s_i)$, in which $M(\cdot)$ is the aforementioned outcome function. In the k-th iteration, we define loss $\mathcal{L}_{PA}$ as the expected distance between the current prediction p_i^k and the running average prediction q_i^k for $\boldsymbol{x}_i$, i.e.,

$$\mathcal{L}_{PA} = \sum_{(\boldsymbol{x}_i, t_i, s_i) \in \tilde{D}} d(p_i^k, q_i^k), \quad (8)$$

where $d(\cdot)$ calculates the distance of p_i^k and q_i^k, which is selected as L_2 distance in our work. The running average prediction q_i^k at the k-th iteration is as

$$q_i^k = \eta q_i^{k-1} + (1 - \eta) p_i^k, \quad (9)$$

in which η is a balance parameter. With this module, the model can be further improved, as exemplified in the orange line for P in Fig. 3b.

3.3 Further Extension and Discussion

This work aims at exploring an effective framework to leverage the cross-domain data for treatment effect estimation. We focus on addressing the issue of *data sparsity* by introducing beneficial modules into existing backbone networks. Importantly, we do not impose any constraints on the backbones. We believe that these modules can be easily incorporated into most available backbone networks. In our experiments, we mainly focus on the basic architecture of TAR-Net [18] and the extended one with disentangled representations [6] to validate the effectiveness. However, other variants, such as the one with local similarity preservation [24], can be also explored by substituting the backbones. We acknowledge that there is still room for further exploration in future work.

4 Experiments

4.1 Datasets

Synthetic Dataset. Assuming a two domain scenario with a binary treatment condition, we set $d = 4$, so $x = \{x^1, x^2, x^3, x^4\}$, and x^j denotes the j-th feature of x. For both domains, we first generate some original samples from the normal distribution, i.e., $x \sim \mathcal{N}([0,0,0,0]', I)$, where $\mathcal{N}$ is the normal distribution and I is the identity matrix.

To simulate the *distribution shift* between domains, we further resample the above-generated samples for both domains. Each sample is assigned a value of r, where $r = 1$ indicates that the sample is reserved during resampling. The domain-related probability function for r can be defined as follows:

$$P(r = 1|x, s) = \epsilon(0.6 - x^2)(1 - s) + 0.05\epsilon(x^2 - 0.6)(1 - s) \\ + \epsilon(0.5 - x^1)s + 0.05\epsilon(x^1 - 0.5)s, \tag{10}$$

where $\epsilon(z)$ is the unit step function, returning 0 if $z < 0$, and 1 otherwise. To simulate the *selection bias* in each domain, we assign the treatment as follows:

$$P(t = 1|x, s) = 0.01\epsilon(-x^3) + 0.99\epsilon(x^2 - 0.5)s + 0.99\epsilon(0.5 - x^1)(1 - s). \tag{11}$$

To simulate the discrepancy for the *behavior patterns*, the control and treated response functions are varied across domains, which can be written as $\mu_0(x, s)$ and $\mu_1(x, s)$. The effect function is $\tau(x, s) = \mu_1(x, s) - \mu_0(x, s)$. Specifically,

$$\mu_0(x, s) \sim \mathcal{N}(g_1(x), 0.1) + s\mathcal{N}(g_2(x), 0.02), \\ \tau(x, s) \sim \mathcal{N}(g_3(x), 0.05) + s\mathcal{N}(g_4(x), 0.02), \tag{12}$$

where $g_1(x) = \sin x^1 + \cos(x^2 + x^3), g_2(x) = -0.5\sin x^1 + \cos(x^1 + x^2), g_3(x) = 0.3x^4 + \cos x^1 + \sin x^2, g_4(x) = 0.1x^4 + 0.1\sin x^1 + 0.1\sin x^2$.

2000 samples are simulated for training, and 10000 samples are used for test in each domain. For test, the treatment assignment is randomized, and the distribution of x remains the same as the original one without distribution shift.

ACIC Dataset. ACIC dataset is introduced by [5]. We cut 4,000 samples for training and 800 samples for testing. To create a domain distribution shift, the whole dataset is split into two domains by the value of a binary feature x^{38}. The treatment assignment is defined as:

$$P(t = 1|\boldsymbol{x}, s) = sx^{20}x^{43} + 0.5x^{24} + (1 - s)x^{45}x^{20} \tag{13}$$

Also, we add two terms to make the outcome functions different between domains. Assume the original outcome in the dataset is $Y_0(t)$, we define the modified outcome as:

$$\begin{aligned} \mu_0(\boldsymbol{x}, s) &\sim Y_0(0) + s\mathcal{N}\left(g_5(\boldsymbol{x}), 0.1\right) \\ \tau(\boldsymbol{x}, s) &\sim Y_0(1) - Y_0(0) + s\mathcal{N}\left(g_6(\boldsymbol{x}), 0.1\right) \end{aligned} \tag{14}$$

where $g_5(\boldsymbol{x}) = 0.5x^8 x^{43} + 0.5\sin(x^{57} + x^{43}) + 0.1\cos(x^{54} + x^{27})$, $g_6(\boldsymbol{x}) = 0.5x^{27} + 0.5\sin x^{40} + 0.5\sin x^{30} + 0.5\cos(x^{10} + x^{20})$.

4.2   Setup

Compared Methods. Our framework is not limited to specific backbone. This section reports the experimental results with two different backbones to validate the effectiveness. We first adapt the extensively used TARNet [18] for our method by incorporating domain information into the outcome prediction network. For comparison, we first adapt the widely used meta-learners [11], including **T-Learner (TL)**, **S-Learner (SL)**, by utilizing s as a feature.

Some other NN-based methods, such as the ones adapted from TARNet and CFRNet [18], are also compared, which can be found in the following competitors. The basic architecture of them is illustrated in Fig. 4. Specifically, (i) We denote the basic version of the adapted method for the studied setting as **TAR**$_\alpha$, which does not explicitly address distribution shift issues and is similar to TAR-Net [18]. (ii) To tackle the selection bias issue between the control and treated groups, the re-weighting based strategy [15] is extended based on **TAR**$_\alpha$, i.e., the $\mathcal{L}_{DL}$ term in Fig. 4, and we denote it as **TAR**$_\beta$. (iii) To handle the distribution shift issue between domains, we extend another two versions based on **TAR**$_\beta$ by employing MMD and Wasserstein distances [18] to perform the representation matching for the two domains, i.e., the term $\mathcal{L}_{DG}$ in Fig. 4. We denote them as **TAR**$_{\gamma M}$ and **TAR**$_{\gamma W}$. Our method, which uses the TARNet backbone and includes the PA and CA modules, is named **TEC**. It can be seen as an extension of **TAR**$_\beta$. Furthermore, We replace the TARNet backbone with a recent one that incorporates disentangled representations [6] and obtain another set of compared methods, i.e., **DR**$_\alpha$, **DR**$_\beta$, **DR**$_{\gamma M}$, **DR**$_{\gamma W}$ and **TEC**^{DR}.

Moreover, for **TAR**$_\beta$, **TAR**$_{\gamma M}$ and **TAR**$_{\gamma W}$, we switch from the re-weighting-based strategy to the MMD-based strategy to handle the selection bias issue, and the corresponding variants are denoted as **MTAR**$_\beta$, **MTAR**$_{\gamma M}$ and **MTAR**$_{\gamma W}$. Similarly, for the **DR** related methods, we also test the corresponding variants, denoted as **MDR**$_\beta$, **MDR**$_{\gamma M}$ and **MDR**$_{\gamma W}$.

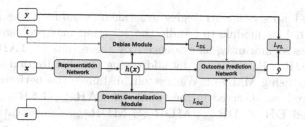

Fig. 4. The fundamental architecture of the compared NN-based methods.

Implementation Details. These meta-learners are implemented by the third-party library EconML [2] with 200 trees in ensemble models, a learning rate of 0.1 and a max depth of 6. All the compared NN-based methods were trained using ADAM optimizer [10] with an initial learning rate of 0.0005, a learning rate decay of 0.96 every 3000 step, and a batch size of 512. For the remaining hyperparameters, we conducted preliminary experiments to search for the best combinations, and repeated the experiments 10 times to obtain the final reported results with the identified hyperparameters.

Evaluation Metrics. We use the Precision in Estimation of Heterogeneous Effects (PEHE) [18] and the Absolute Error of Heterogeneous Effects (AEHE) to estimate the performance. Assume the treatment effect estimation of unit i is $\hat{\tau}(x_i)$, PEHE and AEHE can be written as:

$$\text{PEHE} = \frac{1}{N} \sum_{i=1}^{N} (\tau(x_i) - \hat{\tau}(x_i))^2, \qquad (15)$$

$$\text{AEHE} = \frac{1}{N} \sum_{i=1}^{N} |\tau(x_i) - \hat{\tau}(x_i)| \qquad (16)$$

Goals of the Experiments. Through our experiments, we aim to answer the following questions:

- **RQ1** How do our proposed methods compare to other baselines in terms of performance?
- **RQ2** How do our proposed methods perform when different backbone networks are used?
- **RQ3** How does each module in our framework affect the performance?
- **RQ4** How do the compared methods perform in real-world tasks?

4.3 Experimental Results and Analysis

Main Results (RQ1). The main results are summarized in Table 1. Regardless of the backbone network used, the methods of the basic version (**TAR**$_\alpha$ and **DR**$_\alpha$) perform far from satisfactory. This is expected, as distribution shifts exist and no specific strategies are taken to address the issue. As we introduce the debias module to handle the selection bias problem, the performance improves significantly, as evidenced by the results of $\mathbf{TAR}_\beta$, $\mathbf{MTAR}_\beta$ and $\mathbf{DR}_\beta$, $\mathbf{MDR}_\beta$. Furthermore, by addressing the distribution shift problem across domains using MMD or Wasserstein distance, the performance can be further enhanced, as observed in the results of $\mathbf{TAR}_{\gamma M}$, $\mathbf{TAR}_{\gamma W}$, $\mathbf{MTAR}_{\gamma M}$, $\mathbf{MTAR}_{\gamma W}$, and $\mathbf{DR}_{\gamma M}$, $\mathbf{DR}_{\gamma W}$, $\mathbf{MDR}_{\gamma M}$, $\mathbf{MDR}_{\gamma W}$. Consistently, our methods outperform the corresponding compared ones. These results provide a rough answer to **RQ1** and demonstrate that our methods can achieve more competitive results compared to the corresponding baselines.

Table 1. The main results of the compared methods.

Dataset Metrics	Synthetic Dataset PEHE S_0	S_1	AEHE S_0	S_1	ACIC Dataset PEHE S_0	S_1	AEHE S_0	S_1		Synthetic Dataset PEHE S_0	S_1	AEHE S_0	S_1	ACIC Dataset PEHE S_0	S_1	AEHE S_0	S_1
SL	0.136	0.166	0.289	0.324	0.796	1.066	0.692	0.780	TL	0.160	0.180	0.306	0.322	0.702	0.929	0.610	0.668
TAR_α	0.124	0.119	0.229	0.237	0.769	1.065	0.638	0.730	DR_α	0.093	0.091	0.201	0.205	0.718	1.030	0.620	0.719
TAR_β	0.089	0.097	0.204	0.216	0.628	0.963	0.570	0.687	DR_β	0.076	0.070	0.179	0.184	0.614	0.888	0.556	0.639
$TAR_{\gamma M}$	0.072	0.072	0.176	0.187	0.589	0.876	0.553	0.636	$DR_{\gamma M}$	0.063	0.060	0.164	0.166	0.533	0.801	0.524	0.608
$TAR_{\gamma W}$	0.078	0.080	0.187	0.194	0.581	0.824	0.537	0.628	$DR_{\gamma W}$	0.069	0.063	0.172	0.178	0.528	0.734	0.504	0.584
$MTAR_\beta$	0.089	0.088	0.205	0.214	0.618	0.930	0.563	0.672	MDR_β	0.072	0.070	0.175	0.184	0.595	0.849	0.553	0.634
$MTAR_{\gamma M}$	0.077	0.071	0.186	0.187	0.586	0.866	0.550	0.627	$MDR_{\gamma M}$	0.064	0.060	0.163	0.174	0.519	0.761	0.519	0.599
$MTAR_{\gamma W}$	0.083	0.081	0.191	0.207	0.563	0.796	0.533	0.626	$MDR_{\gamma W}$	0.071	0.075	0.174	0.193	0.522	0.706	0.501	0.583
TEC	0.050	0.052	0.151	0.158	0.538	0.689	0.529	0.581	TEC^{DR}	0.049	0.047	0.149	0.149	0.409	0.601	0.465	0.531

Results with Extended Backbones (RQ2). In Table 1, the results with different backbones can be also compared. Compared to the **TAR** versions, the **DR** versions are with the extension of learning disentangled representations. As we can see, the latter is consistently more effective. We conjecture this phenomenon may be related to the underlying outcome functions. If the underlying outcome functions conform to the disentangled assumption, the **DR** versions are more likely to perform better. Additionally, the **DR** versions offer better flexibility due to the disentangled representations. Furthermore, when combined with our proposed modules, the performance of both **TAR** and **DR** versions is significantly improved compared to their respective competitors. These results provide a preliminary answer to **RQ2** and demonstrate that our proposed method can achieve consistent superiority with varied backbones.

Ablation Study (RQ3). To evaluate the impact of the CA and PA modules, we further conduct some ablation studies. The results are presented in Table 2, which includes the performance of **TEC** and two variants where either the CA or PA module is removed. We also included the results of $\mathbf{TAR}_\alpha$ and $\mathbf{TAR}_\beta$ for comparison. The results show that removing either of the two modules leads to a decrease in performance, although the models still outperform $\mathbf{TAR}_\alpha$ and $\mathbf{TAR}_\beta$. These findings provide a preliminary answer to **RQ3** and indicate that both the CA and PA modules play critical roles in improving the overall performance of the framework.

Table 2. The results of the ablation studies on synthetic dataset.

Dataset	Synthetic Dataset			
Metrics	PEHE		AEHE	
	S_0	S_1	S_0	S_1
$\mathbf{TAR}_\alpha$	0.124	0.119	0.229	0.237
$\mathbf{TAR}_\beta$	0.089	0.097	0.204	0.205
w/o PA	0.066	0.064	0.170	0.170
w/o CA	0.060	0.059	0.161	0.164
TEC	**0.050**	**0.052**	**0.151**	**0.158**

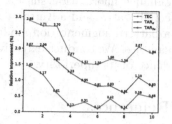

Fig. 5. Relative improvement (%) of the compared methods.

4.4 Online Application (RQ4)

This subsection presents the performance of **TEC** in handling real-world applications, using a recommendation task as an example. In this task, we have two distinct user groups with significantly different distributions and behaviors. We consider the modeling for these two groups as a cross-domain treatment effect estimation task, aiming to assign recommendations to users effectively to improve user experience and increase click-through rate.[2]

All users were randomly divided into multiple groups, and each compared method was applied to one group. The control group, which received no specific recommendation, served as the baseline. We report the relative improvements of other models compared to the control group. We included a basic model that did not address the distribution shift issue, similar to $\mathbf{TAR}_\alpha$. Additionally, we compared with $\mathbf{TAR}_{\gamma_M}$. For our method, we compared with the **TEC** version. The backbones of the compared models were identical for fair comparison.

The average relative improvements of these methods compared to the control group, measured by the click-through rate (CTR) metric over a period of 10 days, are shown in Fig. 5. As observed, all model-based methods achieved better CTR compared to the control group, with our proposed method consistently demonstrating superiority over the other models. Specifically, the average relative improvement of **TEC** compared to the control group was 2.03, which is significantly better than the results of $\mathbf{TAR}_\alpha$ (0.54) and $\mathbf{TAR}_{\gamma_M}$ (1.21). These results demonstrate the effectiveness of **TEC** in handling real-world tasks, providing an answer to **RQ4**.

5 Conclusion

This paper presents a novel framework named **TEC** for leveraging cross-domain data to benefit treatment effect estimation. We address not only the *missing counterfactual* and *selection bias* issues but also the *data sparsity* problem. To

---

[2] Due to commercial confidentiality, we omit some details here and below.

enhance information sharing and transferability across domains, we propose a collaborative and adversarial generalization module. Additionally, we introduce a pattern augmentation module to address the underrepresentation of few-shot samples. We conduct extensive experiments, and the results provide strong evidence of its superiority.

References

1. Athey, S., Imbens, G.: Recursive partitioning for heterogeneous causal effects. PNAS **113**(27), 7353–7360 (2016)
2. Battocchi, K., et al.: EconML: A python package for ml-based heterogeneous treatment effects estimation (2019)
3. Bica, I., van der Schaar, M.: Transfer learning on heterogeneous feature spaces for treatment effects estimation. In: NeurIPS (2022)
4. Dahabreh, I.J., et al.: Study designs for extending causal inferences from a randomized trial to a target population. Am. J. Epidemiol. **190**(8), 1632–1642 (2021)
5. Dorie, V., Hill, J., Shalit, U., Scott, M., Cervone, D.: Automated versus do-it-yourself methods for causal inference: lessons learned from a data analysis competition. Stat. Sci. **34**(1), 43–68 (2019)
6. Hassanpour, N., Greiner, R.: Learning disentangled representations for counterfactual regression. In: ICLR (2020)
7. Hatt, T., Berrevoets, J., Curth, A., Feuerriegel, S., van der Schaar, M.: Combining observational and randomized data for estimating heterogeneous treatment effects (2022). arXiv:2202.12891
8. Huang, Q., Ma, J., Li, J., Sun, H., Chang, Y.: SemiiTE: semi-supervised individual treatment effect estimation via disagreement-based co-training. In: ECML PKDD, pp. 400–417 (2023)
9. Johansson, F., Shalit, U., Sontag, D.: Learning representations for counterfactual inference. In: ICML, pp. 3020–3029 (2016)
10. Kingma, D.P., Ba, J.: Adam: A method for stochastic optimization (2014). arXiv:1412.6980
11. Künzel, S.R., Sekhon, J.S., Bickel, P.J., Yu, B.: Meta-learners for estimating heterogeneous treatment effects using machine learning. PNAS **116**(10), 4156–4165 (2019)
12. Kyono, T., Bica, I., Qian, Z., van der Schaar, M.: Selecting treatment effects models for domain adaptation using causal knowledge. Health **4**(2), 1–29 (2023)
13. Pearl, J.: Causality. Cambridge University Press, Cambridge (2009)
14. Powers, S., et al.: Some methods for heterogeneous treatment effect estimation in high-dimensions. Stat. Med. **37**(11), 1767–1787 (2018)
15. Rosenbaum, P.R.: Model-based direct adjustment. JASA **82**(398), 387–394 (1987)
16. Rosenbaum, P.R., Rubin, D.B.: The central role of the propensity score in observational studies for causal effects. Biometrika **70**(1), 41–55 (1983)
17. Rubin, D.B.: Causal inference using potential outcomes: design, modeling, decisions. JASA **100**(469), 322–331 (2005)
18. Shalit, U., Johansson, F.D., Sontag, D.: Estimating individual treatment effect: generalization bounds and algorithms. In: ICML, pp. 3076–3085 (2017)
19. Shi, C., Blei, D., Veitch, V.: Adapting neural networks for the estimation of treatment effects. In: NeurIPS, vol. 32 (2019)

20. Sun, Y., Zhang, Y., Wang, W., Li, L., Zhou, J.: Treatment effect estimation across domains. In: CIKM, pp. 2352—2361 (2023)
21. Tang, C., et al.: Debiased causal tree: heterogeneous treatment effects estimation with unmeasured confounding. In: NeurIPS, vol. 35 (2022)
22. Tarvainen, A., Valpola, H.: Mean teachers are better role models: weight-averaged consistency targets improve semi-supervised deep learning results. In: NeurIPS, vol. 30 (2017)
23. Yao, L., Chu, Z., Li, S., Li, Y., Gao, J., Zhang, A.: A survey on causal inference. TKDD **15**(5), 1–46 (2021)
24. Yao, L., Li, S., Li, Y., Huai, M., Gao, J., Zhang, A.: Representation learning for treatment effect estimation from observational data. In: NeurIPS, vol. 31 (2018)
25. Zhang, Y., Zhou, J., Shi, Q., Li, L.: Exploring the combination of self and mutual teaching for tabular-data-related semi-supervised regression. ESWA **213**, 118931 (2023)
26. Zhang, Y., et al.: A backcasting framework for approximating macro-outcome via micro-treatment. SSRN 4494664 (2023)

Author Index

© The Editor(s) (if applicable) and The Author(s), under exclusive license
to Springer Nature Singapore Pte Ltd. 2024
D.-N. Yang et al. (Eds.): PAKDD 2024, LNAI 14645, pp. 379–380, 2024.
https://doi.org/10.1007/978-981-97-2242-6

Printed in the United States
by Baker & Taylor Publisher Services

Printed in the United States
by Baker & Taylor Publisher Services